NASA
SP-483

S0-EST-332

LIVING ALOFT
Human Requirements for Extended Spaceflight

Mary M. Connors
NASA Ames Research Center

Albert A. Harrison
University of California, Davis

Faren R. Akins
University of Santa Clara

Prepared at Ames Research Center

Scientific and Technical Information Branch   1985
National Aeronautics and Space Administration
Washington, DC

Cover artist: Jack Perlmutter

Library of Congress Cataloging in Publication Data

Connors, Mary M.
   Living aloft.

   (NASA SP ; 483)
   Bibliography: p.333
   Includes indexes.
   1. Manned space flight. 2. Space flight—Psychological aspects. 3. Human engineering.
I. Harrison, Albert A. II. Akins, Faren A.
III. Title. IV. Series.
TL873.C66 1985   302   85-5149

For sale by the Superintendent of Documents, U.S. Government Printing Office
Washington, D.C. 20402

... the early phases of a science require ... a certain disregard for the formalisms and pedantic, creeping construction of the ultimate scientific edifice. Perhaps what is essential is a foundation of sensible, if vague, ideas and orienting attitudes — correct in their broadest sweep if not in their precise predictions.

Jack P. Hailman
Science, 168, 701 (1970)

TABLE OF CONTENTS

ACKNOWLEDGMENTS xi

CHAPTER I – LIVING IN SPACE 1
   BACKGROUND 1
   A FRAMEWORK FOR FORECASTING 3
      Guiding Assumptions 3
      Theoretical Orientation 4
      The Available Data 5
   SPACE ENVIRONMENTS 7
      The Physical Environment 7
      The Social Environment 9
      Basic Reactions to Space-like Environments 11
      Temporal Fluctuations 14
   SUMMARY AND CONCLUSIONS 15

CHAPTER II – BEHAVIORAL AND SELECTION
IMPLICATIONS OF BIOMEDICAL CHANGES 19
   PHYSIOLOGICAL DECONDITIONING 20
      Simulation Studies 20
      Resistance to Deconditioning 25
      Countermeasures 30
   VESTIBULAR ALTERATIONS 35
      Manifestations and Theory 35
      Resistance to Vestibular Effects 41
      Countermeasures 46
   VISUAL CHANGES 51
   SUMMARY AND CONCLUSIONS 53

CHAPTER III – HABITABILITY 59
   BACKGROUND 59
   THE PHYSICAL ENVIRONMENT 60
      Interior Space 60
      Food 63
      Hygiene 65
      Temperature and Humidity 66
      Decor and Lighting 67
      Odor 69
      Noise 70
   HEALTH AND LEISURE 75
      Recreation 76
      Exercise 80
   PRIVACY 82
      Meaning and Functions 82
      Theory 83
      Bases of Needs 84
      Mechanisms 86
      Crowding 87
      Territoriality 89
      Privacy in Space 90
   COMPLEX EFFECTS 94
      Multiple Stressors 95
      Aftereffects 96
   SUMMARY AND CONCLUSIONS 97

CHAPTER IV – PERFORMANCE 107
   DESCRIBING PERFORMANCE 108
      Work Requirements in Space 108
      Human Performance Abilities 109
   ASSESSMENT OF HUMAN PERFORMANCE 110
      Discrete-Task Assessment Techniques 110
      Multiple-Task Batteries 114
      Partial- and Full-Scale Simulation 115
      In-flight Performance Assessment 118
      Future Focus of Research on Performance
         Assessment 120
   ISSUES IN ASTRONAUT WORK REGIMES 121
      Factors Affecting Work Capacity 121
      Factors Affecting Work Schedules 123
      Factors Affecting Workload 125
      The Effects of Desynchronosis 132
      Sleep Disturbances 136
   SUMMARY AND CONCLUSIONS 140

CHAPTER V — SMALL GROUPS 145
   INTRODUCTION 145
   INDIVIDUAL CHARACTERISTICS AND CREW
   COMPATIBILITY 146
      Gender 147
      Age 150
      Culture 151
      Personal Attractiveness 152
      Emotional Stability 153
      Competence 154
      Cooperativeness 155
      Social Versatility 156
      Similarities and Complementarities 157
      Group Homeostasis 160
      Crew Size and Social Compatibility 160
      Assembling Groups 161
   INTERPERSONAL DYNAMICS 162
      Leadership 162
      Cohesiveness 169
      Compliance, Conformity, and Independence 173
      Group Performance 175
      Temporal Dynamics 177
   SUMMARY AND CONCLUSIONS 180

CHAPTER VI — COMMUNICATION 187
   INTRODUCTION 187
   DIRECT INTERPERSONAL COMMUNICATION 188
      Verbal Communication 188
      Nonverbal Communication 192
   MEDIATED COMMUNICATION 194
      Planning Considerations 194
      Systems Requirements 195
      Systems Effects 196
      Application to Space 204
   COMMUNICATION NETWORKS 205
      Internal Communication 206
      External Communication 209
   SUMMARY AND CONCLUSIONS 212

CHAPTER VII — CRISES 217
   INTRODUCTION 217
   EXTERNALLY PRECIPITATED CRISES 217
      Experience in Space 218

Individual Response to Threat 220
  Group Processes 228
  Implications for Space 233
 INTERNALLY PRECIPITATED CRISES 234
  Psychological Episodes 234
  Transcendant Experiences 243
  Substance Abuse 245
  Grief 247
  Crisis Intervention 252
  Implications for Space 253
 SUMMARY AND CONCLUSIONS 253

CHAPTER VIII — ORGANIZATION AND MANAGEMENT . 259
 INTRODUCTION 259
 SPACECREW STRUCTURE 261
  Power Structures 262
  Work Roles 269
  Normative Structures 277
 MOTIVATION 278
  Rewards 280
  Sanctions 284
 EXTERNAL RELATIONS 287
  Boundary Roles 288
  Interorganizational Conflict 289
  Models of Conflict Management 293
  Reassimilation 298
 SUMMARY AND CONCLUSIONS 299

CHAPTER IX — SUMMARY AND RECOMMENDATIONS .. 305
 PURPOSE 305
 CHAPTER SUMMARIES 306
 DIRECTIONS FOR FUTURE RESEARCH 312
  General Research Issues 314
  Extended Spaceflight Variables 319
  Competing Perspectives 324
  Neglected Research Areas 327
 RESEARCH OPPORTUNITIES 328
  Situations and Environments 329
  Methods and Approaches 331
 CONCLUSIONS 332

REFERENCES 333
AUTHOR INDEX 397
SUBJECT INDEX 413

ACKNOWLEDGMENTS

We wish to express our deep appreciation to R. Mark Patton, Special Assistant to the Director of Life Sciences, NASA Ames Research Center, who provided both initial support and continuing encouragement throughout this study, and who offered helpful comments on each and every chapter of the manuscript. We also gratefully acknowledge the improvements made to individual chapters by the comments and suggestions of Earl A. Alluisi, Frederick E. Guedry, Robert C. Haygood, J. Michael Lentz, Joseph E. McGrath, Ronald E. Rice, Don A. Rockwell, Everett M. Rogers, Daniel S. Stokols, and John L. Vohs.

Artist: Mitchell Jamieson

I
LIVING IN SPACE

BACKGROUND

Since the earliest days of spaceflight, substantial concern has been expressed regarding the physical needs of astronauts, including any biological damage that might result from exposure to radiation or from reduction in gravitational forces. In contrast, relatively little concern has been directed towards people's psychological and social adjustment to space. At one time this difference in emphasis was justified. The Mercury, Gemini, and Apollo flights were measured in hours and days and it could be reasonably assumed that astronauts would be able to withstand certain deprivations for these brief periods. The longer flights of Skylab presented a different picture. Early in the development of Skylab, it was recognized that steps would have to be taken to accommodate a wider variety of human needs. However, the needs that were addressed remained narrowly defined and centered primarily on habitability considerations.

We are now at the point in the development of spaceflight where the range of psychological and social requirements of the human participant must be given full consideration. There is hope that NASA soon will move ahead with its long-awaited space station. As this is written, information from the Russian space program suggests that the Soviet Union will soon launch a space vehicle capable of carrying a crew of twelve, possibly as a prelude to a manned Mars mission. Whatever the specific projects or time frames, it seems clear that tomorrow's manned spaceflights will involve large numbers of people living and working together under close confines and in "unnatural" environments for long periods of time. Adjustment to such conditions has important implications for mental health, for social organization, and ultimately for mission success. Because tomorrow's astronauts are likely to expect, and even demand, greater autonomy in living and working arrangements, the planners' perspective must extend beyond concern for effective functioning within the space community and encompass the relationship between the space community and the home planet.

In this book we attempt to identify and assess, in a serious and systematic fashion, the psychological and social problems that may be associated with future space missions, and to explore some possible solutions. This task involves establishing both a structure in which relevant issues can be considered and a level of analysis that can contribute to a scientifically based understanding of human adaptation to space.

Several authors, expressly or tacitly, have affirmed the need for the integration of behavioral and social science methods and findings into space mission planning and management (e.g., Leonov and Lebedev, 1975, 1972; Berry, 1973a; Kubis, 1972; Sells, 1966). In 1972 the Space Science Board of the National Academy of Sciences issued a publication entitled *Human Factors in Long Duration Space Flight,* which attempted to approach the question of the human in space in a comprehensive fashion. Through a series of essays, the contributors to this volume looked at various aspects of human adjustment to space, placing particular emphasis on the neglected behavioral, psychological, and sociological factors of this "microsociety in a miniworld." The present volume attempts to build upon and broaden this past work, and, on occasion, to challenge the assumptions upon which it and other earlier discussions rest.

Much of the writing on man in space has focused at one extreme on narrowly defined experiments, usually involving basic biomedical processes, and at the other extreme on highly speculative and even Utopian views of potential social arrangements. The present work is an attempt to fill the gap between these two approaches by examining the behavioral science literature which has either direct or indirect application to the space environment. From this examination we hope to draw conclusions about adaptation to space and identify those areas where further psychological and social research is needed.

Space missions have generally involved small crews, drawn from highly homogeneous pools (such as white, educated, young adult males) and functioned for limited periods of time. We anticipate that future missions will involve large crews drawn from diverse populations, and that these missions eventually could last years. As crew size, crew heterogeneity, and mission duration continue to increase we enter the realm of "extended spaceflight." The general question we attempt to answer in this book is "What are the psychological and social issues of future spaceflight?" Reformulated in terms of the variables of extended spaceflight, the question becomes "What will happen when space crews become larger, more heterogeneous, and as space missions become substantially longer?"

A FRAMEWORK FOR FORECASTING

Our attempts to forecast the psychological and social dynamics of extended spaceflight have proceeded within a particular framework. This framework has been imposed by a set of assumptions, by a theoretical orientation, and by available data. Since this framework has left an imprint on each chapter that follows, we consider it important to make this framework explicit at the outset.

Guiding Assumptions

Three assumptions have guided our efforts. The first assumption is that psychological and social factors will become increasingly important determinants of the success or failure of future space missions. Indeed, we shall review some evidence indicating that our understanding of spaceflight technology has already outrun our understanding of technology's human users. A goal of the present effort is to draw the reader's attention to the kinds of psychological and social issues which may prove critical during the coming generations of manned flight.

Our second assumption is that it is essential to avoid a premature commitment to a narrow perspective. Instead, it is necessary to entertain as many potentially viable alternatives as possible. The old ways do not necessarily remain the best ways, given rapidly changing technology, societal change, and likely shifts in mission specifications and goals. For example, early flights were modeled along military or paramilitary lines, and all astronauts to date have been highly trained professionals. Military models of organization may not prove desirable when astronauts are selected from a heterogeneous pool of applicants, and not all members of large crews need be rigorously trained professionals. Future flights thus may require a careful weighing of model alternatives. Our corollary assumption is that no single plan will be desirable, or even workable, for all future missions. Optimal crew selection and training methods, habitability and communication specifications, and viable social structures will all depend upon such variables as crew size, crew heterogeneity, mission duration, and mission objectives.

Our third and most important assumption is that some of the uncertainties regarding life in space can be reduced through careful and rigorous behavioral and social science research. As we shall see, extreme care must be taken in extrapolating past findings to future spaceflight. Yet, however tentative, there does exist a research base that can provide useful information to the manned space program. Although this base may not always provide definitive answers to space-related questions, it can move us far beyond the point where we must rely entirely upon speculation.

Theoretical Orientation

Following Sells (1966) and the Space Sciences Board (1972), we have adopted a systems perspective on spaceflights. That is, missions are viewed as comprised of highly interdependent components (e.g., technical, biological, and social), such that variations in one component typically have repercussions in one or more of the others. We have attempted to expand this conceptualization by incorporating elements of open-systems theory as devised by J. Miller (1960, 1955) and advanced by Katz and Kahn (1966) and J. Miller (1978). Two important features of open-systems theory should be stressed. First, since open-systems theory can be applied to biological and social units of varying sizes, large-scale missions fruitfully can be analyzed in terms of component systems or subsystems. Second, by viewing systems as open, full acknowledgement is accorded the importance of the surrounding environment. Although space missions are separated from Earth by immense distances, mission-Earth transactions remain frequent and critical. For instance, a high rate of information exchange occurs during the course of all missions, whereas personnel rotation and resupply are likely on missions involving large, orbiting satellites. In addition, the preparation, launch, and recovery phases of spaceflight will involve intense spacecraft-Earth exchanges.

Although no attempt will be made to formulate data in systems terms, the interrelationships which are fundamental to the open-systems approach will be basic to all discussions. Equally important, it is our intention to *think* in systems terms, i.e., to avoid categorizing and instead to search for "connectedness principles, isomorphisms, interrelationships — in short, a holistic approach" (Fisher, 1978, p. 96).

The Available Data

At present, relatively few data are available from space itself. In attempting to forecast some of the dimensions of life in space, behavioral and social scientists have concentrated on other environments with elements similar to those of space, i.e., those marked by isolation, confinement, deprivation, and risk (Kubis, 1972; Haythorn, McGrath, Hollander, Latané, Helmreich, and Radloff, 1972; Sells and Gunderson, 1972; Kanas and Federson, 1971). Thus, much of the data that has particular relevance to space comes from experiences in such settings as polar camps, underwater habitats, and space-capsule simulators of varying degrees of verisimilitude.

Since these environments approximate certain aspects of space conditions, relevant data will be given considerable attention both in this chapter and in the chapters that follow. We will refer to these environments as "space-like" or "similar to space." However, it should be remembered that they are not truly space-like; they merely approximate certain aspects of space. Indeed, a disclaimer underscoring the tenuousness of offered findings to the space environment is *de rigueur* in most original research reports, and in all serious literature reviews. There are two major shortcomings associated with much of this research. First, many studies have involved settings which lack the degree of isolation, confinement, and risk typical of long-duration spaceflight. Studies which have involved a high degree of isolation, confinement, and risk have often lacked methodological rigor.

Earth-space discontinuities— There are important discontinuities between isolated and confined environments that presently exist on Earth and those that are expected to predominate in space. Many of the studies cited in discussions of long-duration spaceflight involve subjects, tasks, and settings that bear little or no correspondence to those likely to be encountered in space. The results of such studies are not necessarily inapplicable to space, and repetitive findings that point to the same general conclusion may extrapolate quite well. Nonetheless, considering such studies, it is necessary to keep in mind the kinds of variables that could render generalization questionable.

First, most studies of isolated and confined groups involve a very restricted range of subjects (all males, all college sophomores, all naval personnel, etc.). There may be appreciable differences between these subjects and the people who will participate in future space missions. In addition, most experimental (as compared with naturalistic) studies of isolated and confined groups involve subjects who are basically unacquainted with one another prior to the experiment. There may be substantial differences between such assemblages and the preformed groups that are likely to be sent into space.

Second, most studies of isolated and confined people involve shorter periods of time than those anticipated for many future space missions. For example, the longest space-simulator studies terminated between 90 and 105 days. "Wintering over" at a polar camp requires, at the outside, a year's commitment, and most studies deal with much shorter periods of time. However, an interplanetary mission will consume the better part of two years. The available data, therefore, may not reflect some important temporal variables.

Third, interplanetary missions, orbiting laboratories, and other space missions will involve a degree of isolation, confinement, and risk rarely equalled by any environment on Earth. Although submarines and Arctic bases may effectively be cut off from civilization by seemingly large distances, such distances are small compared with those that future space travelers can anticipate. The degree of confinement imposed on a space mission may be approximated in a few Earth locations (e.g., in subaquatic dwellings and at polar stations during winter months), and occasionally a very high degree of risk may be found (e.g., in subaquatic dwellings). For the most part, however, Earth environments, by comparison, are tame and hospitable.

Data collected in exotic environments— It is always difficult to conduct good applied research, but the difficulties are multiplied when the research is carried out in an exotic environment. The obstacles exotic environments impose are not insurmountable, but they have handicapped researchers and occasionally forced them to abandon preferred procedures and techniques. To the extent that investigators are forced to eliminate control conditions, to use gross rather than refined measurements, or to rely on impressions and memories, there is increased latitude for error in their observations.

It has often proved difficult for an investigator to be present when a group is undergoing conditions of isolation, confinement, and risk. There may be severe constraints on the number of people who can participate, and inquisitive individuals who are not essential to getting the primary job done may have to be left behind. As a result, some researchers have been forced to rely on data gathered before and after the mission. Researchers who do accompany a mission may find it difficult to build the necessary rapport with the other members of the expedition. Unless a researcher is making some very clear contributions to group welfare (e.g., by serving as a cook), he or she may be seen as a drain on the group's resources and become a target for resentment. In addition, certain research activities, particularly those examining the relationship among members, may threaten the stability of the group.

In some remote environments, electronic surveillance can be used instead of participant observation. The underwater-habitat research shows that devices such as video monitors can be put to good use (Radloff and Helmreich, 1968). However, the same research shows that electronic surveillance has certain limitations. Much of the activity may take place out of the range of the surveillance gear and the quality of signals may be poor. Such gear may have to be installed in inaccessible locations, making it difficult to adjust or repair. In some remote environments, electronic surveillance equipment may be inadmissible because of space and weight restrictions.

Reviewing the literature, one cannot help but be impressed by the ingenuity, and frequently the bravery, of the researchers who have ventured into the realm of exotic environments. Nonetheless, it is important to remember that not all of the barriers to good research have been overcome successfully, and the results of such studies must be interpreted with these limitations in mind.

SPACE ENVIRONMENTS

Space environments are characterized by isolation (a separation from the normal or daily physical and social environment), confinement (restriction within a highly limited and sharply demarcated physical and social environment), deprivation, and risk. Such environments may be expected to place heavy demands on astronauts' psychological and social resources.

The Physical Environment

Space itself is totally inhospitable to life as we know it. Penetration of the wall of a spacecraft or a major equipment malfunction could lead to instant — or lingering — death. Although concern for astronauts' welfare will continue to be primary, there is a limit to the amount of safety redundancy that can be built into any system. Malfunctions or mistakes, should they occur, will have to be corrected by a relatively small group of people who have limited access to tools and supplies. Because space environments are so removed, the chances for outside rescue are slim. In addition to the high risk of the space environment, space travelers must contend with limited facilities, equipment, and supplies, and with a low level of perceptual stimulation.

Economic and engineering considerations limit both the size and the weight of a spacecraft. Like most sea and air vessels, a considerable portion of the spacecraft's total weight must be devoted to propulsion mechanisms, fuel, and navigational equipment. In addition, spacecraft require elaborate machinery to ensure an adequate supply of air and water and are likely to carry considerable scientific baggage. All such items use internal space that could otherwise be devoted to living quarters and recreational facilities.

Equipment transported to space needs to be carefully designed or selected, and thoroughly tested. In the future it will become increasingly important that equipment not only fulfill its intended function, but also operate in a way that does not yield dysfunctional social consequences. For instance, equipment that must be used by many people should have a reset requirement built into its operation. This would avoid the interpersonal frictions that have occurred in space simulation, when people fail to leave things as they find them (S. Smith, 1969). However, equipment that is completely satisfactory during pretesting on Earth may not meet expectations *in situ*, as has been shown in space flights from Mercury (Wolfe, 1979) through Skylab (Cooper, 1976) and Shuttle (Stockton and Wilford, 1981).

Within the space environment, life proceeds without abundant provisions or supplies. Even in the future, when space settlements may have certain renewable resources and when shuttling craft may supply orbiting laboratories, goods will be limited. Although it may be possible to guarantee adequate reserves of life-sustaining items, many of the common luxuries that can be found in the smallest U.S. communities will be either unavailable or extremely scarce.

Even the simple pleasures associated with mealtimes may be muted in space. Providing food that is tasty and otherwise appealing has been a particular problem in space and in certain subaquatic environments that require artificial atmospheres. At the same time, foodstuffs and other oral supplies have been found to offer special gratification to isolated and confined people (Shurley, 1973; Berry, 1973a; Earls, 1969; S. Smith, 1969; Serxner, 1968; Mullin, 1960). It is unlikely that alcohol as a social beverage will find its way into space, at least until relatively large and stable settlements are established. Alcohol, as a recreational drug, may be keenly missed by space travelers, since there is evidence that alcohol plays an important social role in exotic environments (Mostert, 1974; Shurley, 1973).

On many future missions, the perceptual field will be relatively unchanging. Inside the vehicle, the walls and furnishings must be functional. Machinery will hum monotonously, and the temperature and humidity will remain within narrow limits. Outside the spacecraft, the view may be spectacular; however, this view may be highly repetitive (as in the case of an extended orbital mission) or shift at a barely perceptible rate (as in the case of an interplanetary mission). Furthermore, since viewing ports are structurally costly, they are likely to be only sparingly available.

Early fears that space missions would produce hallucinations and other effects associated with prolonged and pronounced sensory deprivation (Zubek, 1969; Flinn, Monroe, Cramer, and Hagen, 1961; Heron, 1957) have been allayed (Kanas and Fedderson, 1971), but boredom remains a likely concomitant of extended-duration spaceflight. There are various techniques, involving judicious selection and arrangement of interior design and personal articles, that could raise the visual stimulation inside the space vehicle. However, the physical environment, when not terrifying, is likely to be bland.

The Social Environment

Adapting to unusual and frustrating physical conditions in space is but half the battle confronting today's astronauts and tomorrow's space settlers. Challenges are also posed by the social environment. Participants on long-duration space missions will be temporarily, or perhaps permanently, extracted from the ongoing relationships that are important to them, and will be thrust into a microsociety which will impose its own social deprivations and hardships.

Withdrawal from the home community— People under normal circumstances are embedded in a complex social matrix that links them with family members, friendship groups, one or more large-scale organizations, and society. Space-mission participants are withdrawn from this social matrix. Space travelers are likely to be separated from loved ones and friends, with the concomitant loss of reassurance, affection, and respect that flow in such relationships.

A second effect of withdrawal from the larger society is a loss of variety in social relationships. A corollary of this reduced variety is a lessened opportunity to exercise one's typical social roles. In the course of daily life, people enact a variety of different roles (teacher, husband, father) with those who enact interlocking or reciprocal roles (student, wife, son). In addition to providing stimulation, such diversity allows the person to exercise different skills and talents, and may be important for a complete sense of identity (D. Miller, 1963). Variety in relationships also provides the opportunity to formulate a balanced response, as when a dispute at work is placed in perspective in a session with a spouse or a friend.

Withdrawal from preexisting relationships removes known comparison points and limits the self-evaluation process (Festinger, 1954). For this reason, withdrawal from one's usual relationships may make it difficult to maintain a sense of personal continuity or identity.

Challenges of the microsociety— At the same time that astronauts must cope with the effects of withdrawing from the macrosociety, they will be confronted with social challenges within the microsociety. Unlike the usual acquaintance process through which people get to know one another gradually and against a backdrop of other developing and continuing relationships (Altman and Taylor, 1973), space missions will involve intense contact with very few people. Acquaintance processes are thus likely to be forced, rapid, and relatively unbalanced by alternative relationships. Also, prolonged isolation and confinement appear to magnify the effects of attitudinal dissimilarities, need incompatibilities, annoying habits, irritating mannerisms, and other sources of interpersonal friction (Haythorn et al., 1972), while reducing the opportunity to express dissatisfaction. For example, under conditions of isolation and confinement antagonisms are frequently suppressed, apparently because of fear that their expression could spark a conflict with unacceptably costly consequences.

Basic Reactions to Space-like Environments

We have considered some of the difficulties in extrapolating from data gathered on Earth to problems that could occur in space. We must also be aware that certain factors cloud the interpretation of the overall pattern of results among Earth-based studies. First and foremost is the noncomparability of studies themselves. Rarely have two studies involved similar degrees of isolation, confinement, and risk, and only occasionally do different studies involve similar dependent measures.

Second, there are many possible causes of most of the reported effects. The field settings in which these studies are conducted typically contain a number of elements, any one or a number of which might provoke psychological or social dysfunction.

Finally, most naturalistic studies of isolated and confined individuals do not include controlled observations of people who are living and working together in nonisolated, nonconfined locations. Although some informative data from natural settings may be available, these data rarely offer the degree of comparability desired.

Given these limitations in comparing data from Earth-based studies, certain behaviors occur with sufficient consistency in the various studies to command the attention of space-mission planners. (For a consideration of how data from isolated and confined groups compare with experiences on the Russian Salyut 6, see Bluth, 1982).

Impaired intellectual functioning— There is some evidence that isolation and confinement lead to impaired intellectual functioning. Several investigators have reported decreases in alertness, concentration, and memory among polar expeditioners (Natani, Shurley, and Joern, 1973; Strange and Klein, 1973; Mullin, 1960), with need-induced fantasies sometimes distorting perceptual judgments (S. Smith, 1969). In a carefully controlled laboratory experiment, Taylor, Altman, Wheeler, and Kushner (1969) found that isolation and confinement led to impoverishment of ideational and imaginative output, as evidenced by responses to the Rorschach Inkblot Test.

The declines observed thus far have not appeared among all, or even the majority of participants. Also, the declines that have been reported are not necessarily severe, and many studies report no intellectual decline at all (see, e.g., Cleveland, Boyd, Sheer, and

Reitman, 1963). A continuing concern, however, is that prolonged isolation and confinement can lead to cognitive impairment of sufficient magnitude to prevent satisfactorily coping with emergency situations.

Motivational decline— Many studies suggest that prolonged isolation and confinement results in a loss of motivation. A common report is that study participants begin confinement with every intention of engaging in creative activities such as writing, completing a project, or accomplishing some serious reading. These worthwhile goals rarely are achieved, or even attempted. Instead, most hours of confinement are spent in time-marking activities such as solitaire.

Gunderson and Nelson (1963) observed a general decline in work satisfaction and group accomplishment among polar expeditioners, and Natani and Shurley (1974) have reported fatigue, inertia, and general apathy among a similar group. In the course of an Arctic trek, scientists interviewed by Pope and Rogers (1968) reported increased feelings of helplessness and worthlessness, feelings which are hardly conducive to goal-directed activity. Cleveland et al. (1963) report a "diminution of vitality" among fallout-shelter confinees, and Earls (1969) has found that the high motivation shown by submariners during the first few days of a cruise dissipates as the cruise progresses.

Somatic complaints— Relatively common correlates of isolation and confinement are insomnia, headaches, digestive problems, and other somatic complaints. For the most part, it is felt that these findings do not simply reflect unappetizing rations, cramped sleeping quarters, or extreme conditions; there is presumed to be a fairly strong psychological component as well.

Insomnia and other sleep disturbances are the most prevalent complaint. Sleep disturbances have been noted among members of polar parties (Natani and Shurley, 1974; Strange and Klein, 1973; Gunderson and Nelson, 1963; Law, 1960), among submariners (Earls, 1969; Weybrew, 1963; Mullin, 1960), and among experimental subjects (Hammes and Osborne, 1965).

Headaches are a common problem in various confinement settings (Natani and Shurley, 1974; Earls, 1969; Kubis and McLaughlin, 1967), and upset stomachs, constipation, and other digestive complaints also appear with considerable frequency (Natani and Shurley, 1974; Hammes and Osborne, 1965). In general, sick calls increase over time in confinement, with the highest incidence occurring during the third quarter of the expedition's term.

Psychological changes— Studies to date do not support the notion that isolation and confinement result in severe psychiatric disturbances (Lugg, 1973; Gunderson, 1963). However, mild psychiatric symptoms have been reported with considerable consistency. A large number of incidences involve depression (Natani and Shurley, 1974; Natani et al., 1973; Strange and Klein, 1973; Serxner, 1968; Gunderson and Nelson, 1963; Palmai, 1963; Law, 1960) and anxiety (Pope and Rogers, 1968; Serxner, 1968; Ruff, Levy, and Thaler, 1959). Other psychological changes that have been noted in confinement are regressive, highly polarized and immature attitudes (Lugg, 1973; W. Smith, 1966), increased defensiveness or belligerance (Palmai, 1963; Ruff et al., 1959), and a high emphasis on oral satisfaction (Serxner, 1968; Mullin, 1960). Ruff et al. (1959) offer a general explanation for these observations by suggesting that isolation and confinement are conducive to the "eruption of the unconscious."

Social tensions— It has been suggested repeatedly that prolonged isolation and confinement impairs people's ability to get along with one another (Kubis, 1972; Haythorn et al., 1972; Kanas and Federson, 1971). Touchiness or social irritability has been observed among polar expeditioners (Natani and Shurley, 1974; Gunderson and Nelson, 1963; Palmai, 1963; Law, 1960), submariners (Rohrer, 1961), fallout-shelter confinees (Cleveland et al., 1963), and space-simulator subjects (George Washington U., 1974). It appears that the confinement environment magnifies incompatibilities which might otherwise be overlooked (Haythorn, 1970, 1968; Haythorn et al., 1972).

One method by which confined individuals control interpersonal tensions is by shunning those activities, such as competitive games, which might lead to conflict. More broadly, interactions among confined individuals tend to lack affect. For instance, Flinn et al. (1961) found that most of the social acts performed by subjects in a space simulator fell into the interpersonally neutral categories of Bales' (1950) Interaction Process Analysis.

A second method by which confined individuals avoid conflict is by withdrawing from one another. Haythorn and Altman (1967) have termed this withdrawal behavior "cocooning." However, unlike a physical cocoon, the barriers which confined individuals establish are primarily psychological. Although confinees become disinclined to interact with one another, they do not necessarily wish to be physically separated. The result is an aggregation of people engaged in individual activities in what might be thought of as an adult version of parallel play. Rogers (1978) has found that this preference for shared space over private space extends to mixed-gender crews. Being in line-of-sight contact with others seems to comfort the confined individual, probably by keeping him current on what is happening in the group.

Another commonly reported method for dealing with interpersonal tension in confinement is to direct it to outsiders. Numerous studies show that confinees direct their hostility towards those who maintain communication with the confined group, or to others associated with the confinement project. Space travelers are not immune to this phenomenon; members of certain Apollo and Skylab crews have expressed peevishness and defiance towards ground-control personnel.

Temporal Fluctuations

Almost any condition, note Kanas and Fedderson (1971), can be tolerated for a brief period of time. It is with the passage of time that the deleterious effects of isolation and confinement gain prominence. In the Antarctic, work satisfaction, social relations, and group accomplishment have been found to deteriorate with time (Gunderson and Mahan, 1966; Gunderson and Nelson, 1965). In a Boeing 30-day simulation, increased time was associated with (1) increased annoyances, interpersonal conflicts, and hostility; (2) decreased feelings of being happy, comfortable, and satisfied; and (3) increased dislike for the experimenter (George Washington U., 1974).

Although some studies show a general relationship between time in confinement and negative feelings, other studies indicate that mood and morale do not deteriorate monotonically. Rohrer (1961) has identified three broad stages of reaction to prolonged isolation, confinement, and stress. The first stage is a period of heightened anxiety brought about by the perceived dangers in the situation. Moderately heightened anxiety may improve alertness and performance, and is hence not necessarily detrimental. The second stage, which emerges as the crew settles down to a routine day-to-day existence, is marked by depression. Moods during this stage are likely to result in regrets about having joined the mission. The third stage is a period of anticipation which occurs as the end of the mission draws nigh. During this period emotional outbursts, aggressiveness, and rowdy behavior are likely. This expressiveness can be dangerous because at the end of a mission many complicated operations may have to be performed (Kanas and Fedderson, 1971).

Departing one stage and entering another appear to be more dependent upon the relative than the absolute passage of time. Substantial evidence suggests that whether the mission lasts days, weeks, or months, morale reaches a low ebb somewhere around the one-half to two-thirds mark. For example, Palmai (1963) reports that morale reached its lowest ebb somewhere around the second third of a 1-yr Antarctic stay. In 30-day submarine missions, morale reached nadir at about day 15, and in 8-wk missions, morale reached nadir during the fourth and fifth week (George Washington U., 1974). In the McDonnell-Douglas 90-day simulation study, crew morale was rated as good except for a period of 10 days which occurred about two-thirds of the way through the mission. One- and two-week confinements in fallout-shelter simulations undertaken by the American Institute for Research also show a pattern of depression toward the midpoint of the stay (George Washington U., 1974).

These evidences of temporal mood shifts are not perfectly in accord for, as already noted, some reports make no mention of an upswing in morale toward the end of a mission, whereas in the Sealab II study, morale did not decline over time (Helmreich, 1973). However, many studies do suggest that mood is related to those psychological anchor points which help the individual mark his confinement.

Although morale may recover as the end of a mission comes in sight, there is no evidence that the withdrawal behavior just described similarly abates. Confinees have been reported to remain withdrawn throughout confinement and to show increasing signs of proprietary attitudes towards possessions as the confinement progresses (Grumman Aerospace Corp., 1970; S. Smith, 1969; Haythorn and Altman, 1967).

SUMMARY AND CONCLUSIONS

For a long time to come, people who enter space will be exposed to high physical risk. They will live in restricted, stimulus-impoverished environments, and will have to make do with severely limited equipment and supplies. They will be cut off from family and friends, and face many challenges within their microsocieties.

Numerous researchers have pursued the hypothesis that isolation and confinement adversely affect psychological and social functioning. The literature reports impaired intellectual functioning; motivational decline; somatic complaints; depression, irritability, and other psychiatric symptoms; and impaired interpersonal relations. These difficulties are not omnipresent in isolated and confined settings, and when they are present, they are not usually severe. However, they must be considered possible and potentially dangerous concomitants of extended spaceflight.

Tomorrow's spaceflight can be expected to pose different challenges to human participants than have previous missions. Judging from Earth-based studies, future missions involving large crews, more spacious environments, and more sophisticated communications with the outside world, might be expected to result in diminution of somatic complaints (Gunderson, 1968), reduction of emotional disturbances (Nardini, Herrmann, and Rasmussen, 1962), and an increase in group harmony and efficiency (Gunderson and Nelson, 1963). On the other hand, large and heterogeneous crews present the opportunity for misunderstandings, value conflicts, clique formation, etc., and long-duration flights can be expected to introduce stresses beyond those experienced in Earth-based studies.

In the chapters that follow we will examine a wide variety of research areas that can contribute to our understanding of the psychological and relational aspects of humans in space. Such an endeavor naturally focuses on identifying problems which could confront the space traveler. This approach should not obscure the likelihood that the space environment contains many of its own psychological and social balancing systems. Nor should this problem orientation be interpreted as a rejection of the role of mankind in space. On the contrary, it is precisely because we believe that the human can adapt to extended spaceflight that we seek to explore the potentially deleterious effects of the space environment and those circumstances which might offset these effects. Efforts to counter any adverse effects of space should not be limited to reinstating conditions of neutrality. If large numbers of people are to spend extended periods of time isolated and confined in space, the goal must be to discover or to establish positive conditions under which psychological functioning and social life can prosper and flourish.

Until now, space has been a time-limited challenge — something to survive, to master, and to *have* endured. We are now entering an era when the new astronaut must foresake his short-term approach to space travel and begin to deal with the space environment as a continuing condition. When this occurs, the space traveler will truly begin to live in space.

Artist: Robert Shore

II

BEHAVIORAL AND SELECTION IMPLICATIONS OF BIOMEDICAL CHANGES

Considerable study has been directed toward understanding how physiological functioning is affected by weightlessness (Nicogossian and Parker, 1982) and other concomitant spaceflight stresses such as radiation (Tobias and Grigor'yev, 1975), decompression (Rayman and McNaughton, 1983), acceleration (Waligora, 1979), etc. There has been substantially less concern about understanding behaviors which may change as a direct result of these biomedical alterations. The present chapter will examine the behavioral implications of biological changes associated with space travel. The behavioral implications of various countermeasures used in the treatment of biomedical alterations are also addressed. This analysis focuses on those biomedical topics that have possible behavioral/performance implications: physiological deconditioning, alterations in the vestibular system, and changes in perception and processing of the visual system.

Until recently, spaceflight crews have been a generally homogeneous group, usually male pilots in top physical condition and trained with rigorous endurance tests. For future spaceflight, we need to understand how different groups of humans may adapt to and function in the spaceflight environment. The present chapter will investigate issues of human adaptability to the biomedical stressors of space and, where data exist, will compare demographically distinct subgroups.

PHYSIOLOGICAL DECONDITIONING

A potentially serious consequence of weightlessness is the deconditioning of such physiological systems as the cardiovascular complex. On Earth, the heart must operate against gravitational pressure to sustain blood flow and proper functioning of the cardiovascular system. Under zero-gravity (0-g) conditions, no such hydrostatic pressure gradient exists. Consequently, the heart lessens its pace to achieve an equilibrium appropriate to decreased demands. Reduced output of the heart, decreased heart rate, decreased heart size, and diminished blood volume regulation result (Berry, 1973b).

Diminished loading inherent in weightlessness is also a significant problem as it affects the musculoskeletal system. Reduced weight bearing in space leads to bone "disuse" symptoms including loss of calcium, nitrogen, and phosphorus; decreased bone size and volume; and formation of urinary stones (Berry and Catterson, 1967; Hattner and McMillan, 1968; Biryukov and Krasnykh, 1970; Mack, 1971). Decreased muscle tone and strength, weakened reflexes, and decreased tolerance for physical work are further negative consequences of 0-g.

Simulation Studies

Problems of physiological deconditioning were noted during the 28-, 59-, and 84-day Skylab missions (Johnston and Dietlein, 1977) and during the recent Soviet missions of 175, 185, and 211 days (Toufexis, 1983). Given that significant alterations in biomedical functioning occur during spaceflight, it is crucial that, where possible, Earth-based models be developed to study these changes. Current knowledge suggests that a hypokinetic state is the most appropriate ground-based method with which to study many of the effects of prolonged weightlessness. Data have been generated using plaster-cast immobilization (Dietrick, Whedon, and Shorr, 1948; Billman, Teoh, Dickey, and Stone, 1981; Dickey, Billman, Teoh, Sandler, and Stone, 1982); chair rest (Lamb, Johnson, and Stevens, 1964; Lamb, Stevens, and Johnson, 1965); confinement (Hanna and Gaito, 1960; Alluisi, Chiles, Hall, and Hawkes, 1962; Lamb, Johnson, Stevens, and Welch, 1964; Kurash, Andzheyevska, and Gurski, 1980); dehydration (Dunn, 1978; Dunn and Lange, 1979; Dunn, Leonard, and Kimzey, 1981); water immersion (Gauer, 1975; Epstein, 1978); "dry" immersion (Gogolev, Aleksandrova, and Shul'zhenko, 1980); partial body support (Morey, 1979; Jordan, Crownover, Sykes, Schatte, Simmons, and Jordon, 1980); and clinostatic rotation (Cogoli, Vallucki-Morf, Mueller, and Briegleb, 1980). The major source of information, however, has come from studies of bedrest (see Birkhead, Blizzard, Issekutz, and Rodahl, 1964; Miller, Johnson, and Lamb, 1964, 1965; Chase, Grave, and Rowell, 1966; Vogt, Mack, and Johnson, 1966, 1967; Lynch, Jenson, Stevens, Johnson, and Lamb, 1967; and the Ames Research Center studies detailed in later sections of this chapter).

The lack of activity imposed by recumbancy effectively simulates many of the effects that weightlessness produces[1] (see Berry, 1973b, for comparative review). Bedrest results in reduced hydrostatic pressure and a relative lack of stress upon the body similar in many ways to the conditions of 0-g. Within 24 to 48 hr following recumbancy, there is a loss of fluid volume when an initial shift of volume from the lower extremities to the chest is interpreted by central mechanoreceptors as a relative increase in volume. This perceived volume increase triggers diuresis, resulting in fluid loss. Fluid loss is one of the reasons for body weight losses observed during prolonged bedrest (and spaceflight). It is also a factor in the deconditioning of the cardiovascular system, since decreased blood volume along with decreased blood pressure results in diminished reactivity of vessels.

Bedrest also simulates the effects of weightlessness on the musculoskeletal system. When the human body is maintained in the horizontal position, gravitational forces do not exert pressure upon the support structure in the same way as experienced in the vertical position. Consequently, muscles tend to atrophy from disuse and bones tend to demineralize from lack of stress. These factors combine with cardiovascular and metabolic alterations to decrease work tolerance and capacity.

Since recumbancy produces many of the biomedical changes which occur in space, bedrest would appear to be a valid model for the study of behavioral changes likely to occur in flight. Unfortunately, relatively few experiments have explored this possibility. Furthermore, existing studies have produced contradictory data and have experienced procedural problems. For example, Ryback, Trimble, Lewis, and Jennings (1971a) were among the first investigators to publish performance data obtained from prolonged (5 wk) bedrest. They employed four basic test units to assess psychomotor

---

[1]Calcium loss, however, appears to occur at a faster rate in space than during bedrest simulations.

performance during pre- and post-recumbancy. The tasks employed measured such factors as hand-eye coordination, reaction time, vigilance, short-term visual memory, hand and limb steadiness, auditory coding, etc. Hand-eye coordination was the only factor that decreased significantly during post-bedrest recumbancy relative to pre-bedrest values. Following bedrest, subjects were not as proficient in inserting a stylus into various hole sizes while attempting to avoid touching the edges of the holes. However, in a second study (Ryback, Lewis, and Lessard, 1971b) subjects tested in a reclining position (rather than in the standing position used during the first study) did not demonstrate decrements in reaction time and hand steadiness as found in the original experiment. Apparently, instability in the vertical position, rather than any biomedical or psychological effects, was the main reason for the decrements in performance observed in the first study. Additional tasks used in the second study involving measures of speed and accuracy of movement also failed to show consistent changes. The only significant differences found involved measures of neuromuscular strength of handgrip. Handgrip for the nondominant hand declined significantly from pre-bedrest levels among subjects not exercising during bedrest. This increment was most pronounced after the third week of hypokinesis. However, an earlier study (Bourne, Nandy, and Golarz de Bourne, 1968) has shown that little change in handgrip strength occurs even after 60 days of bedrest, at least when the dominant hand is measured.

More recent studies of bedrest subjects conducted at Ames Research Center have also failed to demonstrate consistent group changes in performance. For example, Winget, De Roshia, and Sandler (1979) studied 35- to 55-yr-old females exposed to 9 days of bedrest. Performance was assessed using an ATC-510 flight simulator modified to permit operation while the subject was in the bedrest position. Daily test scheduling was arranged so that trials occurred both during "peak" and "trough" portions of heart-rate and body-temperature cycles. Although decrements in performance were observed, there were no statistically significant group differences between pre-bedrest, bedrest, or post-bedrest periods. Apparently, while some individual subjects exhibited marked changes during the study, considerable variability in performance across subjects tended to mask group effects. Other studies reported by Winget and his colleagues (Chapman, Winget, Vernikos-Danellis, and Evans, 1975; Chapman, Winget, and Vernikos-Danellis, 1976) have noted similar difficulties for other age groups in isolating performance decrements.

Data from these studies are still under analysis and few detailed reports are available in the literature. It is hoped that more thorough statistical assessment may yield significant predictors of changes in bedrest performance.

Other investigators have failed to demonstrate significant changes in psychomotor performance following bedrest. Jex, Peters, Di Marco, and Allen (1974) tested eight carrier-qualified Navy Reserve A-7 pilots before and after 10 days of bedrest. Using controlstick tracking tasks, no differences in performance were observed when pre- and postrecumbancy measures were analyzed. Likewise, Storm and Giannetta (1974), using tests of complex tracking performance, eye-hand coordination, and problem-solving ability, found no significant effects due to hypokinesis among airmen subjected to 2 wk of absolute bedrest.

If performance does not deteriorate at the psychomotor level, alterations could still occur at the perceptual or cognitive level. Thus far, only limited research has been devoted to these topics. Ryback et al. (1971b) did observe a reduced range of ideas on a measure taken during bedrest compared with a 2-month followup retest. However, a test of associational fluency and a test to measure capacity to shift the function of an object and use it in a different way did not reveal performance changes. Rothstein and Kikoshima (1978) studied 45- to 55-yr-old males exposed to 10 days of bedrest. Four different tasks, including item recognition, visual search, category recognition, and analogy recognition were used to measure various aspects of the subjects' cognitive processing abilities. Although no significant differences between experimental and control group performance were discovered, Rothstein and Kikoshima comment that this finding may have been due to methodological and procedural difficulties rather than the lack of genuine performance changes.

Current data concerning how bedrest (and assumedly the space environment) affects performance have yielded either negative or inconsistent results. Given the significant biological alterations that occur during bedrest, it is difficult to imagine that performance would not be negatively affected in some way. There is considerable evidence that subjects experiencing bedrest do not find it a particularly pleasant condition. In fact, just the anticipation of starting the process is sufficient to increase depression, anxiety, and hostility. Such effects have also been noted throughout the period of recumbancy (Ryback et al., 1971a, 1971b).

There appears to be a number of reasons why many bedrest studies fail to demonstrate any substantial change in performance. These include the length of the simulation, variability of results, and the identification and development of appropriate testing instruments. In all cases of bedrest simulation, the period of recumbancy has been brief (at least in terms of the months and even years that future space missions may entail). We might speculate that a longer period of accumulated stress and biomedical alteration would be necessary before marked performance changes would be evident for all subjects during bedrest. However, researchers at Ames Research Center (personal communication, Charles Winget, August 1983) have noted that performance decrements are most noticeable during the first few days of recumbancy, at least for studies up to 21 days in length. It remains to be determined whether more pronounced deterioration would occur over very long-duration simulations.

High levels of inter- and intra-individual variability appear to be other factors that substantially reduce the chances of observing statistically significant group differences in bedrest studies. Certain individuals are more susceptible than others to performance decrements associated with bedrest. One factor contributing to both inter- and intra-individual variability could be the level of personal motivation. During an otherwise monotonous period of inactivity, performance tasks may be viewed as a welcome challenge and relief to the prevailing boredom. Increased effort derived from heightened motivation could overshadow otherwise expected performance decrements.

The development of appropriate test measures appears to be another stumbling block. Many tasks shown to be effective assessors of decremental performance under other adverse conditions cannot be effectively modified for use by subjects confined to a horizontal position. Further, because of the expense involved in bedrest studies, it is not always possible to have subjects available for lengthy pre-bedrest performance-task training. Performance tasks used during recumbancy usually must be fairly simple and quickly learned. This requirement excludes the use of sensitive but complex tasks that require extensive pretraining. If bedrest studies are to enhance our understanding of the behavioral as well as the physiological aspects of weightlessness, methodological and procedural issues must first be addressed.

The possible use of more frequent performance samplings, of additional performance tests to evaluate different levels of performance (i.e., psychomotor, cognitive, perceptual, etc.), and perhaps bedrest regimens of longer durations are avenues to be explored. Further, elucidation of the individual characteristics influencing performance changes might be obtained by concurrent psychological testing and EEG measurements. Concurrent testing could help evaluate the contributions of subject mood, motivation, sleep quality, fatigue, anxiety, and general well-being to performance in simulated weightlessness.

Resistance to Deconditioning

Although we cannot, at present, draw conclusions regarding how physiological deconditioning affects performance, there is certainly clear biomedical evidence that deconditioning can be dangerous to the safety and survival of the astronaut, if measures are not taken to limit or reverse the deconditioning process. The successful completion of two 6-month-duration missions within an 18-month period by a Soviet cosmonaut (Gazenko, Genin, and Yegorov, 1981) does allay many of the fears for future missions. However, the capacity of different groups of humans to resist pressures will continue to be an area of obvious concern.

Traditionally, our approach to reducing physiological deconditioning in space has been to select men in top physical condition and maintain this condition through exercise. Rigorous preflight and in-flight conditioning programs have been maintained under the assumption that the better the astronaut's physical condition, the greater is his overall resistance to the stresses of spaceflight. In-flight exercise does appear valuable in reducing muscle deconditioning. The comprehensive exercise program used during Skylab missions was effective in preventing loss of weight, maintaining leg strength and leg volume, and maintaining the integrity of muscle systems in general (Thornton and Rummel, 1977). However, in-flight exercise by no means offers complete protection. Cosmonauts Berezovoi and Lebedev returned from their recent 211-day flight aboard Salyut 7 in obviously debilitated condition (Toufexis, 1983). Although they had exercised daily, their muscles were so flabby that they were barely able to walk for a week, and for several weeks afterwards required intensive rehabilitation.

Although in-flight exercise has been shown to benefit external muscles, the benefits to other physiological systems must be questioned. The Skylab exercise program did not deter decalcification or related problems of the skeletal system. Also, its usefulness in preventing cardiac deconditioning and anomalies is suspect (Miller, Johnson, and Lamb, 1965). These data certainly suggest the need to reevaluate the role and value of in-flight exercise.

Many studies suggest that there may be little advantage to rigorous pre-flight exercise for well-conditioned individuals. For example, when athletes (male subjects, ages 20 to 34 yr, who regularly exercised in endurance sports) and nonathletes were exposed to $+G_z$ forces on a centrifuge, no significant differences on measures of acceleration tolerance were found between groups (Klein, Bruner, Jovy, Vogt, and Wegmann, 1969). Again, during a 20-min head-up 90° vertical tilt test, no significant differences between groups with respect to orthostatic tolerance was found (Klein, Backhausen, Bruner, Eichhorn, Jovy, Schotte, Vogt, and Wegmann, 1968). Other investigators (Luft, Myhre, Leoppky, and Venters, 1976) have extended these studies and have shown that athletes actually have reduced orthostatic tolerance compared with nonathletes. During lower-body-negative-pressure (LBNP) tests, aerobic work capacity was significantly reduced for both groups compared with baseline levels. However, although nonathletes demonstrated a slightly lower absolute decrement in capacity during LBNP testing, the athletic subjects showed a much larger relative loss in aerobic work capacity (almost 50%). Luft et al. also found that athletes show a greater tendency than nonathletes to accumulate fluid in the legs during LBNP.

Athletes also appear to be more susceptible to the effects of altitude than nonathletes. Rimpler (1970) found that the aerobic work capacity of athletes compared with nonathletes declined at a significantly faster rate during simulated altitude increments. Accordingly, athletic subjects seem to be more adversely affected by a reduction in environmental partial oxygen pressure than their nonathletic counterparts.

Athletic and nonathletic populations compared during water immersion show similar differential tolerance (Stegemann, Meier, Shipka, Hartlieb, Hemmer, and Tiebes, 1975). Athletes showed a significantly greater relative reduction in aerobic work capacity following 6-hr immersion than did nonathletes. Furthermore, all athlete subjects fainted during a 10-min vertical tilt test following immersion, whereas none of the nonathletes experienced this problem. One other aspect of Stegemann et al.'s experiment is intriguing. Intermittent swimming exercise during immersion was found to improve circulatory responses during subsequent tilt table tests.

However, this aid was more effective for nonathletes than for athletes.

Data obtained in-flight appear to support the findings so far discussed. For example, using the aerobic capacity of athletic and nonathletic populations for comparison, Klein, Wegmann, and Kuklinski (1977) classified the preflight physical-fitness status of the Skylab 4 Scientist Pilot and Pilot as "good" to "excellent." The Commander was classified as "fair" in fitness. During the mission, the two crewmen in better physical condition clearly demonstrated poorer responses to provocative gravitational stress than did the Commander. This effect was reflected in a more pronounced increase in heart rate and calf volume, and by a greater reduction of pulse pressure during LBNP (Johnson, Hoffler, Nicogossian, Bergman, and Jackson, 1974). Also during postflight LBNP stress, the Scientist Pilot and the Pilot showed a higher degree of orthostatic intolerance than the Commander, even though they exercised in flight more than the Commander.

From the data available so far, it appears that there is a need to reevaluate the role of physical fitness in the astronaut program. This is true particularly in light of the inclusion of mission and payload specialists whose physical capabilities may differ from those of the traditional astronaut pilot (NASA, 1977a, 1977b, 1977c). We need to better determine the degree of physical conditioning most suitable for astronaut crews (and their respective specialist members) as well as the type, amount, and scheduling of exercise to be used.

Age— Several studies have found that older, less physically active individuals adapt better to certain stresses of 0-g than their younger, more athletic counterparts. For example, Hull, Wolthuis, Gillingham, and Triebwasser (1978) found that +G_z acceleration tolerance was slightly greater among healthy male subjects between 40 to 55 yr old than among comparably tested younger subjects. Sandler, Goldwater, Rositano, Sawin, and Booher (1979) found comparable +G_z acceleration tolerance results for individuals (ages 46 to 55) following bedrest. Similarly, Convertino, Olsen, Goldwater, and Sandler (1979) found that reductions in functional working capacity following bedrest as measured by cardiorespiratory responses was less for older (age 50 ±1 yr) than younger (age 21 ±2 yr) individuals. However, there was a greater increase in the maximal heart rate observed after bedrest among the older men.

Older subjects also appear to withstand orthostatic stress better than their younger counterparts. Goldwater, Montgomery, Hoffler, Sandler, and Popp (1979) found that men aged 46-55, when compared with men aged 35-44, were more resistant to the effects of LBNP following bedrest. Older subjects demonstrated greater preservation of leg blood flow, less leg and pelvic blood pooling, and smaller decreases in end diastolic volume and stroke volume.

These studies, combined with those examining the adaptability of various athletic populations, suggest the intriguing possibility that normally active older subjects may have greater tolerance to weightlessness deconditioning, at least cardiovascular deconditioning, than the more highly conditioned astronauts we have selected in the past. If so, the mechanisms which confer protection with aging need to be understood. One possibility is that a body that is not "well tuned" does not discriminate as well as a conditioned body and therefore does not respond as markedly to environmental changes. An important question is whether older subjects may be more or less tolerant of 0-g stresses with respect to physiological systems other than the cardiovascular system. There is evidence from animal research suggesting that older individuals may be at a disadvantage in tolerating weightlessness as it affects the skeletal system. Novikov and Il'in (1981) have shown that among rats of various ages subjected to immobilization simulating null gravity effects, there is an age-dependent variation in the relative rates of bone formation and resorption. Older animals show the highest net rate of bone loss during immobilization. It is important that such effects be investigated in humans along with a more comprehensive evaluation of age-dependent biomedical changes in other systems as affected by space-flight conditions.

In discussing the effects of aging on the adaptability of individuals exposed to simulated weightlessness stressors, we might do well to consider also the other end of the age scale. As we anticipate an expanded age range of crews and especially of passengers in space, we will need to know what effects weightlessness will have on younger populations. Since ethical considerations severely restrict the inclusion of infants, children, and adolescents in experimental investigations, other methods of assessing the likely impact of weightlessness on these populations need to be identified.

Gender— Current space-related research with subject populations radically different from those historically considered appropriate for space travel leads to a discussion of the biomedical implications of women in space. During the early 1960s, the Lovelace Clinic in Albuquerque, New Mexico, conducted tests to determine the qualifications of women for spaceflight (Lovelace, Schwichtenberg, Luft, and Secrest, 1962). These tests showed women to be as suitable as men for space travel, especially so when one takes into consideration the fact that physical strength is much less important in space than on Earth. Further, women may be more suitable for space missions than men in some ways. Women generally weigh less and therefore consume less food and oxygen than men (Boyle, 1978). Also, it has been reported that women are more radiation-resistant than men (Rowes, 1982).

Recently, in light of the decision to include women crew members in the Space Shuttle, research has been reinitiated to further extend our understanding of female responses to the space environment and particularly to weightlessness. This renewed investigation has come in the form of bedrest studies involving comprehensive analyses of female responses to hypokinesis and related stressors.

These studies, as reviewed by Sandler and Winter (1979), have shown that women do demonstrate some significantly different responses to bedrest than men. Tests during LBNP have demonstrated greater change in cardiovascular dynamics among women as evidenced by a higher rate of fainting from lowered blood supply to the brain than observed among men. Significant sex differences exist also in postsimulation centrifugation tolerance. Mean tolerance losses among women exposed to bedrest is about 50% greater than among men, although the range of values is quite similar across sexes. However, there is also evidence that women may recover faster from centrifugation than men.

In most other areas investigated, only minor differences appear in the response of men and women to hypokinesis. Although the absolute physical work capacity of women is normally much lower than that of men, relative losses due to deconditioning from bedrest are about the same for both sexes. Also, no significant differences in biochemical changes between men and women exposed to hypokinesis have been demonstrated. Finally, the rhythmicity of various internally regulated biological functions appears to be only superficially different between the sexes. Some phase differences in the peak and trough values of hormonal and metabolic circadian rhythms do exist, but they are of minor consequence.

Although the results of studies investigating gender differences in response to simulated space conditions are still sparse and tentative, the overall conclusion is that men and women differ little in their ability to adapt to the biomedical stresses of space travel. Further, Sandler has indicated that women may be superior to men at withstanding the combined stresses of space (Rowes, 1982). According to a 5-yr study conducted at Ames Research Center, 27 women studied did a better job than their male counterparts in adapting to the physical and psychological challenges of bedrest and to the strenuous biomedical testing used to assess the effects of bedrest.

Given the lack of available in-flight data on women in space, many research questions remain regarding the biomedical response of females to space travel. For example, the question of the relationship between monthly cycles and performance is one topic that has been extensively researched in mundane environments (Redgrove, 1971). Sandler and Winter (1979) found no significant physiological changes during menses as a result of bedrest. Nonetheless, it is known that women experience shifts in blood volume and fluid balance as a result of their monthly period. How these shifts interact with shifts in blood volume and fluid balance produced by weightlessness is at present unknown. It is hoped that data from Soviet flights and data resulting from the inclusion of women in the American space program will provide further insight to the various aspects of female adaptation to space.

Countermeasures

Several alternatives have been explored as possible countermeasures to the physiological deconditioning that occurs in space. Exercise and selection procedures are options discussed in an earlier section of this chapter. Diet and nutritional supplements (Ushakov, Myasnikov, Shestkov, Agureev, Belakovsky, and Rumyantseva, 1978) have also been suggested. Two other alternatives, pharmacological manipulations and artificial gravity regimes, are discussed below.

Drugs— The possibility of using drugs to treat biomedical alterations produced during spaceflight has been discussed for some time (Parin, Vinogradov, and Razumeev, 1969). Soviet scientists have employed a wide range of pharmaceuticals to treat cardiovascular deconditioning in space (Shashkov and Yegorov, 1979) and to normalize water-sodium metabolism. Parin et al. (1969) and Parin, Kosmalinskiy, and Dushkov (1970) have discussed the use of pharmaceuticals in the treatment of various cardiovascular changes produced by weightlessness. They have also considered the possible pharmaceutical treatment of metabolic anomalies associated with weightlessness, radiation dangers inherent in space, and biomedical alterations associated with launch and landing accelerations. More recently, investigators have considered the use of certain drugs to compensate for bone-demineralization problems in weightlessness. For example, certain drugs show some promise as countermeasures for the effects of bedrest on the skeleton and may be effective for spaceflight (Nicogossian and Parker, 1982; Dietlein and Johnston, 1981).

Although the possibility of using drugs to treat spaceflight physiological alterations offers promise, there are some important research questions which have yet to be addressed adequately. One general question concerns whether or not a given drug will affect the body in space in a manner similar to its action on Earth. Due to changes in blood pressure, fluid volume, and metabolism produced by weightlessness, the time course of drug action and its overall effect may be altered significantly. Since major biomedical alterations occur in weightlessness, one cannot assume that the particular system to be treated will react to the drug in the same way as on Earth.

No major research has been conducted in space to examine these possibilities. However, we can gain some insight as to possible changes in space by examining how drug action is altered during exposure to other types of extreme environments. Aviation studies of drug action under varying altitude conditions demonstrate that this type of environment does produce significant changes in drug action effect. Drug toxicities are known to change, often increasing, for various drugs ingested under high-altitude conditions (Margolis, Bernheim, and Hurteau, 1951; Nedzel, 1955; Baumel, Robinson, and Blatt, 1967). High altitude may also potentiate drug action at nontoxic levels. For instance, Sparvieri (1960) reported that increased altitude potentiated the tranquilizing action of chlorpromazine, and altitude has been found to increase the morphine pain threshold (Truchaud, 1966). We also know that drug action can be altered in high altitudes due to metabolic changes. For example, altitude increases metabolism of some drugs, but decreases metabolism of other drugs (as reported by Medina, 1970).

It is not entirely clear which factors in high altitudes produce these changes. Probably, it is a combination of (1) reduced oxygen (although many studies have found comparable results with and without ground-equivalent oxygen), (2) pressure changes, and (3) nonspecific stress factors. Depending on the characteristics of a particular mission, spaceflight may or may not mimic these effects. However, the results from high-altitude studies do indicate that environmental changes which alter normal ground-based physiological functioning can significantly influence the course of drug action. This implies that drug action under spaceflight conditions could be different from Earth. Fortunately, a slight intensification or reduction in the effect of most drugs usually does not produce great problems. However, the importance of such an effect depends upon the situation and the medication. Small but adverse changes in biomedical responses to potentially toxic chemicals could prove problematic during critical mission phases. As a basic research question, it seems important that we consider the interaction of drugs and physiology under conditions of weightlessness.

A second question involving the use of drugs in space concerns their interaction with biological rhythms. Recently, chronobiology, or the study of biological rhythms, has emerged as a distinct field of research (Brown and Graeber, 1982). A related field, chronopharmocology, or the study of drugs administered at different times of the day, is also gaining interest. It is known, for instance, that an oral dose of antihistamine given at 7 p.m. will last 6 to 8 hr, whereas the same dose given at 7 a.m. will last 15 to 17 hr. On the other hand, digitalis (a heart medication) has twice the effect when given at night as when administered during the day. One implication of these findings is that drugs can be given in effectively lower doses if properly timed (NASA Activities, vol. 13, no. 10, 1982, p. 10). Extending these considerations to the space environment raises questions of the interaction effects of weightlessness, biological rhythms, and the administration of medication, and the possible behavioral implications of such interactions.

A third question concerns how to administer drugs in space. A new method has recently been developed which has the potential of offering greater control over absorption, and therefore over the effects of medication (Hackler, 1982). One device, called a Transderm-V System, consists of a patch worn just behind the ear. Whatever chemical it contains is absorbed directly into the bloodstream through the skin. Medication can be dispensed at a continuous and adjustable rate for up to 3 days. Such a system has distinct potential advantages. Since orally administered chemicals must pass through the gastrointestinal tract where much of the dosage is lost, large concentrations must be taken in order to ensure that a therapeutic amount reaches the bloodstream. Gradual absorption directly into the bloodstream could permit the use of much lower dosages, minimizing the risk of side effects. Another advantage is that administrations do not have to be scheduled into an already hectic workday.

The Transderm-V System was employed recently by astronauts aboard the Space Shuttle Columbia. Although the system appears to have advantages over oral administrations of prophylactic drugs, it is not clear that the transdermal system is any more effective than oral ingestion. Wurster, Burchard, and von Restorff (1981) found that the effect of an anti-motion-sickness medication administered transdermally could not be distinguished from that of a placebo; Homick, Degioanni, Reschke, Leach, and Kohl (1982) report similar findings. Both reports note great inter- and intra-individual variations in drug action, which hindered statistical analyses. Further tests are needed to detail the effectiveness of routine and/or long-term use of such a system for spaceflight.

Artificial gravity— One possibility that has been suggested as a countermeasure to the effects of weightlessness is the use of artificially produced gravity aboard spacecraft. Artificial gravity could be accomplished either through rotation of the entire vehicle or by the inclusion of an on-board centrifuge.

Rotation of the vehicle has a high apparent validity, and to the degree that it can be developed to approximate a normal linear gravitational environment, would produce the more comfortable living arrangement for long-duration spacecrews. Unfortunately, a rather large vehicle is necessary to produce axial rotation simulating Earth-type gravitational conditions. A short-radius craft gives a poor approximation to linear gravity and may yield a number of undesirable phenomena, including locomotor difficulty, spatial disorientation, and motion sickness. Because of the design constraints in producing 1-g artificial gravity through vehicle rotation, it is important to determine how little a g-load is sufficient to maintain adequate protection against deconditioning.

An important issue regarding the use of any rotational-vehicle artificial-gravity system is how adaptation to such a system would proceed. Earth-based rotational studies using vertical axis rotation have shown that adaptation can occur within about 24 hr for angular velocities at least as great as 6.0 rpm (Graybiel, Guedry, Johnson, and Kennedy, 1961; Newsom, Brady, and Goble, 1965; Newsom, Brady, Shafer, and French, 1966). However, the rotation in a weightless spacecraft is significantly different from that in an Earth-based room (Guedry, 1965). In a rotating spacecraft, pitch and roll head movements, angular limb movements, and horizontal translation will have dramatically different effects from those found on Earth, depending upon the crewmembers' orientation. Adaptation to rotation in space may therefore occur at a slower rate than in Earth-based rotation studies.

An alternate approach to the negation of 0-g effects through artificial gravity is the use of an on-board centrifuge. Much of the centrifuge work has been done in connection with the development of the Manned Orbiting Research Laboratory (Singer, 1968; Stapp, 1969). Researchers have shown that as few as four 7.5-min exposures to +1.7 G_z at the heart largely prevents orthostatic intolerance (W. White, 1965b; W. White, Nyberg, White, Grimes, and Finney, 1965; P. White, Nyberg, Finney, and White, 1966). However, adverse bedrest-induced changes in heart rate and blood pressure were not improved.

The most recent attempts to test the centrifuge system in space produced some promising results. In the Cosmos 936 flight of 1977, rats were exposed to centrifugation during 18.5 days of weightless spaceflight. Flight results showed that the lifespan of centrifuged animals was significantly greater than that of noncentrifuged control animals (Leon, Serova, and Landaw, 1978). Hemolysis increased three-fold in non-centrifuged flight animals, but was significantly less in those subjects exposed to centrifugation.

The bone system also seems to be positively affected by centrifugation in space (Holton, Turner, and Baylink, 1979). Bone mechanical properties including torque, stiffness, and energy were shown to significantly decrease among noncentrifuged flight animals, but were normal among flight subjects exposed to centrifugation. Also, although centrifugation did not correct loss of bone formation during flight, it did speed up recovery following return to Earth. Furthermore, centrifugation helped to prevent changes in calcium and phosphorus content in rat long bones (Gazenko, Il'in, Genin, Kotovskaya, Korol'kov, Tigranyan, and Portugalov, 1980) and helped to prevent osteoporosis (Stupakov, 1981).

Although these results are based on a small sample of subjects, they are encouraging and illustrate the need for further understanding of the effects of on-board centrifuge systems. It remains to be determined whether engineering innovations and the use of artificial gravity systems will prove more advantageous than alternative physiological countermeasures such as drugs, diet, exercise, etc. in countering the deconditioning effects of weightlessness.

VESTIBULAR ALTERATIONS

Manifestations and Theory

Another system adversely affected by the conditions of spaceflight is the vestibular complex. Two categories of vestibular side effects result from weightlessness. One category includes a variety of vestibular reflex phenomena such as postural and movement illusions, vertigo, and dizziness; the second category is space motion sickness. These two categories of response are believed to be closely tied; motion sickness often follows vertigo and postural illusions, and there is evidence to suggest that as vestibular reflex phenomena disappear with adaptation, the risk of motion sickness subsides (Reason and Brand, 1975).

Vertigo and spatial disorientation— The first reported instance of spatial disorientation in spaceflight was experienced in 1961 by the Soviet pilot aboard Vostok 2 (Graybiel, Miller, and Homick, 1974). For a brief period immediately after exposure to 0-g, the pilot felt that he was flying upside down. Also, he experienced vertigo during his fourth orbit. During the Voskhod I flight of 1964, the scientist-cosmonaut and the on-board physician both reported the inversion illusion (one spaceman imagined he was half crouched and facing downward, and the other felt he was hanging upside down). These illusions persisted regardless of whether the cosmonauts' eyes were open or shut. Yakovleva, Kornilova, Tarasov, and Alekseyev (1982) report that 21 of 24 cosmonauts who made flights in Soyuz-type spacecraft and served in the Soyuz/Salyut orbital complex reported postural illusory reactions. During the Apollo and Skylab programs, several American astronauts reported similar sensations. There were also reports of movement illusions: feelings of tumbling and sensations of spinning (Berry and Homick, 1973; Graybiel, Miller, and Homick, 1977).

Symptoms of vertigo and spatial distortion in space have so far posed no serious hazards to crews. They have been relatively short-lived and at worst an annoyance. As reported by the Skylab 2 pilot, although he was aware of illusory phenomena, their intensities made little impression (Kerwin, 1974). Nonetheless, it is important that we understand the nature and control of these phenomena because of their potential impact on crews under emergency conditions. If unplanned tumbling, rolling, or spinning of the spacecraft should occur (for instance, during docking), the consequent vestibular reflex responses could seriously hinder regaining control of the vessel.

Space sickness— The second category of vestibular side effects reported in space includes symptoms resembling Earth motion sickness. These symptoms range from stomach awareness and nausea to repeated vomiting. Symptoms also include pallor and sweating.

Space sickness[2] has been a recurring problem in the history of the space program. While this syndrome appears to decline within 3 to 5 days, in some cases the degree of illness has hindered work capacity and disrupted the scheduling of important mission activities. Nine of the 25 Apollo astronauts suffered some degree of sickness, while five of nine Skylab crewmen experienced symptoms (Berry, 1970; Graybiel et al., 1977). Soviet cosmonauts have reported similar experiences (Yuganov, Gorshkov, Kasian, Brianov, Kolosov, Kopanev, Lebedev, Papov, and Solodovnik, 1966; Volynkin and Vasil'yev, 1969). Four of the nine crewmembers of the Vostok and Voskhod missions suffered some degree of illness. There were also suggestions that illness suffered during the Soyuz 10 flight may have been a reason for the premature termination of this mission.

It is interesting to note, however, that despite the presence of intense space-sickness symptoms, task performance during flight has been only moderately affected. This observation meshes well with our understanding of the effects of motion sickness upon performance under Earth-based simulation conditions. In a series of studies conducted at Wesleyan University (Alexander, Cotzin, Hill, Ricciuti, and Wendt, 1945a, 1945b, 1945c, 1945d, 1955), subjects exposed to wave-like motion showed only slight deficits in performance on eight kinds of tasks, despite suffering nausea and vomiting. Only performance on a tracking device was significantly reduced among experimental (e.g., sick) versus control (e.g., nonsick) subjects. No statistically significant decrements were found in running through sand

---

[2] Traditionally, the terms "motion sickness," "space motion sickness," and "space sickness" have been used interchangeably to describe the illness experienced by space travelers which mimics the symptoms of motion sickness on Earth. More recently "space sickness" has emerged as the preferred term, although "space adaptation syndrome" also has been used (Toufexis, 1983).

and around various obstacles, a 660-yard dash, dart throwing, speed and accuracy of rifle shooting, code-substitution testing, or mirror drawing (tracing the outline of a star when pencil and paper are seen only as mirror images). Clark and Graybiel (1961) found similar results among subjects exposed to 2-day periods of constant slow rotation. Tasks measured included strength of grip, standing on one or two feet, walking a straight line, ball-tossing, dart-throwing, a steadiness test, card-sorting, opening combination locks, dial-setting, and arithmetic computations. Although the absolute levels of performance, on the average, matched performance under normal conditions, the pattern of performance differed. There were frequent occasions when severe sickness was associated with poor performance or failure to perform. There were many cases also in which high performance occurred despite symptoms of nausea, dizziness, general malaise, and vomiting. These results support the notion that motion sickness primarily affects motivation rather than performance potential per se. Birren (1949) incorporated this idea in an early study of seasickness and crew performance. He found that "peak efficiency" is likely to be unaffected by all except the most severe forms of motion sickness. High levels of motivation engendered by a crisis can completely overshadow the physical discomforts of motion sickness. However, "maintenance efficiency" may suffer severe decrement as a result of sickness because of loss of motivation.

Clearly, motivation plays a very large part in determining the level of performance in a motion-sick individual. The superior motivation of spacecrews has no doubt been a reason for the continued high level of performance achieved in flight despite the presence of space-sickness symptoms. As we anticipate future, more routine flights, crew motivation may not be maintained at this high level and decrements in performance could result.

Sensory conflict— As a means of summarizing and integrating information on vestibular effects, it is important to draw upon a theoretical base. While both spatial disorientation and space sickness pose potential threats to the conduct of a mission, space sickness has been the more obvious and disturbing factor in spaceflight to date. As a result, theoretical considerations, as well as methods of averting or controlling problems associated with vestibular changes, have tended to focus on the issue of space sickness. However, spatial disorientation effects and space sickness are believed to be interrelated phenomena. Understanding the mechanisms underlying space sickness is an important step in understanding the general nature and effects of vestibular alterations.

A number of biomedically oriented theories have been proposed for use in understanding space sickness (see Reason and Brand, 1975), but let us here focus on a fairly well acknowledged psychophysiological model — the sensory conflict theory. Stated briefly, this theory postulates that motion sickness occurs when patterns of sensory input to the brain from the vestibular system, other proprioreceptors, and/or the visual system are markedly rearranged, at variance with each other, or differ substantially from expectations of stimulus relationships in a given environment. In space, sensory conflict can occur in several ways. First, there can be conflicting information transmitted by the otoliths and the semicircular canals of the ear. It is postulated that the otoliths are functionally deafferented due to the effects of weightlessness, whereas the semicircular canals are not radically altered by 0-g and continue to send impulses regarding angular accelerations. When the astronaut moves about in space, the semicircular canals signal that movement to the brain, while the otoliths remain inactive, resulting in conflicting sensory input to the brain. Sensory conflict may also exist between the visual and vestibular systems during motion in space. The eyes transmit information to the brain indicating body movement, but no corroborating impulses are received from the otoliths.

A third type of conflict may exist in space because of differences in perceptual habits and expectations. On Earth, we develop a neural store of information regarding the appearance of the environment and certain expectations about functional relationships (e.g., the concepts of "up" and "down"). In space, these perceptual expectations are at variance with the new functional relationships associated with weightlessness. This new environmental orientation is succinctly expressed in the report of Astronaut Gibson (1974) of the Skylab 4 crew:

> Being upside down in the wardroom made it look like a different room than what we were used to. When I started to rotate back and go approximately 45° or so off the attitude which we normally call "up," the attitude in which we had been trained, there was a very sharp transition in my mind from a room that was sort of familiar to one which was intimately familiar. It all of a sudden was a room in which we felt very much at home and comfortable with. It wasn't a gradual thing, it was a sharp transition.

It is important to note that no single source of sensory conflict appears to account entire for the symptoms of space sickness.

Rather, it is the combination of these conflicts which somehow produces sickness, although the exact physiological mechanisms remain unknown.

The sensory conflict hypothesis has proved very useful as a structure in which to organize the available empirical data. It clarifies why a discrepancy in symptomatology between astronauts and cosmonauts existed early in the space program. Initially, Soviet crewmen reported considerable difficulty related to space sickness, although no reports of these symptoms occurred for American crewmen prior to the Apollo missions. The American Mercury and Gemini capsules were considerably smaller than the Soviet Vostok and Voskhod vehicles, thus restricting movement to a greater degree. With less movement capability, there was less potential for sensory conflict between visual and/or canal impulses and otolithic input. In the Apollo vehicles (which were comparable in size to the earlier Soviet vehicles) considerably more opportunity existed for movement-generated sensory conflicts, and space sickness did occur.

The experiences of Skylab missions are of interest here. While aloft, Skylab crewmen moved between the Command Module and the larger Workshop. In this situation, two separate adaptations may have been required. Of the nine Skylab crewmen, two astronauts first experienced space sickness inside the Command Module. After adapting to this situation, one of these individuals again became sick upon entering the Workshop. Two other crewmen, who had evidenced no symptoms inside the Command Module, developed their first symptoms upon entering the Workshop. These findings highlight the importance of a large area in providing more opportunity for movement and for sensory conflict and therefore a higher incidence of sickness. These results can also be interpreted to support the research finding that adaptation to one situation does not necessarily provide protection in another. Adapting to the sickness-provoking conditions of the Command Module was not a satisfactory prophylactic for preventing sickness inside the larger Workshop.

Given the dramatic adaptation which must occur in space, we may ask whether there are specific training and design considerations which would help reduce the feelings of unfamiliarity with the environment, and also perhaps reduce the problems of motion sickness and disorientation. Although there are design considerations favoring a 360° orientation, the sensory conflict theory suggests there might be advantages associated with an environment more Earth-like in orientation. Even a simple feature, such as strips of material attached to the walls to create the impression of "up" and "down" was found helpful by Soviet cosmonauts (Leonov and Lebedev, 1975). However, it is possible that improving orientation within the vehicle could exacerbate the problem if the crew member then had to reorient to a conflicting scene outside the spacecraft. In addition to design aids, we might consider providing crews with preflight experiences in their craft simulators at orientations other than horizontal and Earth-based "up" and "down." This could be accomplished through water-immersed simulators, where the environment could be rotated away from the horizontal (Abbott and Duddy, 1965; Adams and Bulk, 1967).

It seems logical that our most appropriate strategy for understanding motion sickness in space would be to analyze similar conditions on Earth. Following sensory conflict theory, several models can be suggested. Visual-vestibular conflict in space occurs when motion is seen in the absence of vestibular corroboration. On Earth, a similar conflict can be approximated through the use of fixed-based simulators equipped with appropriately moving visual displays (Barrett and Thornton, 1968; Reason and Diaz, 1971). This type of equipment has been shown to produce motion sickness in a significant number of subjects, with the incidence of illness increasing with the experience of operators in the real-world environment. For example, Havron and Butler (1957) found that the greater the subjects' previous experience flying helicopters, the more susceptible they were to sickness when exposed to visually simulated helicopter flight. These results suggest that a profitable ground-based training procedure for astronauts would be to examine exposure and adaptation to visually moving environments simulating those to be encountered in space, with the astronaut in training remaining in a motionless position.

A second type of sensory conflict experienced in space, canal-otolith, also can be simulated on Earth. By arranging conditions such that the canals suggest motion in the absence of otolith input, we can replicate the essential conditions hypothesized to exist in 0-g. As one example, caloric stimulation of the outer ear fits these requirements. When the outer ear is irrigated with water which is either hotter or colder than blood temperature, convection currents are set up in the semicircular canals causing activation. When the subject is stationary there is no otolith signal corresponding to the canal signals and a situation results which in some ways is analogous to that accompanying head movement on 0-g. In the study of motion sickness, considerably less research has been devoted to these types of provocative stimuli than to those associated with rotation, but caloric stimulation studies are planned for future Space Shuttle flights. Based on the explanatory power of sensory conflict theory, it appears that more research is now warranted.

The sensory conflict theory suggests that just as a period of adaptation is required to adjust upon entry into 0-g, a process of readaptation on Earth may be required upon return to 1-g. This is a factor of concern for space crews returning to Earth. As examples, Apollo 16 crewmen indicated some decrement in postural equilibrium 3 days after their return to Earth (Homick and Miller, 1975). The greatest effects were observed when they were deprived of a visual reference. Following Skylab, postural instability measured by requiring astronauts to walk beams of various widths, was substantial even up to 10 days following return to Earth (Homick, Reschke, and Miller, 1977). All crewmen reported that rapid head movement produced a sensation of vertigo. One crewmember was unable to accurately sense small displacements of his head and body, and another crewman fell down in his house when the lights were extinguished unexpectedly. Similar readaptation problems have been reported by the Russians (Gazenko, 1979).

As perceptual habits are altered to adjust to the new orientations of space, these new expectations will be in conflict with Earth-based orientations upon return. An important issue for the future is to determine whether, as some researchers suggest (Kornilova, Syrykh, Tarasov, and Yakovleva, 1979), these return-to-Earth symptoms increase in severity as the length of stay in space increases. Changes in the gravity level, whether from 0-g to 1-g or the reverse, could also produce altered perceptual relationships, sensory conflict, and therefore motion sickness. It should be remembered that one need not travel from Earth to space or from space to Earth to trigger perceptual alterations. If various parts of the spacecraft are subject to varying degrees of gravity, transitions across those parts may require separate adaptations at each sector.

Resistance to Vestibular Effects

One approach to the vestibular problems in space might be to develop instruments effective in distinguishing susceptible individuals from nonsusceptible individuals. By selecting only those individuals with high tolerance to vestibular changes on Earth, the chance of problems in space might decrease.

Although Earth-based research assumes that susceptibility to motion sickness in 1-g and space sickness in 0-g are related, there are significant differences. Parabolic flight research has shown that an individual may have normal vestibular responses on the ground and show markedly greater or lesser susceptibility to vestibular stimulation in weightlessness (Miller, Graybiel, Kellogg, and O'Donnel, 1969). Similarly, a number of astronauts who had considerable jet pilot flying experience and had no history of motion sickness were the most seriously affected in space, whereas some astronauts with definite motion sickness histories challenged weightlessness with no difficulty (Homick and Miller, 1975). Nevertheless, at present it seems reasonable to assume that the conditions of space produce sickness symptoms via the same general mechanisms that influence motion sickness of Earth, albeit via different stimulus conditions. If so, understanding motion sickness and its correlates could enhance our understanding of space sickness. Several characteristics have been found to be predictive of motion sickness, as discussed in the following sections.

Age— There is evidence that motion sickness susceptibility fluctuates with age (Tyler and Bard, 1949; Chinn and Smith, 1953; Money, 1970). Susceptibility appears to be at its highest level between the ages of 2 and 12. There is a highly significant decline in vulnerability between the ages of 12 and 21, with the decline continuing throughout the life cycle, and it is relatively unusual to find motion sickness among individuals over the age of 50 (Reason, 1968). The suggested decrease in susceptibility to motion sickness associated with increasing age could be attributable to a simple avoidance of motion-intense situations. However, researchers have demonstrated that differences in susceptibility to motion sickness vary with age even when subjects cannot avoid provocative stimuli. Chinn (1956) found that among 5,000 individuals on transatlantic troopships, persons between 17 and 19 years of age showed a high incidence of sickness, whereas persons between the ages of 30 and 39 yr became ill at a low rate. A similar age-related decline in motion-sickness susceptibility has been observed among civil aircraft passengers (Lederer and Kidera, 1954).

A possible explanation of these findings is that the sensitivity of the vestibular system is reduced with age. This suggestion gains support from the work of Preber (1958), Kennedy and Graybiel (1962), and Lidvall (1962), who found vestibular sensitivity to be greater in motion-sickness-prone individuals compared with motion-sickness-resistant individuals. However, overall, it does not appear that vestibular sensitivity can reliably predict susceptibility to motion sickness (Reason and Brand, 1975; Lentz, 1976), and it seems unlikely that a decline in susceptibility correlated with age is in any way related to a decline in the reactivity of the vestibular system. An alternate explanation holds that the increased resistance to motion sickness associated with advancing years is somehow linked with experience with the environment, and perhaps a diminution in the cortical or cognitive factors involved. This hypothesis, if supported, could help explain the difference found between Earth-based susceptibility and space susceptibility. Experience with the environment may render individuals almost immune to symptomatology on Earth, yet in the new environment of space, they must learn to adapt again. Little research has been brought to bear on the environmental hypothesis and thus no conclusions can be drawn.

As we anticipate crews, and particularly passengers, with wider age ranges than previously employed, the possibility that some age groups may be more susceptible to space sickness than others should be considered. In designing crew work schedules and assessing who will conduct vital tasks during the first weeks of the mission, information regarding age-related susceptibility may prove important. Further research is warranted defining the relationship between age and motion sickness and the mechanisms involved.

Gender— Research to date suggests that women are more susceptible to motion sickness than men. Reason (1968) found that women students reported a significantly greater incidence of motion sickness than men of comparable age and travel experience, both before and after the age of 12. These results have been confirmed by Lentz and Collins (1977). An unresolved question is whether this gender-linked susceptibility. is based on physiological differences, psychological differences, or are reflective of a socialization process in which it is more acceptable for women than men to report illness. Bakwin (1971) and Abe, Amatomi, and Kajiyama (1970) have suggested that susceptibility to motion sickness may be genetically determined. They report that at age 3 significantly more girls suffer from motion sickness than boys. Also, in partial support of the genetic argument, Lentz and Collins (1977) found that, compared with nonsusceptibles, susceptible individuals more frequently indicate that their parents or siblings are also susceptible. However, such reports could reflect a learning process as well as a genetic inclination.

Given the growing involvement of women in the space program, the extent and degree that space sickness will affect female personnel needs to be determined. Also, understanding the factors which underlie gender-related differences in motion sickness might help explain the general process of motion sickness.

Personality— A review and study by Collins and Lentz (1977) suggests that certain personality factors distinguish individuals who are susceptible to motion sickness from those who are resistant to motion sickness. Using a self-report test battery (composed predominantly of paper and pencil tests), Collins and Lentz confirmed several significant differences between susceptible and nonsusceptible individuals. Motion-sickness-susceptible individuals showed a high degree of anxiety and scored higher on tests assessing general medical and psychiatric problems and on tests measuring potentially serious neuropsychiatric and psychosomatic disturbances than did nonsusceptibles. These authors also confirmed previous work with the Eysenck Personality Inventory which showed that both the introversion and the neuroticism scales are correlated with motion-sickness history (Guedry and Ambler, 1972; Reason and Graybiel, 1972; Wilding and Meddis, 1972; and Kottenhoff and Lindahl, 1960). Collins and Lentz also found that nonsusceptible individuals had higher scores on the masculinity scale of the Guilford-Zimmerman Temperament Survey than did susceptibles, regardless of their gender.

The results of the 16 Personality Factors Test employed by Collins and Lentz provide a summary of those traits descriptive of susceptibles and nonsusceptibles. Nonsusceptibles can be characterized as emotionally stable, venturesome, self-assured, relaxed, adjusted, thinking-oriented, not neurotically disposed, and good leaders. Susceptibles are defined as emotionally oriented, tender-minded, and subjective, and as possessing traits generally the opposite of those defining nonsusceptibles.

Our understanding of the behavioral correlates of motion-sickness susceptibility has been further expanded by research conducted at Ames Research Center by Cowings and her associates (Cowings, 1977a, 1977b; Cowings, Billingham, and Toscano, 1976, 1977; Cowings and Toscano, 1977, 1982; Stewart, Clark, Cowings, and Toscano, 1978; Toscano and Cowings, 1978). These authors confirmed that susceptible individuals demonstrate higher levels of anxiety than nonsusceptibles. However, the most intriguing findings reported in the Ames studies come from autonomic-functioning profiles. Here, subjects were divided into two groups: "sympathetic dominant" and "parasympathetic dominant," based on autonomic-balance measures developed by Wenger (1941). Individuals with sympathetic-dominant autonomic profiles showed less anxiety, were less able to perceive their own autonomic functioning, and, under conditions of unusual vestibular functioning, showed fewer symptoms of motion sickness than did parasympathetic-dominant individuals. These same sympathetic-dominant individuals were shown to be much more susceptible to hypnosis than were their parasympathetic-dominant counterparts. Parker and Wilsoncroft (1978) also examined the relationship between autonomic dominance and susceptibility to motion sickness. Like Cowings, these authors found sympathetic-dominant subjects to be more resistant to motion sickness than parasympathetic-dominant subjects. However, these authors conclude that resistant subjects are more (rather than less) anxious than susceptible subjects, and that anxiety acts to protect against symptomatology.

In comparing the results of these studies, several questions emerge. Although the personality patterns of those found to be motion-sickness-resistant and those found to be motion-sickness-susceptible are generally consistent between the Cowings and the Collins and Lentz studies, it is not obvious that the "extraverted, venturesome, emotionally stable, good leader" (Collins and Lentz) would be highly susceptible to hypnosis (Cowings). The more significant question, however, involves the tie-in between autonomic dominance and motion-sickness susceptibility. The Collins and Lentz description of personality traits associated with resistance to motion sickness suggests an individual usually thought of as parasympathetic dominant. Yet Cowings has identified individuals with traits overlapping those described by Collins and Lentz to be sympathetic dominant. The relationship between autonomic dominance, personality traits, and motion-sickness susceptibility needs to be more fully understood if these variables are to contribute to selection strategies.

One might suspect that motion-sickness-susceptible (compared with -resistant) individuals would be more aware not only of those functions associated with their motion sickness, but also of their internal functions generally. Cowings (1977b) used an autonomic perception questionnaire (Mandler, Mandler, and Uviller, 1958) to assess the ability of subjects to perceive the processes of their own viscera. It appears that susceptibles are much more aware of their own bodily functions than are nonsusceptibles; this is perhaps one reason they respond to the disruptive effects of unusual vestibular stimulation upon their bodily functions.

A related finding has been made by Barrett and Thornton (1968), who examined the relationship between motion-sickness susceptibility and field dependent/independent perceptions.[3] During visual motion simulator training, field-independent subjects experienced significantly greater discomfort and were more prone to motion sickness than field-dependent counterparts. Although it is logical to assume that those who rely on their internal sensations are aware of them, it does not follow that those who operate primarily on external cues are necessarily unaware of internal cues. Studies to clarify these findings would further elucidate the area of motion-sickness susceptibility. Field-independence has also been associated with introverted personality styles, while extroversion is typical of field-dependent subjects (Witkin, Lewis, Hertzman, Machover, Meissner, Brettnall, and Wopner, 1954). These relationships suggest that introverts may be more susceptible to motion sickness than are extraverts. This possibility deserves further attention.

Countermeasures

Drugs— One approach to the treatment of space sickness is the use of medications to help reduce symptomatology (Wood, Kennedy, and Graybiel, 1965). Unfortunately, anti-motion-sickness pharmaceuticals have not been completely satisfactory. Although these pharmaceuticals have reduced the severity of symptoms for some astronauts at some times, they have been ineffective on other astronauts. The pilot on Skylab 3 took an anti-motion-sickness preparation, but vomited anyway. The Skylab 4 commander and pilot continued to experience space sickness periodically during their first 3 mission days, despite the administration of pharmaceuticals on a regular basis (Graybiel et al., 1977).

There are other limitations to the use of drugs as a prophylaxis against motion sickness. To produce the maximum benefit, the chemicals have had to be taken before the symptoms occur. Once illness is established, the drugs are not altogether helpful in alleviating symptoms. Furthermore, medications may have side effects or

---

[3] It has been found that, when placed in a situation with conflicting cues, some individuals rely on the external stimulus field (field dependence), whereas other individuals base their judgments primarily upon internal kinesthetic sensations (field independence) (Witken, 1949a, 1949b).

aftereffects that can potentially influence performance (Wood, Manno, Manno, Redetzki, Wood, and Vekovius, 1984), hence astronauts have been reluctant to ingest the medications regularly. Unfortunately, changes in physiological functions resulting from medications can interact with and overshadow biomedical alterations produced by weightlessness, and interfere with the evaluation of data (Vernikos-Danellis, Winget, Leach, Rosenblatt, Lyman, and Beljan, 1977). As with other drugs, we do not understand adequately what effects usage may have in space.

Head movement schedules— The use of self-induced head movement on planned schedules has been suggested as a way to speed up the process of adaptation to unusual vestibular stimulation (Graybiel and Wood, 1969; Homick and Miller, 1975). This procedure is based on a principle analogous to vaccination. By exposing the subject to a mild level of provocative (i.e., response-producing) stimulation insufficient to result in sickness, it is hoped that the process of adaptation to greater levels of stimulation will be accelerated. Earth-based studies of head-movement adaptation schedules have shown encouraging support for this vaccination approach (see Graybiel, 1975, for review). For example, Graybiel and Wood (1969) showed that subjects required to execute experimenter-paced head movements during rotation adapted significantly faster to the provocative condition than when head movements were not required. Unfortunately, in-flight data have not been as encouraging. The astronauts of Apollo 10 were instructed to carry out a series of slight head movements paced at about 2-sec intervals during the early stages of the mission, but because of busy schedules, only the lunar module pilot actually carried out the exercises. On the first day, movement-induced stomach awareness and nausea developed within 1 min and movement had to be discontinued. The same result occurred the second day and again the head movements had to be stopped.

Controlled head movements have also been used in combination with medications. During Skylab 3, all three crewmen suffered space sickness before treatment. On the second mission day, they were instructed to take an anti-motion-sickness preparation and to execute, after a 1-hr delay, a series of head movements. Although remaining ill, they did report feeling marginally better by evening.

It would seem from the limited in-flight data available, that adaptation schedules are either not sufficiently understood or are not satisfactory in counteracting the problem of space sickness. Even if perfected, there are other problems inherent in the use of this technique. It requires time away from duties during flight. Also, there is evidence that protection conferred under one set of stimulus conditions does not transfer to another. With some exceptions (see discussion by Reason and Brand, 1975), research indicates that adaptation is normally highly specific to the particular stimulus conditions under which adaptation is acquired. As one example, Guedry (1965) found that in a room slowly rotating in a counter-clockwise direction, adaptation to provocative head movements did not transfer when the subjects were rotated in a clockwise direction.

From the data presented in this section, it seems that neither drug usage nor adaptive head-movement schedules are particularly attractive countermeasures at this time. However, there are other directions of research which may prove practical and effective.

Head restraint— An approach opposite to that of head-movement schedules has also been tried in space. Crewmembers onboard the Salyut 6 mission employed a special head-restraint helmet (Neck Pneumatic Shock-Absorber) designed to reduce space-sickness symptoms by limiting head movement (Matsnev, Iakovleva, Tarasov, Alekseev, Kornilova, Mateev, and Gorgiladze, 1983). A soft cap, secured by unstretchable straps attached to the shoulders, minimizes head tilt and turn. Rubber cords, also attached to the cap, further restrict movement unless the crew member exerts considerable force with his neck muscles. Flight-test results from the Salyut mission indicate that the device was beneficial in controlling the development of space sickness.

The use of such a head-restraint device is a logical extension of the fact that space-sickness symptoms are best ameliorated by lying flat (thus placing the head in a position similar to the erect posture maintained by the head-restraint device) and avoiding all head movements (see Reason and Brand, 1975).

Although this device has proven useful in the limited tests conducted in space, many research questions need to be addressed. Does restriction of head movement pose any problems for performance of tasks requiring rapid or agile shifts in the position of the head? Is hand-eye coordination altered by the fact that the hands can be easily monitored only when the head is in the erect position? Does the device permit adaptation to the stimuli associated with the development of space sickness such that eventually the restraint can be removed without fear of illness? If adaptation does occur, is the period of adjustment prolonged because of the lack of head movement? This latter question points up the paradox of treating space sickness. In theory, more active head movements produce faster adaptation to the conditions of space (Graybiel, 1975). However, increased movement also increases the risk of nausea which could debilitate crewmembers and jeopardize performance. Alternatively, restricting head movements (by lying still) lessens the probability of symptom development, but also prolongs the period of adaptation. It remains to be determined if in fact the Neck Pneumatic Shock-Absorber produces a similar extension of the adaptation period.

Cuban boots— In addition to the head-restraint cap, Soviet cosmonauts have also tested specially designed shoes or "Cuban boots," named after the Cuban cosmonaut Arnaldo Mendez who tested the prototype (Engler and Cheshire-Engler, 1983). These shoes provide a pressure of up to 60 torr to the bottom of the feet, making the cosmonaut feel as if he is standing on solid ground. A Soyuz 38 cosmonaut reported that use of the Cuban boots reduced the severity of spatial illusions and motor disturbances, phenomena thought to be produced by conditions which also produce motion sickness. Although the Cuban boot is not likely to eliminate vestibular problems in space, it is certainly an intriguing compensatory mechanism to simulate 1-g conditions and deserves further attention.

Autonomic response control— Various investigators have begun to explore the use of biofeedback and acquired autonomic response control in the prevention of motion sickness. Cowings et al. (1977) have demonstrated that subjects trained to control heart rate, respiration rate, and blood pressure using biofeedback and autogenic training procedures can volitionally diminish and/or prevent their own motion-sickness symptoms. Among 50 subjects trained in the past few years and tested under conditions of provocative motion-sickness stimuli (rotating chair producing coriolis acceleration), 85% have improved their ability to withstand the test conditions (Rasmussen, 1980). Likewise, Levy, Jones and Carlson (1981) showed that biofeedback treatment of 20 air crewmen, disabled by chronic and severe sickness in flight, resulted in an 84% rate of return to flight duty. In this study, subjects were trained to voluntarily control such functions as galvanic skin response, skin temperature, and muscle tonus.

It would appear that the use of autonomic-response control techniques has several advantages over procedures previously used in the treatment of motion sickness. Toscano and Cowings (1978) found that learned control of symptom suppression could be transferred from one situation to another. Subjects learning control of autonomic functions to suppress illness when exposed to acceleration in one direction showed significant suppression of symptoms when rotated in the opposite direction. Nor has training been found to be differentially affected by initial motion-sickness susceptibility (Toscano and Cowings, 1978, 1982), or by gender differences (Rasmussen, 1980). Also, it appears that autonomic-response control training can be used effectively by subjects for whom traditional countermeasures have proved inadequate (Levy et al., 1981).

Despite the promising results reported by investigators employing biofeedback in the treatment of motion sickness, several major issues remain unresolved. It is not clear at this time why and how biofeedback works. In fact, there is evidence to suggest that biofeedback should not be effective. Graybiel and Lackner (1980) measured changes in blood pressure, heart rate, and body temperature in conjunction with the onset of motion sickness. Among 12 subjects studied, no significant variations in these physiological parameters were observed during progressive stages of motion sickness. Other studies have failed to indicate any consistent relationship between the appearance and severity of motion-sickness symptomatology and changes in blood pressure, heart rate, or body temperature (Money, 1970; Reason and Brand, 1975). Logically, if there is no consistent variation in these physiological parameters as a function of the onset of motion sickness, it is difficult to understand how learned control of these functions could be useful in the reduction or prevention of motion-sickness or space-sickness symptoms. There is insufficient data at present to reconcile the positive effects found by Cowings and her associates and by Levy, Jones, and Carlson with the results of Graybiel and others. However, certain prospects suggest themselves. The observed decline in motion-sickness susceptibility could be related, not to the particular autonomic response controlled, but rather to the cognitive aspect which accompanies the process of control. We know from previous research that concentration upon some difficult or engaging mental task tends to alleviate the symptoms of motion sickness. Guedry (1964) found that only three out of 10 subjects given mechanical-comprehension and spatial-relations problems while being rotated suffered motion sickness, whereas twice the number of control subjects not given cognitive tasks experienced severe symptoms: two individuals in this group failed to complete the experiment and four vomited repeatedly.

Although mental activity may lessen the symptoms of motion sickness, it seems unlikely that the effects that have been reported for biofeedback could be explained fully by simple attention to a task. Indeed, Toscano and Cowings (1982) have addressed this issue.

They demonstrated that subjects given training on a cognitive task do not withstand the stress of coriolis acceleration well at all when compared with biofeedback-trained subjects. There is a compelling need to understand why biofeedback does appear to work and whether it is, in fact, related to the learned control of specific autonomic responses.

Another issue regarding the use of autonomic-response control procedures to treat motion sickness concerns the approximately 15% failure rate found by both Cowings and Levy. It is impossible to say at this time why this procedure works for some and not for others. It could prove valuable to devote investigations specifically to these failures. Perhaps an understanding of why autonomic-response control procedures fail to confer protection from motion sickness in some cases would help us understand why it does seem effective in others.

An additional critical research question concerns the degree of effectiveness of biofeedback training in the specific treatment of space sickness. Although there are similarities between conditions on Earth that produce motion sickness and those prevalent in space that produce space sickness, the situations are not identical. It remains to be determined whether biofeedback training, effective in treating motion sickness on Earth, can also counteract sickness in space.

Another area of interest concerns the possible occurrence of symptoms after autonomic control is relaxed. Is response control exerted in the initial phase of flight sufficient to confer protection throughout the mission? Can the suppression of symptoms result in the delayed occurrence of illness? What is the relationship between intensity of illness and the duration of discomfort? These kinds of questions go beyond the issue of biofeedback and relate to the various methods of countering motion and space sickness.

VISUAL CHANGES

Because of the importance of vision to the conduct of space missions, it was suggested in the early literature that the 0-g environment could alter visual capabilities (Cooper, 1963; O'Lone, 1965; Berry, 1970). During the initial phases of the space program, astronauts demonstrated seemingly high capability to distinguish terrestrial reference points such as rivers, groups of people, and vehicles.

Such observations fueled the suspicion that 0-g might actually enhance perception (White, 1965a).

American studies of visual functioning in space have failed to demonstrate any significant changes in visual perception as a result of 0-g. During the Gemini program, Duntley, Austin, Harris, and Taylor (1968) tested astronauts' acuity during preflight, in-flight, and postflight, using devices calibrated to the sensitivity of a clinical wall chart test. No differences were observed in any phase. In addition, visual acuity of astronauts under ordinary viewing conditions from their capsule window was tested. Rectangular patterns were constructed in large fields in Laredo, Texas, and in Carnarvon, Australia, and the astronauts were asked to identify patterns arranged in a 4 × 3 matrix. Again, visual acuity in space paralleled that of 1-g, and no changes were observed across days of testing.

Observations taken aboard early Soviet spacecraft suggested that 0-g, rather than enhancing perception, might result in visual losses, at least during the early days of a mission. Soviet investigators assessed the ability of cosmonauts to judge the direction and number of certain focusing patterns of dashed lines under standard conditions. Decrements for all crewmembers were reported, with losses averaging 20%. Reductions in color perception during spaceflight have also been reported. Measurements made on Vostok 2 and Soyuz 9 indicated a 25% diminution of color intensity, with losses particularly marked with purple, light-blue, and green (Popov and Boyko, 1967). Contrast sensitivity decrements of as much as 40% have also been reported. In spite of these changes, Soviet investigators have expressed the view that vision in weightlessness is reliable and presents no obstacle to the conduct of a space mission (Nicogossian and Parker, 1982). The recent Soviet-French flight aboard the orbiting station Salyut 7 explored the possibility that the role of vision in human spatial orientation and muscular control is altered in 0-g (Pishchik, 1982). Data from these experiments are still under analysis and no conclusions can be drawn at present.

There are factors other than weightlessness which could influence visual functioning in space. High-contrast effects and the requirement to adapt to rapidly changing brightness levels (as in rapid Earth orbits) render space a special visual environment. Another factor is the presence of high energy, high atomic number particle radiation. Variable form, low level light flashes have been reported by many astronauts following the initial sighting by one of the astronauts aboard Apollo 11. These lights were perceived as "spot or starlike flashes" in 66% of the cases reported, as a "streak, sharp or dashed line" in 25% of the sightings, and as "clouds" in 8% of the reports (Pinsky, Osborne, Bailey, Benson, and Thompson, 1974). Such sightings occurred during periods of darkness in the spacecraft. They reportedly occurred in one eye at a time and even with both eyes closed. It is believed that these perceptions are generated by particle radiation traversing the head or eyes and triggering a retinal response. These findings highlight the question of duration effects in space. Although such light flashes are of relatively little consequence during short-term missions, we do not know what their significance to the visual system might be during longer exposures.

Appreciable vibration can be another deterrent to the optimal functioning of the visual system in space. At least in some cases, vibration can be sufficiently intense to interfere with visual tasks, such as reading instruments (Grether, 1971a). Even though vibration effects may last only a few minutes, they can pose problems for crew performance of critical tasks, such as those associated with the lift-off. On the return flight, vibration during reentry and recovery apparently has not been a problem. However, short-lived oscillation combined with deceleration pulse may be a problem of renewed interest with respect to winged aerospace vehicles of the Shuttle type. Here the aircraft is flown to ground by a human pilot following a ballistic reentry, and the effects of initial vibration could influence the successful performance of landing tasks.

Failure to find consistent perceptual changes in the weightless environment suggests that visual changes in space, if they exist, are not pronounced. However, further data, especially those emphasizing brightness and color sensitivity, dark adaptation, and accommodation, along with assessments of radiation and vibration effects, would help complete the picture of visual perception in the space environment.

SUMMARY AND CONCLUSIONS

In this chapter we have discussed some of the consequences and selection implications of the biomedical alterations that occur in space. Although the physiological changes associated with weightlessness-induced deconditioning have been extensively researched, considerably less attention has been given to the perceptual and behavioral correlates of these changes. As an example, bed-rest studies have been the predominant model for the study of space-related biomedical changes. However, very few of these experiments have included performance measures, and those that have included performance measures have generally experienced procedural difficulties or resulted in inconsistent findings. There is both an opportunity and a need to extend hypokinesia studies to include psychomotor and cognitive performance measures.

Aside from the issue of behavioral changes associated with biomedical alterations in space, there are psychophysiological research questions surrounding various deconditioning countermeasures. In the past we have relied upon the selection of athletic men for the astronaut program and have employed rigorous physical training programs to improve or at least maintain their high levels of conditioning. Research now indicates that extensive pre-mission physical conditioning may be inappropriate or even detrimental to space adjustment. Similarly, the type and amount of in-flight exercise which may be a useful countermeasure for particular physiological systems has received renewed attention. The question of the physical characteristics of selected crews has become an issue in light of recent research suggesting that older and/or less physically fit individuals may have certain advantages in adjusting to some of the biological stressors of space. Studies simulating the capacity of women to tolerate space biological stressors also must be considered. On most measures men and women respond similarly to the stresses of weightlessness; on a small number of responses men appear to have an advantage. However, women consume less food and oxygen, weigh less, and appear to be more radiation-resistant. Moreover, there is evidence that, overall, women tolerate the combined pressures of space better than men. Information on possible gender differences needs to be gathered through in-flight experience.

The use of medication as a countermeasure to the biological stressors of space has important implications for understanding performance. Research conducted in extreme environments indicates that alterations in drug action can occur when normal biomedical functioning is modified. We need to assess whether any changes in drug action occur in space and what the behavioral ramifications of these changes may be. Related questions involve assessing how weightlessness influences rhythmicity and how the time of drug administration may influence drug action and possible behavioral correlates. The new fields of chronobiology and chronopharmacology explore these prospects and need to be integrated with the general question of how drugs operate in space.

Artificial gravity, either in the form of vehicle rotation or on-board centrifuge devices, has frequently been suggested as a means of compensating for adverse spaceflight alterations. Vehicle rotation would require a rather massive structure to simulate completely Earth's gravity. Even then the movement dynamics aboard such a vessel would be considerably different from those on Earth and would require a period of adaptation. The benefits and limitations of such a simulation need to be understood, as does the way gravity simulations of less than 1-g might mitigate the adverse effects of weightlessness. The use of on-board centrifuge devices is probably a more practical near-term possibility for use in space. Current experiments with animals suggest that this technique may be prophylactic for some uses. However, this approach does not appear to prevent bone decalcification and may produce some unpleasant side effects related to the abnormal vestibular stimulation produced by centrifugation. The overall usefulness of this technique for future missions needs to be assessed.

Another question concerns the behavioral correlates of weightlessness effects on the vestibular complex. Most investigators conclude that two distinct problems result from the unusual circumstances of the vestibular system in space. One involves spatial disorientation, illusions, and vertigo, and a second class of responses mimics what on Earth is called motion sickness. Most attention has been directed toward the latter category because of the frequency of its occurrence and the potentially debilitating effects that illness could have upon the flight crew. There is a need to continue development of theoretical models that can link together available data and provide direction for future efforts. The sensory-conflict theory appears to be a promising choice.

Several countermeasures now in use for the treatment of motion sickness are discussed in this chapter. Future possibilities for minimizing space sickness include the use of biofeedback and the identification of more effective selection procedures. The relationship of demographic variables such as age and gender to motion sickness needs to be examined more fully. Also, studies correlating certain personality characteristics and propensities toward motion sickness appear promising.

The third major system discussed is the visual complex. Although it now appears unlikely that weightlessness significantly alters visual acuity, routine studies are needed to document how such functions as dark adaptation, brightness and color sensitivity, accommodation, etc., are affected during spaceflight. Radiation may have an effect on vision as noted in previous missions, leading to perceptions of "light flashes." An important question is whether there is any cumulative effect associated with this irradiation which might jeopardize crews of long-duration missions. Another aspect of space travel that may adversely affect vision is vehicle vibration. With manual landings a routine requirement of Shuttle-like operations, it will be important to continue investigations on how vibration alters vision and what steps can be taken to minimize or compensate for vibration-induced losses.

Artist: Tom O'Hara

III
HABITABILITY

BACKGROUND

Habitability is a general term which connotes a level of environmental acceptability. The requirements for conditions to be "habitable" change dramatically with circumstances. For brief periods, almost any arrangement that does not interfere with the health of the individuals or the performance of their jobs would be acceptable. Over the long term, conditions must support not only individuals' physical, but also their psychological health.

Historically, habitability research has focused on the impact of environmental factors. It was assumed, for instance, that the presence of a noise stimulus or the confinement of large numbers of individuals in a small space was sufficient to explain performance variations. In a second phase of inquiry, attention turned towards personal or psychological factors as also being important in determining how individuals respond to less-than-ideal conditions (see review by Cohen, Glass, and Phillips, 1979). The most recent developments in this research area have shifted the focus away from the components of habitability (i.e., the individual and the environment) to the relationship between them (i.e., the person-environment fit). According to this model, people are motivated towards certain goals which, if achieved, result in a sense of well-being. The ability to reach these goals depends both on characteristics of the individual and on the supplies of the environment. A mismatch between an individual's values and abilities and the offerings of the environment results in strain, and eventually in physical and psychological illness. Evidence shows the person-environment-fit model can have predictive value over and above that of the components (Harrison, 1978; French, Rodgers, and Cobb, 1974).

In this chapter we will consider some of the habitability issues of extended spaceflight. Because of the extreme nature of the space environment, the emphasis here must necessarily be environmental. However, as the chapter progresses, we will attempt to highlight both the personal factors that affect adaptation to space and the interaction between personal and environmental factors.

THE PHYSICAL ENVIRONMENT

In dealing with people who are removed from their normal living situation, one can observe a wide range of idiosyncratic behavior, yet few individuals appear to view their own behavior patterns as anything but the norm. Spaceflight cannot be expected, for a very long time, to cater to unusual individual preferences; to a large extent the individual must adapt to the environment. However, it is necessary, in a general way, to anticipate those habitability issues of the space environment that could prove discomforting to a large number of space travelers. Among the more common disrupters are problems associated with interior space, food, hygiene, temperature, decor, lighting, odor, and noise.

Interior Space

Extra living area, costly even on Earth, becomes prohibitive when it must be blasted into space. Until we arrive at the point where dwellings can be constructed in space, astronauts' habitats must, of necessity, be restrictive. For the near term, the goal is to strike a reasonable balance between interior space which is unnecessary and that which is inadequate to meet the needs of the occupants. A basic habitability question concerns the minimum space individuals require, either alone or with others, to support their physical and psychological needs. Here we are concerned with the narrow question of volume requirements. The more complex questions of different perceptions of, and responses to, crowding will be considered in a later section entitled "Privacy."

Most of the estimates of volume requirements for humans to live and work in space assume that the length of confinement will be an important variable. Breeze (1961) concludes that a minimum of

50 ft³/person (1.42 m³/person) is adequate for 1 or 2 days of confinement, whereas 260 ft³/person (7.36 m³/person) is needed for 1 or 2 months, and 600 ft³/person (17.0 m³/person) for more than 2 months. Fraser (1968a) evaluated the results of 60 confinement studies to determine at what point physiological or psychological impairment occurred which was related to spatial restriction. He found that impairment (which he defined as the demarcation between "no impairment" and "marked impairment") occurred at between 50 ft³ (1.42 m³) for very brief confinement, and 150 ft³ (4.25 m³) for 60-day confinement. He concludes that a volume of 250-700 ft³/person (7.08-11.82 m³) length of confinement, is adequate.[1]

The number of individuals sharing confinement is believed to be an important variable affecting the amount of space needed per individual. Davenport, Congdon, and Pierce (1963) employed a model which assumes that more space per individual is needed as the number of individuals increases. This assumption finds some support from the investigation of Smith and Haythorn (1972), which showed that three-man crews suffered greater stress with approximately 70 ft³/man (1.98 m³/man) than did two-man crews. However, it seems at least as likely that the general relationship is in the opposite direction, i.e., that less space per person is needed as crew size increases. Breeze (1961) implies this latter relationship, at least for small crews, and Fraser (1966) terms the requirements for more space per person with increasing crew size "debatable." Clearly, this relationship requires research clarification.

In addition to the variables of crew size and time in confinement, how the available space is allocated must also be considered. Fraser (1968a) suggests that habitable areas be considered in terms of four kinds of functional units: work unit (operational tasks, vehicle management), public unit (dining, recreation, exercise), personal unit (sleeping, personal privacy, personal storage), and service unit (toilet, laundry, public storage). Very little information is available on the need for functionally distinct areas, or, if such need exists, how to appropriately "mark" the area (i.e., color, style, etc.). One study which deals with this issue has recommended that working and living areas be made sharply distinct from each other by virtue of design

---

[1] Cabin volumes were: on Mercury, 55 ft³ (1.53 m³); on Gemini, 44 ft³ (1.25 m³); and on Apollo, 107 ft³ (3.03 m³) per person (as reported in *Human Factors in Long Duration Spaceflight*, National Academy of Sciences, Washington, D.C. (1972).

features such as furniture, lighting, and acoustics (Jackson, Wamsley, Bonura, and Seeman, 1972).

Another question involving interior space concerns the utilization of the available areas within the spacecraft. In theory, unrestrained in their weightless state, astronauts should be able to utilize all available space and to work from the "walls" and "ceilings" as well as the "floors" of the vehicle. It has been found that weightlessness does permit astronauts to use space more efficiently than on Earth (Berry, 1973a). However, some space travelers have experienced considerable difficulty in trying to orient themselves in the absence of the familiar cues of gravity, showing a clear preference for rooms with a defined "up" and "down."[2] Research on environmental cognition indicates that mental representations help individuals to organize and manage their environment (Evans, 1980). The zero gravity in the space habitat places the individual in the totally unfamiliar environment of ungrounded three-dimensional space; in this environment the normal frame of reference is fundamentally disturbed. Lynch (1960) has used the term "legibility" to describe how parts of a scene can be organized into a coherent pattern. The desire of astronauts to define direction in space can be thought of as an attempt to establish legibility in space. It needs to be determined if space travelers are able to acclimate with time (or familiarity) to a directionless world or if their need for a directional orientation persists. If the latter proves to be true, we must conclude that not all surfaces in the spacecraft can be made equally usable.

A problem in using interior space efficiently concerns mobility within the space vehicle. Each time the astronauts move to a new location they must free themselves from the restraints holding them in one area and attach themselves in the new area. Experience in Skylab indicates that of the devices used so far, the ones that keep an individual securely in place are difficult to engage and disengage,

---

[2] Astronauts of Skylab preferred those areas where there was a local vertical, i.e., a defined "floor" and "ceiling." Astronauts felt least comfortable in the large upper deck of Skylab where, because of size and the lack of architectural cues, orientation was difficult (Life in a Space Station, *New Yorker Magazine*, 8/30/76 and 9/6/76). Story Musgrave, Mission Specialist who participated in an Extravehicular Activity on STS 6 has reported that he felt no need to identify a particular direction as "up" or "down" (personal communication, June, 1983). He further raised the interesting possibility that the presence of a defined vertical within the vehicle might cause conflict with outside perceptions, possibly contributing to space sickness. (For a discussion of the role of sensory conflict in space sickness, see chapter II.)

whereas those that are easy to slip into and out of are virtually useless. Grounding shoes based on a suction principle have been tried on Shuttle, but there is little indication that this mobility problem has been solved.

Another limitation to the full use of available space comes under the heading of etiquette or acculturation. For instance, in Skylab, congestion in the dining area would have been relieved if the astronauts were willing to "float" over the table to their places. Apparently passing over an eating area was perceived as inappropriate behavior and the astronauts chose rather to squeeze past each other or to take turns in the eating area (Johnson, 1974). Issues to be addressed for future spaceflight include determining how long it takes for space travelers to adopt an etiquette appropriate to a weightless environment, and identifying visual cues or other aids that might be furnished to make the transition easier.

Food

The Workshop on Controlled Ecological Life Support Systems (Mason and Carden, 1979) enumerated many of the food-related research needs associated with advanced spaceflight. These include questions of the storage stability of food, analysis and development of feeding systems, nutritional requirements in space, and the development of criteria for evaluating health status in response to diet. From the behavioral perspective, this workshop also emphasized the question of diet acceptability, including the possible altering of acceptability by such methods as behavior modification.

Experiences in space have not provided much information on how food will be used routinely in extended spaceflight because food intake in space generally has been restricted for medical-monitoring purposes. Also, because of the problems involved in preparing and consuming food in a weightless environment, attention has focused on the physical arrangements associated with food consumption, such as packaging, dispensing, scheduling, etc. On the early short-duration flights, food had to be extracted from its container and astronauts gave little attention to mealtime; food and its preparation were viewed as an inconvenience and an intrusion on a very busy schedule (Berry, 1973a). The Skylab flights were the first to demonstrate that food could be eaten from open dishes; Skylab also provided the first opportunity for astronauts to "share a meal." However, even here, astronauts did not always eat together, since there was difficulty in accessing the pantry area when all three astronauts occupied the dining area simultaneously. One somewhat puzzling aspect involving food is an apparent change in taste between ground and space. On Skylab, food which had been judged to be adequately seasoned prior to flight tasted bland. On the Soyuz 26 mission, Cosmonaut Grechko reported that canned ham which had been judged correctly seasoned on the ground was "too salty" in space, and cosmonauts on later flights were reported to have developed insatiable cravings, for instance for apricot juice and honey (Oberg, 1981). These latter findings are particularly notable since the Russians have given considerable attention to tailoring meals to the preferences and tastes of individual cosmonauts. The reported changes may reflect taste-preference shifts related to the general diet in space as compared with Earth, or they may suggest that there are taste or odor threshold changes in weightlessness.

It seems obvious that the view of mealtime as a necessary but annoying interruption in a busy day is bound to change as space travelers become more acclimated to their environment, and as menus offer greater variety and personal choice. On the 15-day Sealab II experiments, food neither offered much gratification nor occasioned much notice (Radloff and Helmreich, 1968). By contrast, on the 60-day Tektite experiments, food took on considerable importance (Berry, 1973a).

In space-simulation studies, responses have ranged from food as a primary source of irritation (McDonnell Douglas Astronautics Company, 1968) to food as no problem at all (Coburn, 1967). Even the effect of time in confinement on the acceptability of the food supplied has not been consistent. Experience on the Ben Franklin submersible showed that complaints about food increased during the 30 days of confinement (Grumman Aerospace Corp., 1970), whereas in a spaceflight simulation of the same duration, it was reported that the men seemed to enjoy their meals more toward the end of the mission than they had at the beginning (General Electric Corp., 1964).

Probably the most significant food-related question for future spaceflight is how food will be used to fulfill psychological and social needs. Clearly, there are advantages to be gained from the social aspects of mealtime. Bluth (1982) suggests that space travelers should plan to share at least one meal a day together in order to help dissipate any tendencies towards divisiveness that might develop. However, there may be special dangers in the social or recreational aspects of food use in space. When confined in an environment which has grown boring, subjects paradoxically tend to underutilize the recreational facilities provided. Yet these same subjects tend to place great emphasis on mealtime (Mullin, 1960). Eberhard (1967), reviewing experiences in the Arctic and Antarctic, at sea, and on missile bases, concludes that men in confinement take almost twice as long to eat as men in the general population. In a confinement experiment (Rogers, 1978) subjects spent a sizeable portion of their incomes on the embellishment of meals. Overall, expenditures for meals were about 30% higher than that required to maintain an adequate diet (Sullins and Rogers, 1975). The challenge for future spaceflight may be not only to make the food sufficiently appetizing and satisfying to meet the needs and desires of the individual, but also to provide sufficiently engaging activity options, so that the individual's psychological and social needs can be met without a harmful over-reliance on food or mealtime.

Hygiene

Americans are accustomed to very high standards of personal hygiene. Probably related is the finding that limitations on bathing facilities and waste management problems have been high on the list of discomforts reported by participants in various confinement experiments (Hammes, 1964, 1965; Hammes and Ahearn, 1966, 1967; Hammes, Ahearn, and Foughner, 1968; Rasmussen, 1963; Strope, Etter, Goldbeck, Heiskell, and Sheard, 1960; Strope, Schultz, and Pond, 1961; Coburn, 1967). In at least two experiments, lack of water for washing was listed as the number one annoyance of subjects confined for a period of 2 weeks (Rasmussen, 1963; Rasmussen and Wagner, 1962).

The hygienic and waste-management facilities of early spaceflight must be considered, at best, primitive. Following Apollo, considerable attention was given to this problem and a newly designed hygienic facility was evaluated during a 56-day simulation of Skylab (Van Huss and Heusner, 1979). Although vastly improved over earlier systems, bathing and waste management facilities on Skylab still did not allow the kind of ease and comfort that long-duration space travelers would require. For instance, mechanical difficulties were encountered with the integrated fecal/urine collector.[3] The

---

[3] For a graphic description of problems of human waste in space, see: There Ain't No Graceful Way: Urination and Defecation in Zero-G, Astronaut Russell Schweickart talking to Peter Warshall, *Space World*, Vol. P-1-181, pp. 16-19 (Jan. 1979).

Russians have also reported difficulties in addressing hygienic problems in space. These issues and measures to deal with them were assigned high priority in the flight program of Salyut 7 (Chernyshov, 1982). Problems in the hygiene and waste-management areas are well known to space planners, and it can be expected that these concerns will receive continued attention.

Temperature and Humidity

Variations in temperature affect human performance in diverse ways. In general, performance begins to deteriorate in any circumstance where heat reaches about 75% of the physiological tolerance limit (Roth, 1968), with high levels of humidity exacerbating temperature effects. Several studies (Pepler, 1958, 1959; Mackie, O'Hanlon, and McCauley, 1974) have demonstrated decrements in cognition and in psychomotor performance at temperatures at or above an effective temperature of approximately 85°F, with accuracy continuing to decrease as temperature rises. Working at a high constant temperature, performance also has been found to decrease with length of exposure (Roth, 1968). Cognitive tasks are more adversely affected than motor performance tasks (Poulton, 1970). The ability to withstand heat is limited by the amount of physical effort required. The length of time that can be tolerated at different temperature levels decreases dramatically as the amount of energy expended increases (Poulton, 1970). There is evidence that women are less able to withstand the effects of heat than men (Mackie et al., 1974), possibly due to less active sweating mechanisms. Uncomfortably high temperatures have also been related to temper outbursts and negative reactions to others (Griffitt, 1970; Griffitt and Veitch, 1971).

Decreases in temperature also can produce efficiency decrements. Most notable is the decline in skilled motor performance with continued exposure to the cold (Dusek, 1957); declines have also been found in visual reaction time (Teichner, 1957). Although skin temperature of the hands would seem to be an obvious factor in decreased motor performance, other factors also appear to be involved. Lockhart (1968) investigated the effects of a cold body, with and without cold hands, on three manipulative tasks. All three tasks were performed inefficiently when both body and hands were cold. Warming the hands improved performance on all three tasks, but did not raise performance to the level of the controls for two of the tasks.

Temperature/humidity problems did develop on Apollo 13 and on Skylab 2 when malfunctions occurred, and apparently the Russians experienced difficulty with humidity control of Salyut 4 (Oberg, 1981). However, with the exception of the Apollo Command Module, which was reported to be too cold for sleeping, the temperature/humidity question has not been a major problem in space operations. Temperature has been reported as a source of annoyance in several confinement studies (Smith and Haythorn, 1972; Rasmussen and Wagner, 1962; Rasmussen, 1963). By itself, this finding is not particularly noteworthy, since it is possible that environmental temperature and/or humidity drifted outside an acceptable range. Of more significance are the findings of Coburn (1967) that temperature/humidity ratio was the most common source of annoyance in a 32-day space simulation study, while at the same time *no single set of conditions* could be identified that would satisfy all six subjects of the experiment. It may be that as space travel expands, and travelers come to expect more amenities, temperature and humidity preferences will need to be taken into account, along with other compatibility considerations. It would also be important to determine if individual differences in temperature/humidity preference have corollaries in performance alterations.

Decor and Lighting

Space vehicles have now evolved to the point where they have something approximating decor to discuss. Rogers and his colleagues have investigated the importance of decor in confinement (Rogers, 1978). These investigators found that, in 10-day confinement, "plush" decor had very little value to all-male crews who were occupied in meaningful work. Attractive surroundings were found to be more important when crews were composed of both men and women, and when all-male crews had no meaningful work to perform. Decor was found to be most important when crews were both mixed and without meaningful work. In terms of extended spaceflight, it is reasonable to assume that it will become increasingly important as spaceflights lengthen.

It is generally accepted that the interior design of a spacecraft should have built-in flexibility. Such flexibility could include the use of movable partitions, removable wall covers, projectible designs, etc. It can be assumed that, in space as elsewhere, there is a general aversion to sameness. Visual variety can be introduced through the judicious use of texture or color. Skylab astronauts reported that the sameness of colors within their vehicle was disturbing (Berry, 1973a). Russian investigators have looked at the visual environment of a spacecraft and have proposed ways that changes in decor could be employed not only to relieve visual monotony but to maintain the space traveler's link to the home planet (Petrov, 1975).

In addition to a general aversion to sameness, there is also a recognized aversion to clashing designs. Terrestrial designs feature variety, but a variety which flows from a theme. Individuals experiencing this theme also have the opportunity of experiencing other themes in the course of a day. In space the number of designs must be limited. We need to ascertain what constitutes acceptable versus unacceptable variety in this closed environment. There is evidence that people prefer greater environmental complexity with time (Dember and Earl, 1957). If so, we should plan for increasingly complex arrangements as spaceflight lengthens.

Since there is no atmospheric absorption in space, the visual environment is marked by higher brightness levels than experienced on Earth and, more importantly, by abrupt contrast effects. Disruptive contrast effects could occur if, for instance, an astronaut suddenly encountered a bright object through a window, or if he or she were exposed to a glare from a reflecting surface. For these and other reasons, there has been concern that spaceflight could cause injury to eyesight, or otherwise alter the visual experience.[4] In this environment lighting becomes an especially important habitability consideration. Proper lighting is important to safeguard vision, to minimize annoyance, and to enhance the visual environment. Petrov (1975) has pointed out the desirability of individual light controls on all principal instruments so that adequate visual acuity can be maintained in the presence of sudden light flashes or extreme vibration. On Skylab, crews found light levels inadequate for the performance of some tasks and offered recommendations that levels be increased up to sixfold. While modifications, at least those affecting work areas, may be required, it should be remembered that many adult Americans today grew up before energy conservation was a widespread concern and are accustomed to higher levels of illumination than are required for good vision. For extended spaceflight, it will be necessary to maintain levels of illumination adequate to the tasks required, while readapting to lower levels of illumination generally.

---

[4] Other visual concerns include the effects of radiation and general effects of weightlessness.

In addition to its direct effect on vision, light levels can have an indirect effect on other behaviors. Changes in illumination levels have been found to influence the motor activity of animals (Alexander and Isaac, 1965) and, for humans, raising the level of illumination has been found to occasion an increase in the sound or noise level (Sanders, Gustanski, and Lawton, 1974). A related concern is how to simulate day/night cycles. It will be important to determine how the use of lighting might impact this area.

Odor

A frequently neglected aspect of habitability is the odor environment. Because particulate matter does not settle out in a weightless environment, odor problems in a space habitat may be more severe than under similar Earth conditions. Although some odors may, on occasion, be considered pleasant, "no odor" is the preferred condition. Odors have been associated with a number of medical symptoms including nausea, sinus congestion, headaches, and coughing (Goldsmith, 1973). They also contribute to general annoyance.

At the simplest perceptual level, an increase in the intensity of an odor stimulus is accompanied by an increase in response. However, adaptation effects are particularly pronounced for the olfactory sense, making the relationship between stimulus and response especially complicated. When odor stimuli are mixed, the perceptual response is further complicated. Jones and Woskow (1964) found that when two odors are mixed, the perceived intensity of the mixture is less than the sum of the subjective intensities of the component mixtures, but more than a simple average of the two. Berglund, Berglund, and Lindvall (1976) have developed a vector sum model to describe the findings that the perceived odor intensities of mixtures containing two to five odor stimuli only slightly exceeds the odor intensity of a single odor. From the standpoint of managing the odor environment, the significance of this model lies in the inverse of this relationship — i.e., that removal of some odors from a multi-odor mixture does not necessarily reduce the intensity of the odor perception.

Responses to odors can be accentuated by the presence of visual cues, and can also be influenced by prior biases, such as favorable or unfavorable attitudes towards the source of the odor (Kendall and Lindvall, 1971).

Materials used in spaceflight are subjected to testing for odor as well as for flammability and toxicity. Odor evaluations are made by a panel of test subjects who rate materials on a scale from 0 (undetectable) to 4 (irritating) with a score of 2.5 (falling between "easily detectable" and "offensive") considered passing (NASA, Office of Manned Spaceflight, 1974). It appears that, so far, the combination of material selection and adaptation effects have kept ahead of odor buildup in spaceflight. This does not mean that odor problems will not arise in the future. During the Spacelab Mission Development Test III (see Helmreich, Wilhelm, Tanner, Sieber, and Burgenbauch, 1979a, for a description of this simulation), participants complained of disturbing odors which they attributed to the primates and test rats which shared their facilities. A significant aspect of this test was that animal cages were in view during the simulation. We need to determine what odor-related problems are likely to occur in space and what steps, such as physical or visual separation of areas, could help address such problems.

Noise

Many of the initial concerns about space reflected the view that the environment would be understimulating to the space traveler, possibly leading to the disorganizing effects associated with sensory deprivation (see review by Suedfeld, 1980). Although the environment of space may prove to be understimulating in some respects, it may prove to be overstimulating in others. Among the latter concerns is the question of noise or unwanted sound. Berry (1973a) has commented that in space (p. 1142):

> The "silence of the void" is replaced by the sound of machinery, which makes the spacecraft cabin at least as noisy as any typical office, and sometimes noisier.

Noise has proven to be a problem in several confinement studies (Farrell and Smith, 1964; Page, Dagley, and Smith, 1964; Grumman Aerospace Corporation, 1970).

In the extreme, noise can cause pain and even damage to the inner ear, resulting in a hearing loss. Even low levels of noise can interfere with communication. Noise can also be a physiological stressor, exerting adverse effects on the cardiovascular system (Cantrell, 1975a), on the autonomic nervous system (Cantrell, 1975b), and on the vestibular system, resulting in disorientation, nausea, and dizziness (Harris, 1972). The clinical symptoms that have been associated with noise are so extensive and nonspecific that these symptoms have been aggregated under the general heading "noise sickness" (Andreyeva-Galanina, Alekeseyev, Kadyskin, and Suvorov, 1973).

The effects of noise on various performance tasks are not fully consistent. However, it is generally found that noise results in a narrowing of attention. Narrowed attention allows simple tasks to be performed, but negatively affects the more demanding tasks. Grether (1971b) reports that noise generally results in decrements in performance when the task involves complex reaction time, two-handed coordination, vigilance, or time estimation. Cohen, Conrad, O'Brien, and Pearson (1974) found decrements in information processing related to noise, with the effects becoming more adverse as the work pace quickens. Broadbent (1979) provides evidence of higher error rates and greater variability of performance in a noisy environment. In addition to its direct effect on performance, noise is also thought to impact performance negatively by lowering motivation and morale. (For a review of the direct and indirect effects of noise on human health and welfare, see National Academy of Sciences, 1977.)

Noise data show some adaptation effects to steady-source stimuli. However, intermittent noise shows reduced adaptation effects. Teichner, Arees, and Reilly (1963) found that a change in noise level, either up or down (from 81 dB), resulted in a decrement in information processing. Decisionmaking (Woodhead, 1959), monitoring of visual displays (Woodhead, 1964a), success on arithmetic tasks (Woodhead, 1964b), and information processing (Cohen et al., 1974) all have shown continuing decreased performance in the presence of intermittent noise.

In addition to its specific effects on performance, noise is generally conceded to be both fatiguing and distracting, disturbing sleep and interfering with waking activities. It constitutes a major source of irritation and has been linked to aggressive behavior. Donnerstein and Wilson (1976) and Geen and O'Neal (1969) have shown that subjects are willing to deliver a higher level of shock in a noisy environment than in a quiet one, and Mathews and Canon (1975) have found that experimental subjects are less altruistic when ambient noise levels are relatively high than when they are relatively low.

Although methods of arriving at acceptable noise standards have been quite diverse, recommendations that have evolved have been remarkably similar. The usual conclusion is that a sound level equivalent to approximately 45 dB indoors is a desirable and safe maximum. This value reflects primarily the requirement that speech not be interfered with. However, it has proven to be an appropriate level generally (Environmental Protection Agency, 1974). It is clear that many people in large and bustling cities or those living close to airports manage to survive at higher noise levels than those recommended. However, it is also clear that even the recommended levels can disturb the sleep of a sizable portion of the population, with women more easily disturbed than men and the elderly aroused more readily from sleep than younger persons (Central Institute for the Deaf, 1971).

One of the consistent problems with noise is that it is plainly annoying. Fidell, Jones, and Pearsons (1973) adopted the methodology of equipping subjects with signalling devices so that they could register annoyance levels in real time. This approach offers an alternative to survey techniques and could aid in the difficult task of quantifying annoyance data. Annoyance in response to noise has been found to be related to various factors tangential to sound. Numerous studies show noise associated with fear to be more disturbing than simple noise. Annoyance with noise stimuli has also been related to the individual's beliefs about the noise source. If the person believes that the noise-causing activity is related to his or her overall well-being, or that the person regulating the noise is concerned with the exposed person's well-being, the individual will be less annoyed than if the reverse is true (Central Institute for the Deaf, 1971). Unpredictable noise (Glass and Singer, 1972) and noise over which the individual has no control (Cohen et al., 1979; Averill, 1973; Lefcourt, 1973) result in particularly high levels of stress.

Although excessive noise is disturbing to everyone to some degree, there is considerable evidence that individuals vary dramatically in their sensitivity to noise (Borsky, 1977; Pearson, Hart, and O'Brien, 1975; Becker, Poza, and Kryter, 1971). Those who are most sensitive to noise become increasingly disturbed over time, whereas the annoyance level of less sensitive individuals remains relatively constant over time (Weinstein, 1978). Several studies have shown these individual differences to be related to personality characteristics (Central Institute for the Deaf, 1971; Becker et al., 1971; Pearson et al., 1975) and to remain relatively stable over time (Becker et al., 1971). Among the traits that have been ascribed to the noise-sensitive individual are reduced intellectual abilities and social skills and increased desire for privacy (Weinstein, 1978).

In spaceflight, potential noise problems are of two general types. One type occurs during the launch and reentry phases, the other during the cruise phase of flight. During the launch and reentry phases, propulsion and aerodynamic noise levels can reach 120-130 dB in the cabin. If unabated (for instance by the helmet-spacesuit system), this kind of auditory assault could be borne for only brief periods of time, and then, only at considerable cost.[5] But of more importance to extended spaceflight are cruise noise levels. Cruise levels of 65-70 dB have been reported for the Apollo flights (Von Gierke, Nixon, and Guignard, 1975). Although this level did not create any serious problems for the crews, it is unlikely that an equally noisy environment would be acceptable on longer flights. Noise levels only slightly higher were found unacceptable in a 90-day simulation test (Langdon, Gabriel, and Abell, 1971). Measurements taken aboard Skylab revealed ambient noise levels lower than those of the Apollo flights, ranging from approximately 43 dB in the sleeping cubicles to 60+ dB in the working area. Although these levels are probably still too high for extended spaceflight, the reported noise problems on Skylab were concerned less with the level of noise than with the medium of transmission. At the reduced pressure of Skylab, sound was considerably dampened, and astronauts separated by only a few meters had to shout at each other to be heard (Johnson, 1974).[6]

Spaceflight raises the spectrum of noise questions: its effect on perception and performance, adaptation effects, the fatiguing and annoying aspects of noise, and individual sensitivity differences. A compounding problem of the space environment is the unrelieved character of the noise environment. In most terrestrial situations, the individual can retreat from a noisy environment for at least a portion of the day. This is true only to a limited extent in space. Although some noise sources can be turned off when relief is needed, others cannot. Jonsson and Hansson (1977) provide evidence that prolonged exposure to a stressful noise stimulus results in a high incidence of hypertension. Similarly, heightened blood pressure has

---

[5] Yuganov, Krylov, and Kuznetsov report that after exposures of 125-130 dB for 20 min, subjects suffered from head pain, and ringing and feelings of stuffiness in the ears for 20-40 min after exposure, and did not recover for 1-2 hr (Yuganov, Ye. M., Yu. V. Krylov, and V. S. Kuznetsov, Study of Features of High Intensity Noise Effects During Spaceflight, in *Problems of Space Biology*. V. N. Chernigovskiy (ed.), Nauka Press: Moscow (1971); NASA TT F-719, Washington, D.C., 29-33, (1973)).

[6] Similar information has also been provided by members of the Skylab habitability team at NASA's Johnson Spaceflight Center.

been found among children living close to a large metropolitan airport (Cohen, Evans, Krantz, and Stokols, 1980). An attempt to reverse aversive effects through noise abatement of classrooms had no significant effect on the hypertension of exposed children, even after one year (Cohen, Evans, Krantz, Stokols, and Kelly, 1981). To the extent that certain minimum noise levels are always present, spaceflight potentially constitutes a more stressful noise environment than a simple consideration of decibel levels would imply.

Although we have been considering noise (i.e., unwanted sound) as a potentially disruptive stimulus, it should be mentioned that the presence of sound is necessary and beneficial. Sound serves as a source of arousal and its absence can be as detrimental to the health of the individual as a noisy environment.[7] Even annoying sounds, when kept within some limits, are not only accepted, but expected (Borsky, 1977). Russian experiments on the use of music in confinement conclude that unfamiliar and unusual music had a positive effect on all confined subjects "causing ecstasy in some, and in others, although even unpleasant, at least activity" (Zarakovskiy and Rysakova, as reported in Leonov and Lebedev, 1975, p. 162).

Music has been found to have a special relationship to behavior. It can aid efficiency when one is required to perform a repetitious task (Fox, 1971) and particular kinds of music can influence mood and behavior (Yingling, 1962). For instance, soothing music has been found to lead to helping behavior (Fried and Berkowitz, 1979).

An issue which will ultimately assume importance in a closed environment is sound preference. Stereos or television, symphonies, country music, and rock are considered necessary background to some individuals. To others, these same sounds are an intolerable cacophony. Even in the dampened acoustical environment of Skylab, the violin selections which enlivened one astronaut's day proved a source of irritation to his two crewmates (Cunningham, 1977). Over the years, various researchers have attempted to understand the correlates of musical taste (Farnsworth, 1950, 1969; Conyers, 1963; Edmonston, 1969). Such understanding could prove helpful in planning for long-duration spaceflight since some accommodation will have to be reached, either through attention to sound

---

[7] Improvements in some aspects of mental arithmetic and in clerical tasks have been reported in the presence of sound (see review by Grether, 1971b, op. cit.).

compatibility of crewmembers, through training, or through the use of earplugs or other sound-dampening mechanisms.

The task, then, is not to eliminate sound, or even to reduce it to a minimum, but rather to control unwanted sound (noise) while using wanted sound as a means of enhancing the total habitability of the space environment.

HEALTH AND LEISURE

Isolation studies present a fairly consistent picture of how time in confinement is spent. First, in situations where there are real mission-related tasks to perform, work dominates the interests of the confinees. When individuals enter such a confinement situation they appear to shift their expectations from the work/leisure/rest cycle of their everyday lives to a focus concerned almost exclusively with work (Parker and Every, 1972). They show relatively little interest in leisure activities and even tend to hoard their workloads (Eberhard, 1967). In confinement situations without specific work requirements, subjects show no special interest in either work or recreation. Although subjects enter confinement with ambitious plans of how to use their time constructively, these plans are rarely carried out. Rather, subjects tend to operate at a low level generally and simply pass the time until the confinement is over.

Since most confined groups emphasize work and deemphasize leisure, the question arises as to why this should be so. We would offer the following explanation. Individuals, on entering confinement, take on challenging new jobs and are committed to succeeding at these jobs. Where work is required as part of the confinement task, they incorporate these assignments into their new emphasis. Where work is not assigned, they concentrate on simply surviving the confinement. Their goals are only for the short term; after reaching these goals, they can resume "living." One might expect that a greater emphasis on leisure, as well as on unassigned work, will occur only when the individuals begin to think of the confinement situation as "normal." This would be expected to occur when the individuals have been (or anticipate being) confined for an extended period of time, or when they have experienced confinement repeatedly and have grown accustomed to its demands.

The physical structuring of the confinement situation may also contribute to an emphasis on work. Most people think of work and recreation as occurring in physically separated locations. In confinement there is no place for the individual to "get away." It will be important to determine if this factor is significant in the preference of confinees for work over leisure activity and if so whether functionally distinct areas can substitute for physical separation.

Recreation

As work loads lighten and as prolonged, uneventful periods become more frequent, the use of off-duty hours in spaceflight will be a major concern. Although recreation does not hold high priority among confined individuals, in some situations, notably long-duration confinements, recreational pastimes have been pursued. In these cases, confinees have shown a clear preference for what might be described as passive/noninteractive recreation, i.e., movies, television, books, music, looking out the windows, etc. Cards, dice, and other games of competition have been used little among confinees (Seeman, Singer, and McLean, 1971). Several interesting exceptions have been noted to the general rule favoring passive/noninteractive entertainment in confinement. A Monte Carlo night was held during a wintering-over at Little America and an amateur concert was staged aboard the "Ra" on its Atlantic crossing (reported by Leonov and Lebedev, 1975, p. 165). Planned activities also marked the fallout shelter experiments conducted at the University of Georgia (Hammes and Ahearn, 1967; Hammes et al., 1968). Here, subjects not only engaged in active and interactive leisure activity, but initiated their own forms of group entertainment, improvising from available materials. Among the activities engaged in by these subjects were group singing, bingo, talent shows, guitar playing, skits, dancing, spelling and history bees, and even a beard contest, as well as games, coloring, letter writing, etc.

The usual selection of passive over active leisure activity in confinement raises some interesting questions. Is this response related to the general tendency to withdraw, found in virtually every confinement situation (see chapter I)? Or does it relate to the kinds of leisure activities to which the individual is accustomed? Each individual entering confinement has a specific (and limited) repertoire of leisure-time activities from which to draw. Assuming for the moment that his or her normal recreational activities are both active and passive, it would be unlikely that the active leisure activities supplied in confinement would correspond to the habits of the confined individual. To engage in active leisure activity, the space traveler would have to substitute a new activity for his usual form of recreation.

Yet Christiansen and Yoestling (1977) have demonstrated the difficulty in substituting one recreational activity for another, even within the same activity type. No such substitution is required for passive activity. Passive leisure pursuits tend to be in the area of the mass media — television, magazines, movies, etc. These activities generally transfer rather easily to a confinement environment. And, as the name implies, mass media are enjoyed by almost everyone, while there is little evidence that scrabble or even card games are enjoyed by a large proportion of people.[8] Also, considerable evidence suggests that our normal leisure activities are more passive than we acknowledge.[9] If so, we should not be too surprised that people in confinement tend toward passive recreation.

In order to foster a healthy psychological environment in space, it would be valuable to know if the choice of spectator diversions such as movies, television, etc., by confinees reflects a desire to be passive, a desire to be noninteractive, a preference for the familiar, or some combination of these factors. An unusual emphasis on passivity

---

[8] Reading and watching television have been found to be popular pastimes among pilots, engineers, and scientists, whereas such activities as games, painting, etc., are relatively low on their preference list (Eberhard, J. W. and F. A. Hooper, Jr.; Off-Duty Activity Equipment and Facilities for Advanced Spacecraft, NAS-9-9338, Serendipity, Inc., Arlington, VA, 1970; and Karnes, Edward W., J. Kirby Thomas and Leonard A. Loudis; Recreational Preferences in Potential Space Crew Populations, *Human Factors, 13(1),* 51-58, 1971). An earlier examination of leisure activities of male professionals in aerospace corporations showed basically the same trends (Eddowes, E. E.; Survey of Leisure Time Activity, Implications for the Design of a Space Vehicle, *Aerospace Medicine, 32,* 541-544, 1961). These findings closely approximate preferences found in the general population (Group for the Advancement of Psychiatry, The Psychiatrist's Interest in Leisuretime Activities, Report No. 5, New York, 1958).

[9] The amount of time spent watching television raises the question of how active the average person is in leisure pursuits. Virtually every home in the United States has a television set, and the television is turned on for an average of 6 hr/day (Television and Growing Up: The Impact of Televised Violence, Report to the Surgeon General, U.S. Public Health Service, from the Surgeon General's Scientific Advisory Committee on Television and Social Behavior, U.S. Dept. of Health, Education and Welfare, Publication No. (HSM) 72-9090, December, 1971). Although youngsters and the elderly tend to be the main consumers of television, most adults report watching television for at least 2 hr daily (LoScinto, L. A.; A National Inventory of Television Viewing Behavior, in E. A. Rubinstein, G. A. Comstock and J. P. Murray (eds.), Television and Social Behavior, Vol. 4, Television in Day-to-Day Life: Patterns of Use. Government Printing Office, Washington, DC, 1971).

would suggest a general morale problem. If selection is based on the desire to be noninteractive, and therefore nonconflicting, this behavior would be expected to change when crews are sufficiently large or flights are sufficiently long. If a desire to avoid conflict is the motivation, then man/machine interactions such as computer games might be well received. By competing essentially against himself, the person could enjoy the challenge of competition without incurring the costs of interpersonal conflict. Similarly, competition with outsiders might be enjoyed. On Soyuz 9 cosmonauts played chess with a ground team, thereby avoiding the in-group competition issue. Active, but noncompetitive activities also offer promise for maintaining a healthy attitude in confinement. For instance, the psychological benefits of planting and tending gardens have been demonstrated in terrestrial settings (Kaplan, 1973; Langer and Rodin, 1976). Russian cosmonauts had an opportunity to tend plants during flight and apparently took great pleasure in this task in spite of the difficulty in completing the plant development cycle in weightlessness (Leonov and Lebedev, 1975, p. 159; Oberg, 1981). It has been reported that cosmonauts on Salyut 7 did successfully complete the cycle of plant growth, from planting seeds to obtaining new seeds, and also harvested onions, parsley, and borage from the vegetable garden onboard (Tass, Moscow, October 12, 1982). This success should heighten the appeal of planting and tending gardens in space. If the choice of passive activity in confinement is based on preference for the familiar, then a careful selection, where possible, of materials geared to the interests of the particular confinees might provide some incentive to break from the passive pattern.

There is some evidence that the use of leisure relates to the composition of the group as well as to the characteristics and inclinations of individuals. A factor which distinguished the more active fallout shelter subjects (Hammes and Ahearn, 1967; Hammes et al., 1968) from other groups was the unusual degree of diversity in their compositions. Each group reflected to some extent the age, sex, education, race, and socioeconomic status represented in the general population. Groups ranged in size from 30 to over 1000 individuals, and in age from 6 months to 79 yr. Apparently in these experiments it was the women who instigated group activities, perhaps through a felt need to entertain the children. The high level of interaction among these confinees is very provocative and suggests that heterogeneous groups, especially those including women and children, may have a completely different approach to the use of leisure than groups of young men, the usual subjects of confinement studies. A better understanding of the dynamics of group leisure could have broad implications for selection practices generally.

An issue that will be of particular interest in extended spaceflight is how the quality of leisure activities changes over time. As Fraser (1968b) points out, data are not consistent. Experience on nuclear submarines shows a movement away from escapist leisure to an educational use of leisure later in the cruise (Kinsey, 1959; Ebersole, 1960); however, experience in the Antarctic has shown a trend in the opposite direction (Rohrer, 1961). These differences could be accounted for by differences in crew morale, with high morale related to a high level of energy and more creative use of leisure. The kind of work performed might also, over time, influence the kind of recreation sought. Assuming a reasonably high level of morale, one might anticipate that a person engaged in demanding mental activity during working hours would seek more escapist recreation than one working at a monotonous task all day. Since work in space is likely to become progressively less demanding, one might expect a shift towards more enriching recreational pursuits as flights become longer and more routinized.

Russian planners have initiated some novel approaches to recreation. Cosmonauts have been able to watch concerts on their "videotheque," and have had music and news broadcasts piped in. They have also had two-way radio and video communication with family members, friends, scientists, actors, musicians, etc. These events were planned giving weight to the individual interests of crew members, the psychological climate in the space cabin, and the flight stage (Gazenko, Myasnikov, Ioseliani, Kozerenko, and Uskov, 1979). Cosmonauts on extended flights also had letters and presents from home, along with special foods and fresh milk, delivered to them in space. An interesting conclusion from a Russian 70-day confinement study was that unfamiliar and eccentric films (in this case horror movies) could be used to relieve a depressive mood (Leonov and Lebedev, 1975, p. 164). Another spaceflight innovation used by the Russians was the grab bag. Cosmonauts on Salyut 6 were surprised by small, but apparently delighting toys, novelties, etc. The unexpected undoubtedly plays a very large role in what we see as the fullness of experience and needs further exploration as to how it can be utilized as a positive force in spaceflight.

Exercise

Until recently it was believed that a proper exercise program could reverse the significant physiological/anatomical changes associated with the body's response to 0-g. However, studies of prolonged bedrest suggest that exercise, by itself, is insufficient to meet these ends. For instance, changes in endocrine and metabolic functions now are believed to result from changes in hydrostatic pressure and from lack of postural cues, rather than from a lack of activity (Vernikos-Danellis, Winget, Leach, and Rambaut, 1974).

If physical activity is less than a perfect solution to the physiological effects of weightlessness, such activity at least slows some of the body's responses to 0-g, and helps maintain the individual at an acceptable level of physical functioning. The Skylab 4 crew were in space for the longest period, exercised the most, and required less postflight recovery time than the crews of Skylab 2 or Skylab 3. Physical activity also confers psychological benefits such as relief from depression or lessening of anxiety (Wood, 1977). In addition, there is a widespread belief that physical activity (eventually) rewards the performer with a general sense of well-being.

To date, the exercise facilities in space habitats have, of necessity, been limited to those which provide the most physical benefit, while weighing as little as possible and occupying the least possible space. Research on exercise conducted at Harding College (C. Smith, Corbin, and Olree, 1976) provided the basis for the exercise regimes adopted for the American Skylab series. Skylab 2 astronauts used both a bicycle ergometer and an isometric device for their 28 days in space. As a result of this flight it was determined that additional exercise times and programs were required. During the 59-day Skylab 3 mission, ergometer time was increased, and a minigym provided additional exercises for the astronauts' trunks, arms, and legs. On the 84-day Skylab 4 mission, ergometer time was further increased, and a treadmill was added. Typically, astronauts used the treadmill 10 min/day (Van Huss and Heusner, 1979).

The Russians used a combination of a treadmill and gravity suits on early Salyut flights, adding a bicycle ergometer on later flights. The cosmonauts followed a compulsory program of daily exercise, for instance, 2.5 hr/day on Salyut 4 (Gorokhov and Stapantsev, 1975).

Related to the question of exercising in space is the problem of limited bathing facilities and the difficulties astronauts have experienced in using them. Even on Skylab with its more advanced hygienic arrangements, astronauts experienced difficulty in rinsing wash rags, collecting water after showers, etc. The success of any serious exercise program in space will require that shower-equivalents be available on demand and easy to use. This means an adequate water recycling system.

The cosmonauts associated with Salyut 4 were reported to have looked forward to their exercise routines. However, reports on Salyut 6 missions present a different picture. On Soyuz 26, cosmonauts Grechko and Romanenko were reported to have taken every excuse to skip the exercises, while on Soyuz 32 Cosmonaut Ryumin reported the exercise program to be "boring and monotonous and heavy work" (Oberg, 1981, p. 213). On Skylab, astronauts were less than enthusiastic about exercising and varied in their commitment to it. When bathing and space limitations are resolved, more flexibility will be available in planning exercise facilities. It is worth considering, at least on a general level, how to improve the likelihood that whatever equipment is provided will be used. A general question concerns just what motivates an individual, in any circumstance, to begin and maintain an exercise program.

In recent years, studies have appeared which demonstrate a longevity advantage to exercise (Leon and Blackburn, 1977; Rose and Cohen, 1977). However, it is doubtful that many people will be moved to maintain an exercise program because it aids their health in some amorphous manner. Unless one can draw an immediate relationship between exercise and a particular health problem which is highly salient to the individual (such as a previous heart attack), the exercise program is likely to be short-lived. Alternatively, we would suggest that if individuals believe that exercise will help them to *look* better (or prevent them looking progressively worse) there is greater probability that the exercise program will be maintained. If the benefits of exercise become internalized, as when the individual feels release of tension or increase in energy, then the chances improve that exercise not only will be maintained, but actively sought.

As noted above, individuals in confinement avoid competitive encounters. This fact alone eliminates much of what is usually thought of as exercise or physical activity. It is likely that the avoidance of competition will diminish as crews become large or as people are in space long enough to risk occasional hurt feelings. However, competition with others is not essential to a physical activity program, as one can work alone or noncompetitively with others.

Two factors appear to us to be common to people who maintain a physical activity. First, they have identified something which particularly interests them, and second, they seek to acquire or improve a skill. People who seek to improve a skill can be reinforced by the progress they make; this gives them a decided advantage over individuals who begin a program for general health benefits or for diversionary purposes. Therefore, some attempt must be made both to match the interests of the crewmembers with the exercise equipment offered and to provide the potential for working towards improvement.

It is likely that individuals in space will develop interests in movements unique to space. For instance, gravity forces less than 1-g constitute a new environment with a new set of physical challenges. Astronauts on Skylab were enthusiastic about the possibilities of tumbling and acrobatics in space, and suggested that all future space stations include a facility for acrobatics. Whether pursuing an old or a new interest, we believe it important to design equipment to aid the individual in setting goals for himself and tracking his progress.

PRIVACY

Confined individuals who report habitability problems generally direct their complaint at a physical aspect of the environment, perhaps because it is more acceptable to complain about equipment than about a fellow confinee. However, as arrangements for living and working in space become better established, habitability issues can be expected to take on a subtler tone and to involve relationships as well as physical conditions. One such relational issue is the need for privacy. This area is not often discussed in terms of spaceflight needs; because of its perceived importance, we shall review privacy research at some length.

Meaning and Functions

The term "privacy" conjures up a variety of meanings. It is used to describe the need for ample space; visual, physical, or psychological separation; low population density; control over space, possessions, or information; freedom of activity; and many other concepts.

Popular responses to the term fall into four broad categories: "aloneness," "controlling access to space," "no one bothering me," and "controlling access to information" (Wolfe and Golan, 1976). Three of these categories involve managing one's direct interactions with others; the fourth involves controlling information about oneself. Privacy in the informational sense reflects a concern not only about immediate events, but also about future events (Laufer, Proshansky, and Wolfe, 1976). It is the informational aspect of privacy that has changed dramatically with the introduction of computers, occasioning much of the recent concern over privacy issues.

Various definitions of privacy have been proffered by researchers and analysts who seek to understand the role of privacy in human development and functioning. A central element in these definitions is the ability of individuals to choose if, when, and to what extent they interact with, and thereby reveal themselves to others. For example (Margulis, 1974, vol. 6, p. 1):

> Privacy, as a whole or in part, represents the control of transactions between person(s) and other(s), the ultimate aim of which is to enhance autonomy and/or to minimize vulnerability, thereby protecting autonomy.

Margulis (1977) outlines three basic functions of privacy. The first and least central function involves the management of the interaction between self and others. This function assumes that roles and relationships are reasonably well understood and that the individual is seeking the most rewarding level of interaction with others. A higher-order function is concerned with defining self in relationship to others. This function involves role definition and the presentation of self, and includes what Goffman (1959) refers to as "image management." The third and most central function of privacy regulation is concerned with self identification. By means of this function, privacy becomes the mechanism for distinguishing between the self and others. This function is demonstrated by its converse. In certain situations the destruction of privacy is used to break down the notion of self identity and individuality (Goffman, 1961) as when prisoners or soldiers have their clothing removed, and after some period, are issued uniforms.

Theory

Theorists of privacy have long admonished that privacy involves having neither too much nor too little interaction. Interaction can expose oneself to others; however, it also exposes others to self. Privacy represents the ideal balance or "homeostasis" between forces to be open versus closed (Altman, 1975), with dissatisfaction resulting when the optimal balance is not maintained (Sundstrom, 1977). However, this balance is not a static one. People seek to withdraw when they feel overly exposed, and seek exposure when they feel excluded, in a constantly correcting process.

Individuals observe others and seek information about others, assumedly because they need to judge the appropriateness of their own beliefs and behavior.[10] People also voluntarily expose aspects of their personality to others. One explanation for this latter behavior is that the individual must be known if he is to be accepted and affirmed. In addition, there is an "expectation of reciprocity" associated with voluntary personal disclosures (Derlega, Wilson, and Chaiken, 1976). A revelation by one individual carries an expectation that the demonstrated trust will be returned by a revelation by the other. In this way, self-disclosure itself becomes a means of access to needed information.

Bases of Needs

The rights of an individual to privacy and the rights of others (frequently the state) to information sometime come into conflict, with legal decisions swinging first one way and then the other (Levin and Askin, 1977). However, the fact that an individual has a right to privacy is well accepted in our culture. Western societies, with their belief in the distinctiveness and unique contribution of the individual, would be expected to value privacy more than societies with different philosophical orientations. This raises the question of whether the need for privacy is a culturally specific phenomenon or whether it is a generalized phenomenon, common to all members of the species.

Altman (1977) has performed an ethnographic analysis of various cultures and has concluded that privacy regulation is a culturally universal process. In those societies where contact among individuals is necessarily high, alternative approaches have evolved to

---

[10] Festinger first focused on the notion that a primary function of group affiliation is the evolution of one's own feelings and beliefs, i.e., self-identification (Festinger, L.; Informal Social Communication, *Psychological Review, 57,* 271–282, 1950). This area, which is known as "social comparison theory" continues to spawn informative research.

counteract these intrusions. The particular practices vary with the culture, but include the seeking out of secret private areas, flexible arrangement of structures, and cultural restraints on social interactions. He cites the Pygmies of Zaire as an example of a society which has developed an unusual method of privacy regulation. This group alternates between periods of very high physical and social interaction and periods of separation, with the periods of separation lasting up to several months. These cycles are repeated year after year. Altman (1978) concludes from the various societies he has studied that (p. 78):

> when social contact is high, or when certain interactions are forced, compensatory behavioral mechanisms are available that permit people to regulate their social contacts, to be open or closed as the situation warrants.

Apparently, privacy in some form is necessary for individual and group survival, and those societies that survive have found their own, sometimes unique, mechanisms for ensuring an acceptable level of separateness.

If privacy is a general need of man, one might ask whether humans are alone in this quest for privacy. Westin (1967) proposes that the need for privacy, rather than being an expression of a human's special ethical, intellectual, and artistic needs, is an expression of animal origins. He observes that virtually all animals seek separation and solitude. Although animals in captivity appear to tolerate repeated invasions of their privacy, reduced fecundity and other aberrant behaviors in such animals indicate a failure in adaptation, and suggest that privacy may be important to the normal physiology of many species, even when particular individuals have never lived in a private environment.

Berscheid (1977) suggests that, since the need for privacy seems to have biological and evolutionary roots, individual variance in the strength of such a need may have genetic, as well as learned components. These genetic components may correlate in some way with other aspects of personality such as a tendency toward extroversion or introversion. Studying large numbers of twin girls, Scarr (1969) found that genetic factors account for more than half the within-family variance associated with introversion/extroversion. How the introversion/extroversion variable corresponds to privacy-related behavior is largely unexplored, although as Berscheid notes, an interesting start has been made by Marshall (1975) who found evidence of a relationship between privacy orientation and scores on an introversion/extroversion measure. Bersheid poses several questions whose answers could be informative to the relationship between personal characteristics and privacy. Do introverts have different and/or more successful ways of protecting themselves from excess social stimulation than do extroverts? How does the interaction between a person's status on the introversion-extroversion dimension and his privacy state affect such dependent variables as aggression, susceptibility to influence, emotionality, group membership, friendship, interaction patterns, etc.? Understanding such relationships, or more broadly the relationship between personal characteristics and privacy needs, could help in predicting an individual's adaptation to closed and demanding environments such as that of space.

Mechanisms

Privacy is controlled by four behavioral mechanisms: (1) verbal content and structure, (2) nonverbal behavior (i.e., body language), (3) environmental mechanisms, and (4) culturally based norms and customs (Margulis, 1977). Verbal and nonverbal privacy in space can be expected to mimic similar behavior on Earth (although as indicated in chapter VII, both verbal and nonverbal communication can be expected to be impeded in space). Culturally based norms will take time to evolve in space. A fruitful avenue of privacy regulation appears to lie in the manipulation of the environment.

Individuals use the environment and the physical objects in the environment to help define themselves as distinct from others. Clothing, personal space, and personal possessions become not just property with a fixed economic value, but extensions of the person. For instance, it is a well-recognized phenomenon that an individual who has had property burglarized experiences a keen sense of personal invasion, whether or not anything is taken.[11]

Privacy regulation through structural arrangements has direct applicability to space. Archea (1977) warns that although most studies of structural arrangements have overemphasized the normative or symbolic association of architectural design, the attributes implicit in the physical arrangement are of primary importance. Archea proposes a model of spatial behavior in which visual access and visual exposure are the two key aspects of privacy regulation.

---

[11] Irving Altman (1975, op. cit.) discusses this phenomenon in terms of the "contamination" of the personal possessions by the intruder.

Visual access is the ability to monitor one's spatial surroundings by sight; visual exposure is the probability that one's behavior will be visually monitored. The crux of Archea's thesis is that the arrangement of the physical environment regulates the distribution of information upon which interpersonal behavior depends. A central element in the relationship of visual access or exposure to privacy is the use of gradients or changes in the visual scene. These gradients can be abrupt (barriers) or gradual (lighting). Archea refers to the process of controlling behavior or adjusting position to achieve a desired level of privacy as "selective conspicuousness." Architectural privacy has been found to correlate with psychological privacy, and both forms have been related to job satisfaction and, for some tasks, with job performance (Sundstrom, Burt, and Kamp, 1980). The efficacy of architectural arrangements in privacy regulation has been shown also by Baum and Davis (1980).

Crowding

One of the environmental limitations of spaceflight is that individuals are confined to a small area. It has long been understood that when individuals are so restricted, both physical and psychological symptomatologies result. At one time the medical problems associated with high-density confinement were thought to be related to an increased opportunity for the spread of communicable disease (as discussed in Loo, 1973; Stokols, 1972; Desor, 1972). It is now known that the physical transmission of disease accounts for only a minor part of the problems of restricted individuals (Dubos, 1970). Of far greater importance to both physical and psychological health is the generalized factor of stress. Experiments with rodents have shown that cramped quarters give rise to reduced fecundity, hormonal pathologies, and reduced life span (Kiritz and Moos, 1974) with very high densities related to prostration, convulsions, and sudden death (Christian, 1963). Humans are rarely subjected to such extreme conditions, and assessment of the effects of high density on people must rely on the observation of more subtle symptoms. Information is based either on relatively intense but brief laboratory experiments, or else on less severe but more prolonged experiences in natural settings.

In the laboratory, the effects of high density on arousal have been somewhat equivocal. However, the preponderance of recent evidence favors the notion that high density does result in stress-related arousal, as measured both by skin conductance (Calhoun, 1963) and by performance measures (Aiello, Epstein, and Karlin,

1975a). In a natural setting, stress arousal has been found to affect health and performance; physiological measures show trends in the same direction (Epstein and Karlin, 1975). In humans, high blood pressure is often associated with stress. Various studies have shown that elevated blood pressure accompanies high density, although in most of these studies other stressors were also present (D'Atri and Ostfeld, 1975; D'Atri, 1975; Harburg, Erfurt, Hauenstein, Chape, Schull, and Schork, 1973; Harburg, Schull, Erfurt, and Schork, 1970; Lang, 1950).

In recent years, researchers have emphasized the importance of the psychological perception of crowding as distinct from the objective condition of high density. (For a review, see Stokols, 1978b.) Crowding is a far more complex variable than density, and a person in a high-density situation may or may not feel "crowded." Stokols (1976) identifies three theoretical perspectives on crowding: stimulus overload, behavioral constraint, and ecological formulations. Stimulus overload relates perceptions of crowding to the excessive levels of stimulation that frequently accompany high density. Behavioral constraint focuses on the stresses which follow from restrictions on freedom. Ecological formulations emphasize the shortages of resources which often accompany high-density living.

Several environmental variables have been found to influence how crowded an individual feels. High density within a dwelling and in private spaces such as in home or working areas are more closely related to stress response than are high densities outside or in public areas. Relational variables also influence the perception of crowding. Friends require less distance than strangers, informal groups less than formal, etc. One's assumptions about the situation also affect how crowded an individual feels. If individuals believe that they are able to control the situation, for instance, by having the option of leaving, their perception of crowding is reduced (Stokols, 1979; Glass and Singer, 1972; Sherrod, 1974). Predictability or expectation about the occurrence or duration of confinement also has an important influence on reducing feelings of being crowded (Glass and Singer, 1972).

There are differences among individuals in their perceptions of crowding. Altman (1975) reports that self-directed, high self-esteem persons have lower spatial needs than individuals searching for identity, and laissez-faire individuals have lower spatial needs than authoritarian individuals. Similarly, subjects who report an external locus of control perceive a greater degree of crowdedness than do internal locus-of-control subjects (Schopler and Walton, 1974).

However, Baron, Mandel, Adams, and Griffen (1976) suggest that the relationship between perceived locus of control and perception of crowding may be quite complex. In their study, internals demanded less distance than externals only when subjects did not feel cramped; when feeling cramped, internal locus-of-control subjects demanded more interpersonal distance than externals. Hackworth (1976) has shown that frustrations concerning crowding translate into other sensitivities. Subjects who reveal a trait preference for large personal space show increased annoyance to an auditory stimulus following high-density confinement; subjects with low requirements for personal space show a slight decrease in sensitivity following the same confinement.

Several investigations have revealed gender differences in response to crowding. In laboratory studies, crowded men were found to respond more negatively, and crowded women less negatively, than their less crowded counterparts (Freedman, Levy, Buchanan, and Price, 1972). Similarly, crowded women rated themselves as less aggressive, and crowded men rated themselves as more aggressive, than similar but noncrowded subjects. These findings and others (Stokols, Rall, Pinner, and Schopler, 1973) suggest that women become more cohesive and cooperative, whereas men become more remote and competitive, under conditions of acute crowding. However, a recent study of long-term crowding indicates that, over time, women pay a high price for their socially adaptive behavior. Aiello, Epstein, and Karlin (1975b) found that chronic crowding results in women perceiving themselves to be more crowded, and being less satisfied with their situation, than were similarly crowded men. Also, crowded women reported more health-related problems, and showed less group stability over time, than did their male counterparts.

Taken together, these studies suggest that there are individual differences in the perception of crowding which may relate to other personal characteristics and which could be predictive of confinement tolerance.

Territoriality

Several studies refer to the "territoriality" of confined individuals in describing their privacy-seeking behavior. However, true territoriality is less obvious in the human than in other species. Sundstrom and Altman (1974) observe that most animals show highly stereotypic territorial behavior, whereas humans show considerable territorial flexibility. These authors conclude that the human territorial response is probably learned and may be just one of many techniques for achieving privacy.

In only a few cases has the specific need to control particular spaces been identified in confinement. In the University of California, Berkeley, "Penthouse" studies, it was reported that subjects staked out territories for their own personal use and became hostile when others violated these rights (Cowan and Strickland, 1965). Several Russian confinement experiences also have concluded that there is a specific need for personal space (Leonov and Lebedev, 1975, p. 86), and in the submarine experience, the lack of personal space was identified as a cause of stress (Earls, 1969). Yet even in these examples it is not clear whether the felt need was for separate space per se, or for control over interactions with others. Further evidence of the chameleon quality of the territorial need in humans comes from studies of Esser and his colleagues (Esser, 1973; Esser, Chamberlain, Chapple, and Kline, 1964). These researchers found that in confined settings with only a few desirable areas, the more dominant confinees exhibited territorial behavior, claiming for themselves the preferred spaces. However, in a uniform setting, the more dominant members did not act in a territorial manner, but rather moved freely throughout all the available space.

In a recent experiment, Rogers (1978) offered confined subjects an opportunity to purchase specific amenities, inluding extra space which could be turned to personal use. He found that although extra space was sometimes purchased, this space was added to the public area and was not used as individual space.

The evidence, then, suggests that for the human, personal and exclusive use of an area is probably less important than the privacy requirement which such space would address. This finding allows us to search for solutions to the privacy problem in space which do not depend on the personal allocation of territories within the habitat.

Privacy in Space

When one looks at space confinement, one is forced to conclude that many of the familiar avenues of privacy maintenance are blocked, while methods that can be used to break down feelings of personal identity and individuality, such as are employed in prison, are naturally present in the environment. In space, personal property is necessarily limited. Physical space is also severely restricted, leading both to forced interactions and lack of opportunity for personal territory. Close contact leads to problems of information management, since the individual can be observed by others almost continuously. This situation increases the probability that the individual will be "caught unaware" in behavior he would rather keep private.[12]

On Earth, lack of privacy resulting from physical constraints would be compensated for by cultural norms. Space has yet to develop these cultural norms. In fact, confinement generally seems to work against privacy maintenance, as individuals engage in greater self-disclosure than they would ordinarily. The limited number of personal interactions available to the space traveler, together with the lack of new inputs, result in compressed familiarity without the opportunity found in more conventional relationships to develop shared values. Self-disclosure leads to reciprocal self-disclosure, which can lead to rejection.

Even in the absence of rejection, it is easy to envision how escalating self-disclosure could leave individuals feeling manipulated and out of control, both of themselves and of their relations with others. The problem is likely to be exacerbated by a strong group commitment. Having volunteered to the project, the participant would feel pressured to accept all aspects of the situation, including violations of privacy. Some even argue that the element of a public, as opposed to a private project, further reduces the individual's right to privacy (Kelman, 1977, p. 180).

It has been found that if individuals will be forced, eventually, to reveal something about themselves that is damaging to their image, they can reduce the negative impact of the disclosure by revealing it quickly at the beginning of a relationship (creating the impression that they relate honestly), and only to selected individuals (thus recruiting these individuals to help keep the secret from others). For a discussion of these points see Jones and Archer (1976) and Jones and Gordon (1972). However, revealing negative information, even quickly and selectively, does not guarantee a positive response. If the disclosure is seen to be appropriate and socially rewarding, it may result in greater attraction toward the discloser (Jourand, 1959, 1971; Worthy, Garry, and Kahn, 1969). However, a disclosure that is

---

[12]Kelman notes that control over our ability to preserve an image is reduced when we find ourselves in a situation where we must respond spontaneously, immediately, or within severely restricted options (Kelman, Herbert C., Privacy and Research with Human Beings. *J. of Social Issues, 33(3)*, 169-195 (Summer, 1977)).

perceived as inappropriate or unrewarding may be unrelated to acceptance (Jourand and Landsman, 1963; Thompson and Seibold, 1978), or a disclosure can relate negatively to acceptance (Kiesler, Kiesler, and Pallak, 1967; Culbert, 1968; Weigel, Dinges, Dyer, and Straumfjord, 1972; Wheeless and Grotz, 1976).

Individuals with limited means of protecting their privacy respond either by "acting out," e.g., by behaving aggressively toward others, or by withdrawing (Wolfe and Golan, 1976). In confinement, acting out has been rejected as a viable coping mechanism. Withdrawal has become the hallmark of the confinement experience, and this withdrawal behavior may have roots in the need for privacy. Although withdrawal in some form will probably be employed by space travelers, there is reason to believe that even this solution will be less advantageous in space than in the usual confinement experience. In space, crewmembers must depend on each other for their day-to-day survival. Individuals in a threatening situation prefer to be in the company of others, not only for the acceptance, verification, and comforts that individuals provide each other, but also for safety. As long as a high level of interdependence exists in space, withdrawal behavior in spaceflight is likely to be either curtailed or highly specialized.

Although spaceflight will limit the use of many privacy mechanisms, the use of other mechanisms could be improved and expanded. Anything that would help emphasize a person's individuality could be expected to help offset privacy loss. If some, even small, area could be designated to an individual, for instance for storage, it could be decorated to reflect private style. Decorating an area seems to have positive connotations; for instance, the quality and diversity of college room decorations have been found to be predictive of commitment to the university life (Vinsel, Brown, Altman, and Foss, 1980). Personally selected clothing could also help the individual maintain separateness. On Skylab, astronauts found particular fault with their clothing (Cooper, 1976). Although this complaint centered on the drabness of the clothing, it should be noted that garments failed to distinguish among individuals, perhaps another basis for the objection. Better use of available space could also alleviate the privacy problem. For instance, Skylab crews suggested that individual sleeping areas be separated further (Bluth, 1982).

Although self-expression can contribute to one's sense of identity and therefore can help protect privacy, not all means of self-expression can be considered desirable for spaceflight. Men on the first Skylab flight were reported to have accentuated their personal idiosynchrasies. Yet personal mannerisms have been found to be a source of great annoyance with increased time in confinement (Grumman Aerospace Corp., 1970). Expressing one's individuality through clothing, decor, or physical separation could help space travelers maintain their identities without the irritation that follows in response to such personal expressions as gum-snapping, knuckle-cracking, or foot-tapping.

One method that has been employed successfully in circumventing privacy loss through self-revelation is replacing depth of disclosure with breadth of disclosure (Jones and Archer, 1976). This technique is often used by individuals wishing to protect their privacy but who feel themselves forced to respond to the disclosure of another. The responding individual reveals something about himself or herself, or perhaps several things, but does not reveal anything that he or she considers sensitive. The demand for reciprocity has been superficially satisfied while avoiding unwanted revelation. Unfortunately, here again confinees may be at a disadvantage. Altman and Haythorn (1965) have found that control subjects (not confined) naturally employed this technique, whereas similar but confined subjects exchanged information at a high level of intimacy.

Friends display less disclosure reciprocity than strangers (Derlega et al., 1976). This may be because the norm of self-disclosure reciprocity is stronger at the beginning of a relationship than at later stages (Altman, 1973). Regardless of the mechanism, the result is that friends regulate privacy more effectively than strangers, and are less likely to engage in destructive self-revelation. If friends tend not to invade each other's privacy, there would be advantages in either selecting crews from among those with established relationships, or in encouraging friendships (not just working relationships) before confinement is attempted.

Perhaps the most important privacy need in space is to familiarize potential crewmembers with the environmental pressures they will encounter, and to help them understand both the value of group cohesion and the necessity of personal separation. As we have seen, privacy is required for the normal functioning of the individual. But individual privacy is also important for the effective functioning of the group. Group pressures can hamper an individual's judgment, even to the point where simple sensory information cannot be correctly perceived (Asch, 1951, 1956). When the individual becomes indistinguishable from the group, he or she is unable to contribute anything further to the group. One method of encouraging continuing group contribution is to permit anonymity in offering suggestions or in voting on issues. However, a better method would be to develop an attitude of cooperation and acceptance within the group that allows individuals to contribute without fear of either retribution or ridicule.

There is one other major aspect of privacy relating to spaceflight which should be mentioned. This concerns the intrusion of outsiders on the privacy of the group as a whole. Several Russian studies have revealed dramatically negative reactions of confined individuals to outside observation (Lebedev, 1975, pp. 83, 85). It is readily understood how such observation would be viewed as particularly offensive, since the person making the observation remains safe and unexposed. This lack of relational symmetry adds the dimension of "subject-object" to an already strained privacy arrangement. Several episodes that have occurred in space may have had some relationship to the perceived intrusion by outsiders on the group. Astronauts on Apollo 7 removed the sensors which were providing the ground with a record of their physiological functions. Also, it has been reported that cosmonauts on Soyuz 36 shut off their radio contact with the ground. A candidate incident also occurred on Skylab 4. The crew had decided among themselves to circumvent established procedures involving space sickness, forgetting that their conversation was monitored by mission control. The encounter which followed is thought to be responsible for a general atmosphere of friction which lasted throughout much of the mission. Although, in this case, the astronauts could not convincingly claim that they had a right to alter procedures, their behavior indicated that they felt their privacy had been wrongfully invaded. Group privacy rights of spacecrews can be expected to be come an increasingly important issue.

COMPLEX EFFECTS

Environmental pressure, in space as elsewhere, can result in complex patterns of responses. Multiple stressors can yield responses which are different in kind from those associated with the component stressors; furthermore, effects may first appear, or become pronounced, after the offending stimulus has been withdrawn.

Multiple Stressors

In assessing how the conditions of spaceflight may affect behavior, it is important to note that stressors are rarely isolated or independent, and that interactional effects may vary in complicated ways. Murray and McCally (1973) provide a review of laboratory studies involving multiple or combined environmental stressors. The measures of interest in these studies were either physiological or perceptual shifts, or changes in one or more performance tasks. These authors note that environmental stressors may, hypothetically, interact in three general ways: by addition, when the physiological effects of multiple stressors are equal to the linear sum of the single effects; by synergism, when the combined effects are greater than the simple sum of effects; or by antagonism, when the total effect is less than the linear sum of the single effects. Complex patterns of response to multiple stressors have also been reported when psychological adjustment and social interactions are observed. The more interesting effects here result from stressors that combine to reduce the response of a single stressor alone. In one study, subjects experiencing the stress of confinement exhibited more hostility when they had sufficient space than when they had the added stress of being crowded (Smith and Haythorn, 1972). Palamarek and Rule (1979) found that men who had been insulted in a comfortable room showed more aggression than men who had been insulted in a very hot room; Baron and Bell (1976) report that high temperatures facilitated aggression among subjects who had received a positive evaluation from their experimental partners, but that high temperatures actually inhibited aggression among subjects who had been negatively evaluated. In a study of groups at the French Antarctic station, it was found that problems of psychological adjustment "paradoxically appear to increase with greater physical comforts" (Crocz, Rivolier, and Cazes, 1973, p. 362).

One explanation of why hostile behavior should decrease with added stress comes from the study of Baron and Bell (1976). These authors hypothesize that, for instance, a reasonably comfortable person, on experiencing a high ambient temperature, reacts with aggression; however, for a person made to feel uncomfortable by a negative evaluation, and then exposed to a high ambient temperature, the drive to escape or to minimize the discomfort may overshadow aggressive inclinations. An alternate explanation for the lessening of aggression following multiple stressors is that a certain energy level is required either to complain or to act out. Below that energy level the person must spend his efforts coping rather than objecting. Whatever the dynamics, multiple stressors are likely to extract a high cost, both physiologically and in terms of reduced performance, even though they may serve to limit the expression of aggression.

Aftereffects

Stressors present in an environment can influence behavior even after the triggering stimuli have been withdrawn. Cohen et al. (1980) found decreased performance on a cognitive task among children tested in a quiet environment but who were chronically exposed to a noisy environment. Examining the literature related to aftereffects, Cohen (1980) concludes that negative effects are not restricted to particular stimuli, but occur as a result of a wide range of both physical and psychological stressors. Although most studies of aftereffects have examined performance decrements, social behavior can also be affected. For extended spaceflight, a particularly worrisome form of poststressor behavior involves the finding of decreased sensitivity to, and increased hostility toward, others following exposure to stress (Cohen, 1980). Glass and Singer (1972) report that the onset of negative responses may not occur until sometime after the passing of the stressful situation (p. 10):

> It is as though the organism does not experience maximal stress until he is no longer required to cope with the stressor. It is only then that the behavioral consequences of the event become evident.

Interventions that increase personal control over the situation or that render the stress more predictable have been found to mitigate poststressor as well as stressor effects (Glass and Singer, 1972; Cohen, 1980). Various hypotheses have been offered as to why environmental stressors should result in later performance decrements. A widely accepted explanation, based on the findings of Selye (1976), is that after a time of coping with stress, exhaustion sets in, with this exhaustion showing itself in various behavioral changes. Another explanation, similar to that offered for multiple stressors above, is that negative behavior requires an expenditure of energy that may be unavailable during stress. If this latter explanation is correct, negative reactions may signal the beginning of recovery.

SUMMARY AND CONCLUSIONS

The habitability challenge of early spaceflight was to sustain the individual for very brief periods of time. More recently, the goal has been to maintain a few individuals for several months within some acceptable limits of discomfort. For the future, the challenge will be to provide not only a sufficient, but an enriching environment for groups of people for periods of a year and more.

Within the vastness of space, areas in which the astronaut can sleep, work, recreate, etc., are necessarily limited. An important question is how limited these spaces can be before they begin to interfere with a human's normal functioning. Since all areas in a spacecraft are "primary" (i.e., areas in which the individual spends a large amount of time), high density in space is likely to result in physiological and psychological effects. These effects could show themselves in specific biomedical responses, in general health problems, and in performance decrements. Therefore, volume considerations must command continued attention from a general habitability standpoint.

Although we now have a reasonable understanding of the spatial needs necessary to sustain a person for brief periods, we have only limited understanding of what constitutes adequate space over long durations. Nor do we understand how the number of people sharing an area affects each individual's requirements. To date, there is not even agreement on whether large crews require more or less area per person than small crews. When the direction of this effect is ascertained, we can begin to explore the details of the relationship between crew size and spatial needs.

Several questions emerge regarding how the space that is available should be allocated. Is private space necessary and if so, how should it be structured? Should public areas be multipurpose, or used for specific activities? If particular areas are used for specific activities, how can these areas be set apart or "marked" in order to provide maximal psychological distance? And, although astronauts are provided with a 360° world, they continue to operate as if they lived in a modified two-dimensional world. If astronauts can learn to use all available surfaces, how does this learning take place and what are the time considerations? Such research would benefit not only the space effort but environmental psychology in general since space, by exposing individuals to the totally new experience of 0-g, offers a unique opportunity to examine how unfamiliar environments are mastered.

There are numerous environmental factors which singularly are of only minor importance, but which together determine the quality of life in a spacecraft. Although few of these factors have yet posed serious problems, certain questions suggest themselves as deserving attention in extended spaceflight.

Food emerges as an important habitability consideration in many confinement situations. Various studies indicate that an even greater emphasis is placed on food and on mealtime generally when individuals are confined. It will be important to determine what role food consumption, as a diversionary or social activity, will play in long-duration spaceflight. Also, it will be important to determine how the use of food can be influenced by other environmental factors. For instance, it is likely that the greater the satisfaction with work and leisure activities, and the more prevalent the opportunity for social interactions, the less will be the interest in food. A psychophysical question was suggested by the astronauts of Skylab. These crewmembers observed that food which had seemed adequately seasoned prior to flight tasted bland in space. This observation raises the question of whether taste and/or odor thresholds shift in weightlessness, a possibility that should be investigated.

Hygienic facilities in space habitats are still only marginally adequate. When designs reach a higher level of utility it will be possible to evaluate hygienic facilities in terms of more demanding requirements such as acceptability over long durations, ability to support physically active space travelers, etc.

Temperature and humidity can also be expected to affect the adjustment to space living. Cold temperatures can influence an individual's ability to perform manipulative tasks as well as cause general discomfort. High temperatures can result in both decreased performance levels generally and aggressive displays. Since individuals differ in what they consider acceptable temperatures, space travelers may have to consider temperature preferences along with other features of compatibility.

We know that decor assumes more or less importance depending on the conditions of confinement. Decor is found to be more important when crews are composed of both men and women and where there is little work required; decor is less important with all-male crews or where there is meaningful work to be performed. Although people differ in what they consider desirable, there seems to be an almost universal aversion to environmental uniformity, along with some evidence that, over time, people seek greater complexity in the environment. In the closed environment of space, it will be necessary to test tolerances to both differences and sameness in design. Flexibility is likely to be fundamental to any spacecraft design. However, even this recognized value will have its limits, since a living environment must provide an opportunity for stability as well as for adventure. Lighting offers an opportunity for rapid environmental shifts. Just how lighting can and should be used, both from a design and a performance perspective, requires further investigation.

To date, olfactory adaptation in spaceflight has apparently kept ahead of olfactory buildup. However, odors may present problems in future flights, especially when humans and experimental animals share the same living area, and when new crewmembers are rotated into the closed environment. We need to determine tolerance levels for various odors and to understand more fully how odors work in concert.

Noise problems are among the more serious habitability concerns of long-duration spaceflight. We have some understanding of the biomedical, performance, and other changes associated with various noise stimuli in a changing terrestrial environment. However, we know little of these effects when, as in space, noise stimuli are virtually unremitting. A complicating factor in understanding the effects of noise is that an individual's response can be influenced by certain mediating factors. For instance, noise can cause little disruption if the individual believes that the noise is necessary, or that the person causing the noise is acting out of necessity or in a considerate manner.

Tolerance levels to various noise stimuli appropriate to space need to be determined, taking into consideration both adaptation effects and individual differences. We also need to establish how spacecraft noise sources can be adjusted or varied to provide maximum relief. Another important issue concerns the relationship of noise to speech communication, especially when atmospheric pressure is low. There are also questions of how sound, and especially music, might be used to enrich the space environment and, perhaps, to increase productivity. As with issues of decor and temperature, there are further questions of personal preferences and of crew compatibility.

The area of recreation and exercise is only now emerging as a research field. Here the approach must be to observe how people behave, both in their normal lives and in confinement; to suggest explanations for these behaviors, testing explanations where possible; and to relate these findings to the needs of extended spaceflight. We know that people in confinement adopt an extreme work orientation and exhibit little interest in leisure activity. It has been suggested that a more balanced approach will occur with time and with experience in confinement. This hypothesis needs testing. If situations of long-duration confinement can be observed, or if confinement studies are conducted in which the same participants are used more than once, special note should be made of any changes in the use of leisure time.

An overall goal for long-duration spaceflight will be to foster the kinds of leisure activities which will contribute to the general health of the individual. Kleiber (1980) has found some support for the hypothesis that individuals who are actively engaged in their free time are psychologically healthier than more passive individuals. Yet, when confined subjects engage in recreational pursuits, they opt for the passive or noninteractive variety, probably to a greater extent than they do in their everyday lives. An important question for long-duration spaceflight is why leisure pursuits tend to the passive and/or noninteractive variety. Once the basis of this choice is understood (e.g., reduced morale, fear of conflict, preference for the familiar, etc.), we can determine whether the choice of pastime is likely to contribute to psychological and relational problems in extended spaceflight, and if so, how alternative activities can be encouraged. For instance, can a matching of the individual's interest to recreational provisions reverse the trend to passive use of leisure? For those activities which cannot easily be transferred to space, what kinds of activities can be substituted satisfactorily? Can a different (more heterogeneous) mix of individuals lead to more active involvement? Can activities such as the tending of gardens or competing with oneself circumvent the occasion for interpersonal conflict while maintaining the individual in an active mode? And, how can the benefits of passive recreation be maintained or increased? Reading is a favorite pastime in confinement. In space, electronic media will replace the written page for leisure reading as well as for work-related tasks. It must be determined how present or planned information systems, and especially visual displays, will meet the demanding requirements of space.

With reference to exercise, we need to assess the kind and extent of activity that is necessary to maintain an individual at peak physical and psychological health during prolonged weightlessness. We must then determine how to motivate astronauts to perform the needed exercise. If an exercise program is externally imposed, it will probably be maintained for only a short period of time. For long-term spaceflight the motivation to exercise must come from the space travelers themselves. We have little data on what motivates individuals to exercise even under ordinary circumstances. The following hypotheses are offered for consideration:

- An individual is generally more likely to maintain an exercise program for reasons of general appearance than for reasons of health.

- An individual who exercises for health reasons is likely to do so in response to a specific and salient health problem, unlikely to do so for diffuse health benefits.

- When, following exercise, the individual begins to feel better (increased energy, sense of well-being, etc.) the exercise program has maximal chance of being maintained.

Some data suggest that cognitive strategies are very important in maintaining an exercise commitment (Morgan, 1978). Although these findings are based on methods employed by extraordinary athletes, the strategies deserve testing on others as well. Space offers new opportunities for space sports, but we don't know if such sports will be of general interest. It is likely that exercise facilities, like recreational facilities, must cater to the special interests of individuals. We need to determine if, by addressing the interests of the individuals, and by providing them with a means of increasing skills and tracking progress, we can improve the likelihood that an exercise program will be maintained.

A critical habitability concern for long-duration spaceflight is the need for privacy. Privacy is a balance between forces to affiliate and forces to withdraw. This balance helps an individual define himself and his relationship to others. An important question is just what this balance implies in space.

In confinement, many of the accustomed mechanisms for privacy regulation are either lacking in the environment, or inappropriate to the situation. Characteristically, confined individuals respond to this unfamiliar circumstance by first exceeding normal privacy limits, and then by withdrawing from others in the group. Although withdrawal is an effective form of privacy regulation, it is destructive to group cohesion and ultimately to mission goals. For extended spaceflight, alternative means must be found both to protect privacy and to foster effective group functioning.

To meet the individual's need for privacy, we must understand what this need entails and which events or conditions most threaten privacy in space. Is there a specific need for physical or psychological separation? Can privacy needs be met through personal possessions (other than space), through individualistic clothing, or by some architectural arrangement/design?

Confined individuals have shown a penchant for oversharing personal information, suggesting that this is an area requiring particular attention for spaceflight. Generally, it has been observed that friends commit fewer privacy violations than new acquaintances. Similarly, we know that when one is with friends, space limits are perceived as less pressing than when one is with strangers. If these findings are verified in the confinement of space, it suggests that established relationships would transfer well to spaceflight. Preliminary data indicate that women adjust more favorably to a crowded environment than men. However, there is also the suggestion that, over time, women suffer more from crowding than men. These studies require verification and extension. Perceptions of crowding have been found to relate to certain personality characteristics. We need to explore more fully just which personality characteristics are most predictive of spatial demands and how these characteristics relate to other requirements of spaceflight. An interesting question concerns the possibility of training individuals for privacy regulation in space. The goal of this effort would be to foster both group cohesion and adequate levels of personal separation. Ideally, privacy violations would be avoided; practically, one would seek to reduce privacy violations to some manageable level and, at the same time, to provide options (other than withdrawal) for dealing with the violations that do occur. We need to determine whether, and to what extent, training could accomplish these ends.

It is necessary to distinguish among the many issues that relate to privacy, such as personal space, control over information about oneself, etc. And, in order to deal with a particular situation, it is important to understand what a person means when he implies that his privacy is violated. For instance, if a person complains of being "crowded," he or she may be expressing a need for personal space (in which case barriers would help) or a need for more space generally (in which case barriers would compound the problem.) Privacy violation can occur as a result of a single event or condition, or because of a compilation of events and conditions. Privacy violations in space are likely to result from multiple events or conditions, and this compounding effect needs to be taken into account.

In addressing privacy issues, it is important to appreciate the entire situation in which the space traveler must function. In attempting to deal with a problem of too much exposure, one must guard against creating a larger problem by cutting off the individual's access to others. One must also not overlook the costs associated with establishing and maintaining privacy. These costs arise because of conflicts between the needs of the individual and the needs of other individuals or of the larger society. If such conflicts occur in space, the victory of the individual will not be taken easily.

Various environmental stressors can combine to trigger behavior which is different in kind from the response to individual stressors. And, in some cases environmental stressors can exert negative aftereffects such as decreased performance or aggressive behavior (Cohen, 1980). For spaceflight, we must consider the environmental effects over long periods of time which include not only the flight itself but pre- and post-flight adjustment.

One of the more promising findings from environmental research is that many of the negative effects of environmentally induced stress can be ameliorated if the stressor occurs in a predictable and controllable fashion. Stokols (1979) points out also the importance of understanding the salience of the event to the individual. A better understanding is needed of how predictability and control counter the effects of stress (Stokols, 1979). However, the identification of mitigating variables such as these does indicate that it is possible to devise methods to help individuals deal with stressful environments, and that research explicating such variables can have an influence on training strategies.

Another method of reducing the environmental stress of space is to match, as far as possible, the individual with the environment. Although the conditions of space will be demanding, they will also present challenges which for some individuals could offset the inconveniences and disruptions inherent in the environment. If conditions do not tax or exceed the individual's adaptive resources, then for this individual they do not constitute an environmental stressor (Stokols, 1978a). When attention has been given to the individual and to the environment, it is probable that further attention to the relationship between the two, i.e., the person-environment fit, will be required.

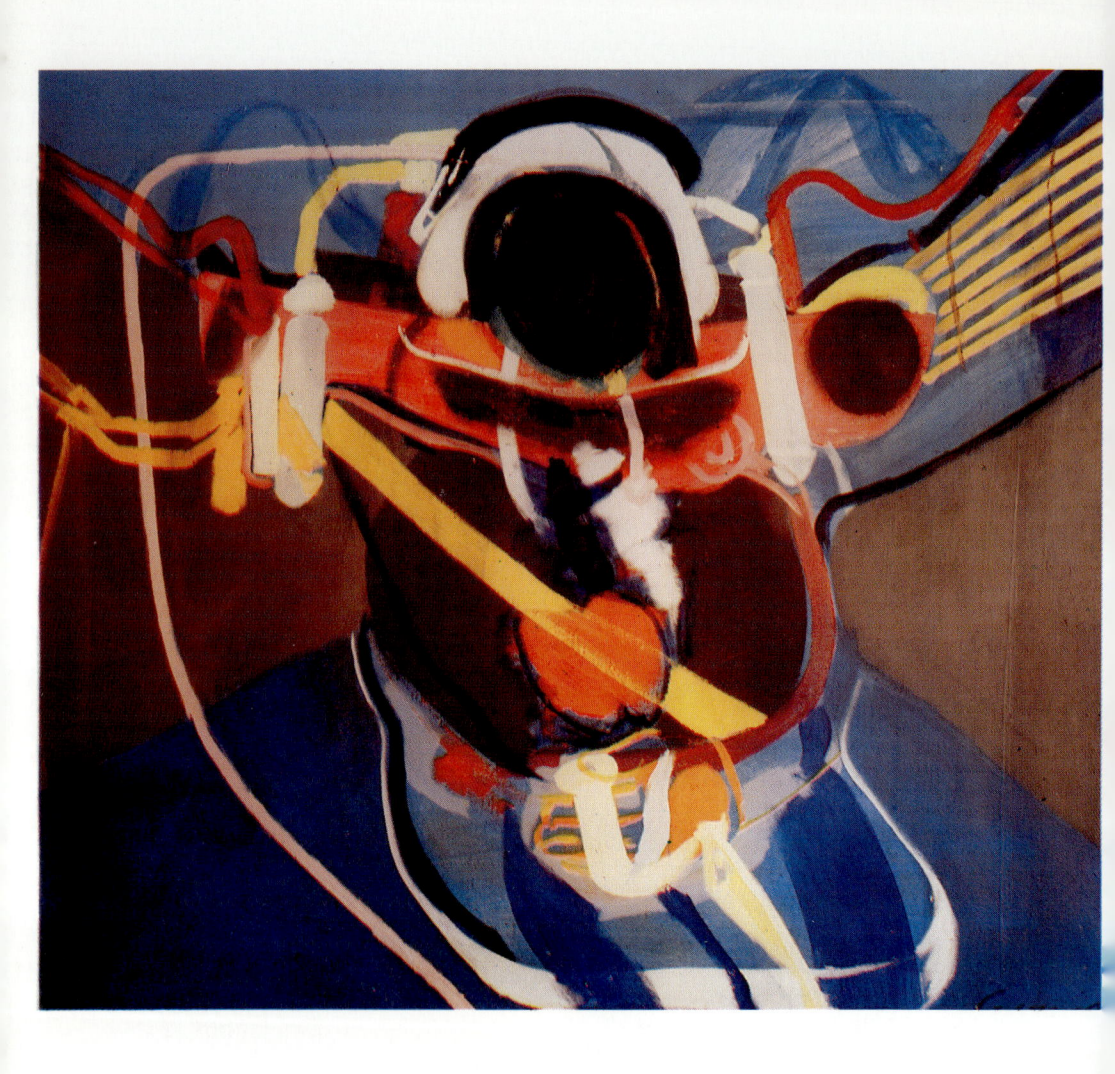

Artist: Hans Cremers

IV
PERFORMANCE

Maintaining skilled performance during extended spaceflight is of critical importance to the health and safety of crewmembers and to the overall success of the mission. During flight it is crucial to know the extent to which the astronauts are operating at levels sufficiently high to ensure safety and mission integrity. Performance in space can also be used to track the effects of weightlessness, sleep deprivation, isolation and confinement, etc.

Performance assessment is also important at other stages of mission development. For example, preflight performance assessment has been a useful tool in understanding engineering and craft-habitability constraints leading to guidelines in the layout and structural composition of spacecraft living and working quarters (Fraser, 1968c; Barnes, 1969), and in instrument and control-panel designs (Shackel, 1959; Woodson and Conover, 1965). Performance assessment has also been used in the selection of astronaut candidates (Wilson, 1959; Lovelace, Schwichtenberg, Lift, and Secrest, 1962; Lamb, 1964), in their training, and in the determination of their readiness for flight (Voas, 1961; Kelly and Coons, 1967; Woodling, Faber, van Bockel, Olasky, Williams, Mire, and Homer, 1973). Indeed, the measurement and evaluation of performance is basic to the role of men and women in space. In this chapter we consider performance, methods for its assessment and prediction, factors that affect performance, and possible means to sustain optimal performance.

DESCRIBING PERFORMANCE

The performance area of special interest is the human in his or her work role. The study of human work capacity, or ergonomics, has much to contribute to our understanding of how to achieve mission objectives. However, before we approach the task of measuring complex operator performance, we should first describe what is to be measured.

Work Requirements in Space

Although specific enumeration of astronaut work units is beyond the scope of this book, it is possible to suggest various broad categories of work requirements. According to Yeremin, Bogdashevskiy, and Baburin (1975) the duties of an astronaut crew have traditionally involved the following tasks:

- Monitoring and controlling the operations of on-board systems
- Controlling spacecraft movements in performing various dynamic operations (orientation, stabilization, approach, docking, orbital correction, descent from orbit, and landing)
- Conducting radio communications and television reporting
- Conducting visual observations, scientific experiments, and investigations
- Assembling and disassembling individual units of the spacecraft, and performing various operations outside the spacecraft
- Operating special gear
- Carrying out onboard documentation

Of course, within each of these categories, there are many individual work units that must be mastered and integrated into a total program in order to carry out a particular mission objective successfully. Also, the division of labor within crews has differed markedly as progressively more complex missions have been employed. In the earlier missions, all astronauts were exposed to similar training programs and performed similar operations aboard the craft. However, as the space program has evolved, the responsibilities of crew members have become more specialized, although considerable cross-training is still given.

Human Performance Abilities

Various attempts have been made to categorize classes of behavior that describe human functioning. For example, Miller (1965) uses such terms as scanning, identification of cues, interpretation, decisionmaking, and short-term and long-term memory. Gagne (1964) suggests discrimination, problem-solving, sequence-learning, identification, etc., as functional categories. Dimensions such as tracking, vigilance, arithmetic, and pattern comparison are used by Alluisi and Thurmond (1965). These categories represent rational, descriptive approaches determined essentially in an armchair fashion. Other investigators have suggested that a small number of descriptive categories is inadequate to systematize the many components that comprise human performance. These investigators argue for an empirically determined taxonomy of human tasks. A refinement of this latter approach would be to identify common elements among the various tasks.

Using experimental-correlational studies and factor analysis, Fleishman (1967) examined a wide range of perceptual-motor performance tasks. His intent was to define the fewest independent ability categories that could be used to describe performance over the widest variety of tasks. The assumption was that complex skills can be described in more basic ability terms. For example, the quality of performance an individual can achieve in operating a turret lathe may depend on the more basic abilities of motor coordination and manual dexterity. Using this logic, Fleishman generated 11 psychomotor factors (e.g., control precision, reaction time, aiming, etc.) and nine factors in the area of physical proficiency (e.g., trunk strength, explosive strength, stamina, etc.) which consistently account for the variance in many performance tasks (Fleishman, 1960, 1962, 1964). While Fleishman's approach has been highly successful in assessing laboratory tasks, the results have been less encouraging when more complex operational settings have been investigated. Since we do not yet have a completely satisfactory method of defining the task abilities which underlie various job requirements, we cannot say with complete assurance exactly what human functions define the job of the astronaut in space. We can, however, approximate these requirements.

One approach to defining the functional requirements of astronauts is provided by Parker, Reilly, Dillon, Andrews, and Fleishman (1965). These authors reviewed and analyzed the technical literature with special emphasis on the factor analytic work of Fleishman. Studies involving regression-analysis models, physical-proficiency studies, and control-dynamics studies were also considered. This process generated a considerable list of basic task dimensions. After making a task analysis of the activities most likely to be required of personnel operating in space vehicles, 18 basic abilities were identified. These 18 abilities were readily classified into six categories: fine manipulative, gross positioning and movement, system equalization, perceptual-cognitive, reaction time, and mirror tracing abilities.

The task dimensions identified by Parker and his colleagues are consistent with similar classification systems suggested by Berlinger, Angell, and Shearer (1964) and by Christensen and Simons (1970). The listing may be taken as one fairly comprehensive representation of the kinds of performance demands made on an operator in space. The next step is to determine how to measure such factors.

ASSESSMENT OF HUMAN PERFORMANCE

A wide range of empirical approaches to the assessment of human performance has been proposed or actually used in astronaut evaluation. These approaches can be conceptualized as lying along a dimension of apparent or face validity. A depiction of how closely the performance-assessment conditions approximate actual mission requirements is given in figure 1. The low-fidelity discrete-task assessment techniques are at one end of the continuum, and measurements of performance during actual missions are at the other end of the continuum. The several approaches to performance assessment are discussed in the following sections, beginning with the low- and proceeding to the high-fidelity end of the continuum.

Discrete-Task Assessment Techniques

Performance analysts have long relied on the use of test batteries consisting of one or a number of discrete, individual tasks to measure such factors as vigilance, reaction time, tracking, limb steadiness, coordination, and perceptual speed. For example, Mackworth's (1950) classic technique for assessing vigilance involves subjects attending to a clock hand that usually moves in single steps, but occasionally gives a double jump. Subjects are instructed to make a

APPARENT VALIDITY WITH RESPECT TO OPERATIONAL CONDITIONS		CATEGORIES OF PERFORMANCE ASSESSMENT
IN-FLIGHT PERFORMANCE ASSESSMENTS	⊤	HIGHEST
FULL MISSION SIMULATION		
PARTIAL MISSION SIMULATION		
MULTIPLE-TASK PERFORMANCE BATTERIES		
SINGLE-TASK ASSESSMENT APPROACHES	⊥	LOWEST

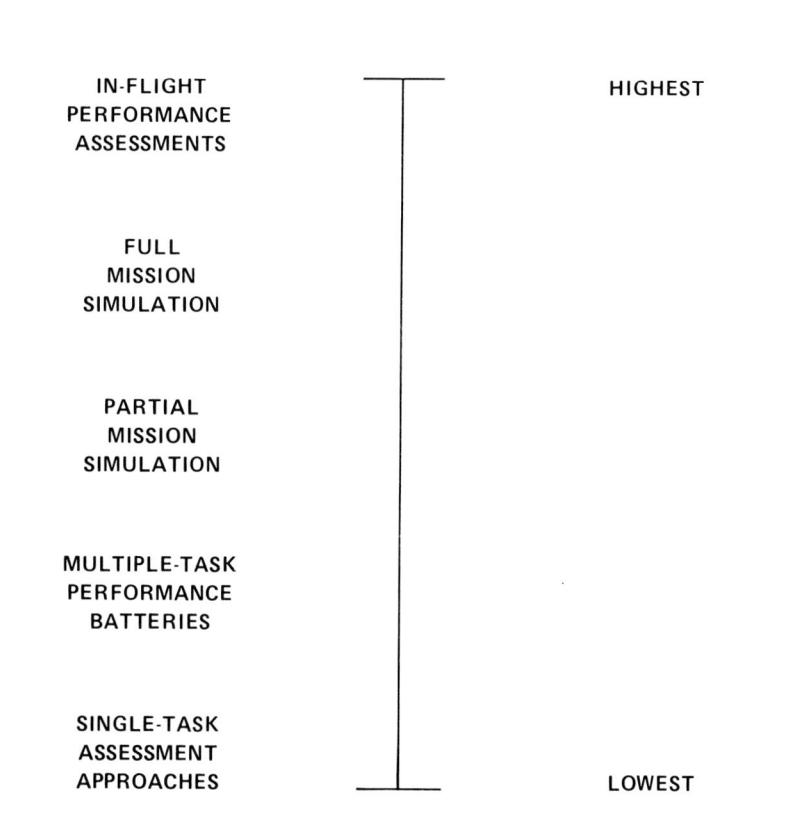

Figure 1.— Comparison of performance-assessment conditions with actual mission requirements.

response via some recording device only when they observe the double jump. This single task has been a standard technique in the assessment of effects of monotony and fatigue associated with the monitoring of radar and control equipment by air crews under operational conditions. A discrete-task technique used for assessing tracking ability is the rotary pursuit task (Ruff, 1963). Here, the subject attempts to maintain constant contact between a stylus held in his hand and a small disk on a rotating turntable.

The essential assumption underlying the use of such tests is that each task measures some basic human ability than can be related directly to the performance of more complex, real-world tasks. By using discrete basic ability tasks under highly controlled conditions, it is assumed that useful information regarding the more complex performance of actual mission tasks can be inferred. The discrete-task-assessment approach has been preferred in many laboratory studies, particularly those assessing the effects of simple factors on a single aspect of performance. Discrete tasks have the advantages of being relatively uncomplicated to use and low in cost. Performance is rather easily defined and quantified with such tests. However, there are several disadvantages in the use of the techniques, especially with highly skilled special populations such as astronaut operators.

First, there is the issue of apparent validity. The tests, even if proven to tap performance on tasks needed in spaceflight, usually do not resemble the actual mission tasks much, if at all. For example, holding a stylus in holes of decreasing diameter without touching the edges may indeed be one measure of limb steadiness, and limb steadiness may be necessary for successful performance of in-flight mission-related tasks; however, that single-task measure of limb steadiness may have little or no direct relation to the in-flight limb-steadiness performance required of astronauts. Also, the point at which a change in task performance occurs in the laboratory may be considerably different from the point at which performance is affected under operational conditions. Therefore, the degree to which the results of such methods can be generalized to operational settings is questionable. These individualized tasks appear best suited for settings where there is a specific question regarding the effect of some isolated variable upon a given aspect of performance. They can help to clarify issues, for example, by identifying questions towards which more sophisticated techniques should be directed; they are of lesser value when used individually in attempts to quantify probable levels of performance under operational conditions.

Fleishman (1967), Parker (1967), and others have argued that discrete-task measures are most useful when factor analyses have identified them as pertinent to features of the more complex operational task system. By selecting those tasks that have elements in common with the operational condition, the likelihood of obtaining relevant, generalizable data can be maximized. This, of course, requires that the investigator be able to define and quantify the specific task demands of the real-world environment.

The usefulness of discrete-task techniques might be enhanced by employing them under conditions that represent the operational setting. One of the important conditions of the operational setting — a condition most often lacking in the use of these sequentially administered tests — is the multitask, concurrent-demand, time-sharing requirements of the work environment. Some investigators have attempted to deal with this issue by combining two or more of these tests into a concurrent-task situation. There is some evidence to suggest that concurrent testing does enhance the sensitivity of the tests (Brown, 1978), but we do not yet know whether it increases the tests' validity, as theory would suggest.

Finkelman and Glass (1970) employed a dual-task methodology requiring simultaneous performance of a primary task (tracking) and a secondary task (recall). They found that unpredictable, uncontrollable noise produced decrements in performances of the subsidiary task, but not in the primary task. Likewise, Bell (1978) found, using a tracking task and a number-processing subsidiary task, that noise and/or heat had detrimental effects on subsidiary-, but not on primary-task performance. The basic notion behind these and other supportive findings is that operators or experimental subjects have only a limited capacity to process information and to respond. When some stressful condition in the environment reduces a person's ability to allocate resources to the tasks, the focus of effort will shift to the primary task, allowing decrements to occur in the secondary task. Thus, performance may deteriorate on what is perceived to be the less important task, while remaining unchanged on the more important task. Such findings represent typical experiences in operational settings. Although the concurrent use of single tasks may increase measurement sensitivity, such an approach still fails to represent the much more complicated real-world work situation.

Multiple-Task Batteries

In efforts to avoid, or at least reduce, the disadvantages of the discrete-task methodology, various investigators have adopted a task-battery approach to the study of human performance. Alluisi (1967) describes such a technique as "synthetic" because it synthesizes various tasks into a general work situation that requires time sharing. Although the tasks individually may have low apparent validity in terms of any specific application, they are selected to represent functions that operators are called upon to perform in a variety of man-machine operational settings and collectively constitute a reasonable and credible "job" to operators and experimental subjects (Alluisi, 1967, 1969; Chiles, Alluisi, and Adams, 1968).

A multiple-task performance battery has been described and an operator's panel pictured by Chiles and his colleagues (1968). Behavioral measures are obtained from the operator's performance in working at the tasks presented with the panel. In the final version of this test battery, six tasks (three passive, watch-keeping tasks and three active, computational/procedural tasks) are displayed at each of the identical work stations — one station for each member of a four- or five-person crew. Communications functions are measured indirectly in the performance of the three active tasks. For further details, including citations of the individual technical reports of research conducted with this approach, consult the three brief summaries (Alluisi, 1967, 1969; Alluisi and Chiles, 1967), the general description of synthetic-work methodology (Morgan and Alluisi, 1972), and the report of the work-rest scheduling research (Chiles et al., 1968).

Using the synthetic-work, multiple-task performance battery, information relevant to a wide range of abilities can be acquired simultaneously, minimizing the difficulties of cross-experimental comparisons often cited as a problem when different subjects in different experiments perform different tasks. The synthetic, multiple-task battery approach has been used by Alluisi and his colleagues to investigate work-rest schedules and also a wide range of variables of potential impact to the crew of spacecraft: desynchronosis (disruption of the body's circadian activity), sleep loss and the recovery therefrom, confinement, illness, etc. (see, e.g., Adams, Levine, and Chiles, 1959; Adams and Chiles, 1960, 1961; Alluisi, Chiles, Hall, and Hawkes, 1963; Alluisi, Chiles and Hall, 1964; Alluisi, Beisel, Bartelloni, and Coates, 1973; Beisel, Morgan, Bartelloni, Coates, DeRubertis, and Alluisi, 1974). Some of this work is discussed in a section of this chapter entitled "Issues in Astronaut Work Regimes."

The synthetic test battery would seem to be the most generally useful approach to the selection and preliminary training of astronauts. It is less expensive than most simulation systems and permits more than a single individual to be measured at a given time. It yields a maximum of information regarding subjects' performance capabilities on several psychophysiological dimensions and can be used to investigate a range of environmental conditions. However, there remains the issue of validity, particularly when mission training is involved. To fully prepare spaceflight crews for the rigors of space, simulators are required.

Partial- and Full-Scale Simulation

Mission simulators provide the highest degree of fidelity possible in the simulation of true operational conditions. They are particularly important in space mission design because there is essentially no opportunity for a graduated series of practice efforts under true operational conditions before the mission takes place. Since space mission crews must be trained and highly proficient in their tasks before the flight, it is imperative that high-fidelity simulator systems be available for training on specific, individual aspects of the mission (partial simulation) and for the completely integrated "dress rehearsal" simulation of the mission (full-scale simulation).

Simulator systems have formed an integral part of our manned space effort. From the beginning, high-fidelity simulators provided much of the information for man-machine design factors and training requirements, and allowed for total system inspection (Link, 1965; Berry, 1967). For example, Mercury astronauts participated in four centrifuge programs to investigate crew capability to control the spacecraft manually during the high-acceleration loads imposed during launch and entry (Link and Gurovskiy, 1975). There was general agreement that the centrifuge technique was the most useful high-g environmental simulation device used during training and that the accurate representation of the actual craft was extremely valuable. Mercury crews used a yaw-recognition simulator during training to familiarize themselves with the wide-angle optics of the periscope they were to use for outside viewing during the actual mission. In addition, a simulation training program using a multiaxis spin-test inertia facility trainer was used to examine reaction to recovery from tumbling flight. In all, some type of simulation was provided for every significant flight phase which required integrating the crew with the flight plan and the ground-support elements of the Mercury project.

As the complexity of spacecraft and missions increased with the advent of the Gemini Program, the importance of simulator training increased. Results from the Gemini Program strongly suggested that crew response during flight was highly dependent upon the fidelity of simulation training received before flight (Kelly and Coons, 1967). During the Apollo Program, simulators were again a significant part of preflight training (Ertel and Morse, 1969; Brooks, Grimwood, and Swenson, 1979). During the Skylab program, simulators depicting diverse visual displays were used routinely in training. Combining such tasks as target recognition and status assessment, complex and time-dependent point operations, malfunction analyses, and rapid response to flare and other transient events, astronauts were able to use high-fidelity simulations to master operations important to the flight and to the experiments conducted aboard the spacecraft (Holt and da Silva, 1977). Simulation continues as an essential preparatory tool for Shuttle (Bilodeau, 1980).

New developments in computer graphics have resulted in significant advancement in simulation experiments. At Johnson Space Center's Graphics Analysis Facility, researchers are able to simulate many of the concerns of man-machine integration without the necessity of hardware development. Full-scale modeling will continue to be required; however, for many applications, computer graphics offer a flexible alternative to this more expensive and time-consuming approach.

While there is no doubt that simulator studies have been useful during all stages of spaceflight, they have been employed primarily for training (e.g., the use of neutral buoyancy simulations achieved through water immersion (Machell, Bell, Shyken, and Prim, 1967) and the Keplerian trajectory aircraft flights (Nicogossian and Parker, 1982) and for evaluating the operation of the entire man-machine complex. In contrast, they have been used only rarely to assess the performance of humans within a system. A partial exception is the work of Milton Grodsky (Grodsky and Bryant, 1963; Grodsky, Warfield, Flaherty, Mandour, and Hurley, 1964; Grodsky, Glazer, and Hopkins, 1966a; Grodsky, 1967), who employed man-machine simulation to assess the reliability of human operators and to investigate performance changes across time.

As one example, Grodsky, Moore, and Flaherty (1966b) compared performance levels during a 7-day, integrated lunar mission simulation with asymptotic performance levels previously obtained during a 5-wk baseline training period. The simulator used was considered to be a good representation of the Apollo command module. Subjects tested included nine Air Force test pilots with pilot qualifications similar to those of the astronauts. The tasks performed during the simulated lunar-landing mission were divided into four categories: switching, flight control, information handling, and malfunction detection. The results of the first two tasks demonstrated some changes from baseline data. Performance was superior to baseline levels for some types of navigational switching tasks, but was degraded for certain other tasks during specific mission phases, and flight-control results demonstrated suggestive, though not statistically significant degradation.

Although Grodsky et al. (1966b) did not delineate any variables (other than the mission time line) that might have influenced performance, there are suggestions that workload, boredom, or task complexity may have been involved. For example, in terms of task complexity, the brake-and-hover requirement of lunar landing was clearly the most difficult, based on the number of trials required during training to achieve an adequate baseline. This was one of the areas that appeared to degrade during the simulated mission. Also, performance during the lunar-orbit insertion was relatively poor (20% to 35% degraded when compared with baseline), although this phase was conceptually and functionally identical to the translunar-insertion and transearth-insertion phases, which showed no such degradations. The authors hypothesize that the observed decrements prior to lunar landing may have been a result of emotional shifts, but they are unable to say if the prevailing mood was one of boredom or excitement.

Although the results of this preliminary study are not particularly striking, they do suggest that important performance effects can be identified and measured during simulation. Performance assessment under such conditions offers certain advantages over the discrete-task and synthetic-work approaches previously discussed. Mission simulation has very high apparent validity. This validity facilitates comparison of performance changes between test and mission settings. Short of actually studying in-flight performance, the simulation approach is our best single indicator of how overall performance under operational conditions may vary as a function of the combined conditions and stresses of the mission. Data obtained during training periods and during mission simulations can be used as a gauge with which to predict, and later to judge the sufficiency of performance observed during actual flights. Future studies should employ simulation as an experimental tool to explore the effects of such independent variables as task scheduling, work-rest cycles, sleep deprivation, and desynchronosis.

There are at least three drawbacks to the use of the simulated mission technique in the study of performance which should be pointed out: (1) some of the conditions of an actual mission, such as weightlessness, simply cannot be adequately replicated, and so their effects cannot actually be measured; (2) although high-fidelity system simulation may permit more direct comparison of specific system performance between test and mission settings, data obtained for one system may not be generalizable to other systems; and (3) the simulation techniques are quite costly in terms of all resources (relative to the two simpler techniques previously described).

In-flight Performance Assessment

In the early days of spaceflight, few quantitative performance data were obtained under operational conditions. During Apollo 15 and 16 (Kubis, Elrod, Rusnak, and Barnes, 1972; Kubis, Elrod, Rusnak, Barnes, and Saxon, 1972) and later on Skylab (Kubis, McLaughlin, Jackson, Rusnak, McBride, and Saxon, 1977), a major source of in-flight data became available through time-motion analyses of image and voice data. For instance, video and auditory recordings from Skylab Missions 2, 3, and 4 were analyzed to determine the amount of time required to complete various tasks. Comparable baseline data had been collected prior to flight. Eight tasks were employed, each subdivided into the many different components required to complete the task. The tasks selected were limited to those with standardized, repetitive maneuvers that would satisfy replication and conformity conditions. Analyses were conducted to determine the degree of performance change (amount of time required in flight versus preflight) and the time to return to baseline (preflight) levels.

The results showed that the first attempts in space to carry out mission tasks usually were inefficient. For example, during Skylab 2, the first in-flight task trial took longer than the last preflight trial in 68% of the cases (Kubis et al., 1977). Similar data taken during Skylab 3 showed 54% of the task elements took longer in space than on Earth, and a comparable figure from Skylab 4 was 58%. However, during each of the three missions, by the end of the second performance trial, approximately half of all tasks were completed within the time recorded for the last preflight trials. This suggests that to facilitate performance adaptation, tasks critical to mission success should be rehearsed early in the flight.

This type of in-flight data collection represents a good first attempt to quantify performance in space. However, the statistical data presented could be further refined. For example, the available reports cite only the number of tasks that took longer to perform in flight than preflight, but give no indication of how much longer. It would be important to determine whether these time periods represent significant delays. Also, the analysis is complicated by the confounding of tasks with the time at which they were measured; different tasks were performed on different days of the missions. In addition, future in-flight measurements should include a specification of the types of errors made. With the time-motion studies presently available, the only inferences that can be made are those involving comparison of fine, medium, and gross motor movements. For example, Kubis et al. (1977) concluded from their Skylab analysis that during adaptation to spaceflight, fine motor movements are affected more adversely than medium or gross motor movements during both intravehicular and extravehicular activities. This result was confirmed in the debriefing comments of the astronauts who reported that they had more difficulty with the control of small objects than of large objects. These findings suggest that extra preflight training should be given to manuevers requiring fine, delicate movements, and that, where possible, these tasks should be scheduled later in the mission, after crews have fully developed their fine motor dexterity.

Work-time ratios constitute another type of in-flight performance analysis which could prove useful in judging the overall work capacities of spacecrews. In one analysis (Garriott and Doerre, 1977), total estimated time associated with tasks accomplished was divided by the total number of hours available to work. This computation was termed the "work-efficiency ratio" and was computed for each crewman of the Skylab missions. (Time estimates were based on the time it took a trained crewman to complete the task on the ground.) A ratio of 0.50 was defined as a normal work day based on "normal" working conditions on Earth, where the hours of useful work (8 hr) is divided by awake time (16 hr). Astronauts maintained or exceeded the 0.50 level on most mission days. However, during the second day of Skylab 3 and the third day of Skylab 4, the work-efficiency ratios were 0.41 and 0.45, respectively. This decrease in work efficiency appears to have resulted from space sickness, as reported by the crewmembers. It might prove useful to combine the analyses of work-efficiency ratios with those from time-motion techniques to determine (1) the extent to which these measures of performance speed may be correlated and how they might be combined to yield maximum information, and (2) the relationship between performance speed and error rate.

Future Focus of Research on Performance Assessment

No single approach to the study of performance variables related to spaceflight can provide all of the resources and data needed. Instead, our future efforts need to focus on the coordinated and integrated use of all the assessment approaches discussed so far (Alluisi, 1975, 1977). Although mission simulations and actual in-flight measurements may be extremely valuable in gauging the overall functioning of crews and in helping to identify problems of concern, these approaches do not support inferences regarding cause-and-effect relationships. For example, there is no way of determining whether the changes in performance observed early in the Skylab missions were due primarily to motion sickness, lack of adaptation to weightlessness, increased arousal resulting from the excitement and novelty of the mission, or some combination of these or other factors. To investigate these variables more directly, we must rely upon more controlled laboratory tests, using the discrete-task and synthetic-work approaches. Important findings from research using such techniques can guide our assessment of performance under more complex, operational conditions.

Other performance-measurement problems also deserve attention. One problem is to define criteria for measuring performance under operational conditions. Until such criteria are defined, there can be no clear direction on how to validate and implement ground-based systems to aid our understanding of work in space. Also, more effort is required to ensure systematic quantitative and qualitative data on in-flight performance. Unfortunately, there are no plans for such performance studies in the Space Shuttle series.

Another problem area concerns task predictive validity under ground-based conditions (i.e., the ability of the ground-based task to effectively predict task performance in space). For research and development purposes, emphasis should be placed on the predictive validity of assessment approaches rather than relying on their apparent validity (i.e., validity based on the extent to which a test seems to measure the variable to be tested because of its similarity to the criterion measure), although for selection and training purposes apparent validity may have to remain a major consideration.

Finally, we need to relate relevant physiological indices to the performance aspects of behavior. Behavioral measures could be useful to determine flight crew status and to indicate changes in their capability to perform mission tasks.

ISSUES IN ASTRONAUT WORK REGIMES

Various aspects of the space environment can adversely affect performance. Many of the environmental, physiological, and interpersonal factors potentially affecting the astronaut, and therefore his or her performance, are reviewed elsewhere in this text. Here, we concentrate on those factors which are directly related to his or her ability to perform work-related tasks.

Factors Affecting Work Capacity

Weightlessness— Several important aspects of work in space are directly affected by weightlessness. The accuracy of psychomotor performance in space is affected by the lack of the gravitational pull that normally weighs down a person on Earth. Simons (1964) has shown that during parabolic flight, the body tends to approach a fetal position as the limbs move toward equilibrium during relaxed posture. Such postural changes have been verified in spaceflight (Thornton, Hoffler, and Rummel, 1977). Because of this postural change, the ability to reach and position the arm and hand accurately can be adversely influenced. Until adaptation occurs, there is a tendency to overshoot a target to be grasped. However, since the phenomenon is recognized, accurate grasping responses are typically not included as mission-critical tasks, especially in the early phases of the mission.

As mentioned previously, fine motor movements may be adversely affected in 0-g. This decrement in dexterity can pose potential problems for the manipulation of control panels, sensitive scientific experimental apparata, etc., although the problem tends to abate when adaptation to the 0-g environment has been achieved. Space system designers must ensure that switches are easy to manipulate, do not require unnecessarily delicate tuning, and are positioned to allow maximum individual access without the chance of accidentally changing one setting while attempting to alter another.

The large metabolic cost required to conduct work in space is also an important aspect in assessing 0-g work capacity (see review by Christensen and Simons, 1970). Experience has demonstrated that metabolic rates during extravehicular activities can be truly dramatic, and in the early Gemini flights, some scheduled maneuvers had to be cancelled because of astronaut fatigue. Even for relatively easy tasks, increased effort is required. This extra effort is related in part to the unique character of working in space where, once momentum has been imparted, new action is required to terminate the original movement. In addition, the astronaut finds it difficult to maintain his or her position in the absence of traction. Problems have also been encountered in moving when wearing a pressurized space suit (Covault, 1983). Although all of these issues have been effectively countered to some degree for short-term missions, the extent to which the increased workload may compound problems in more extended missions has not been determined.

Biomedical changes— Biomedical changes such as reduced cardiac activity and diminished musculoskeletal strength can affect work capacity. As discussed in chapter II, deterioration of these systems can jeopardize the health and functioning of crews and, if unchecked, can reduce the physical capacity of crews for extended high work output. For example, in-flight decreases approaching 10% were observed in exercise (vital) capacity in the Skylab 3 Pilot and the entire Skylab 4 crew (Sawin, Nicogossian, Rummel, and Michel, 1976). The extent to which countermeasures such as in-flight exercise, diet, and supplements will be sufficient to control these changes and to ensure optimal physical work capacity for long-duration missions is not clear.

Space sickness— The problems of space sickness pose an immediate threat to work capacity in space. Valuable mission time has already been lost during several flights because of the debilitating effects of space sickness, making this a high-priority concern for continued research.

One of the models used in space-related work capacity research involves the use of neutral buoyancy (submergence). Performance measurements taken during submergence have demonstrated many of the same difficulties encountered in space. Performance involving manual manipulation (Streimer, Turner, Volkmer, and Guerin, 1970) and force-producing capabilities (Streimer, Turner, and Volkmer, 1966) are degraded in the underwater tractionless mode. Postural and equilibrium changes have also been noted. For instance, Morway, Lathrop, Chambers, and Hitchcock (1963) found that under some conditions subjects exhibited a significant bias in the upward direction when reaching for a target. This tendency may be analogous to the overreaching problem encountered in space (Simons, 1964), as previously noted.

Factors Affecting Work Schedules

During early spaceflight, it was assumed that duty schedules would be long and arduous; it was therefore necessary to investigate how wake-sleep cycles, other than the customary 24-hr day with its uninterrupted 8-hr sleep, might be applied to astronaut crews. This section outlines research pertinent to the issues of work, rest, and sleep cycles.

Altered wake-sleep cycles— When wake-sleep cycles are altered from a 24-hr pattern, this change automatically produces a displacement in the patterning of normal sleep periods — a factor known to adversely affect sleep quality and quantity. In turn, such disruptions in sleep are known to affect performance adversely (as discussed in a later section of this chapter). Considerable research indicates that subjects placed on alternative sleep and wake schedules experience sleep problems (Carskadon and Dement, 1975; Weitzman, Nogeire, Perlow, Fukushima, Sassin, McGregor, Gallagher, and Hellman, 1974). Webb and Agnew (1975) observed subjects placed on 2:1 wake/sleep ratios totaling 9, 12, 18, 30, or 36 hr. Both long days and short ones proved to be inefficient cycles, albeit for different reasons. On the 9-hr day, subjects experienced difficulties falling asleep, and woke up frequently. Under the 36-hr condition, most subjects awoke early and were unable to use the full 12-hr sleep allotted them. Such inefficiencies in sleep patterns as found with abnormal day lengths are known to adversely affect performance when sustained for more than a few cycles (Webb and Cartwright, 1978). But within the desired 24-hr day, the question remains how best to schedule work and rest (or sleep) activities. Much of the work focusing on these issues was done by Chiles, Alluisi, and Adams, who carried out more than a dozen separate investigations over a 10-yr period. Performance was assessed using the multiple-task performance battery described previously in this chapter and outlined by Alluisi (1967, 1969).

Altered work-rest schedules— One question concerns the relationship of work-rest schedules to performance. Two 15-day studies using repeated schedules requiring 4-hr work followed by 2-hr rest (4:2 schedule) and a 30-day study requiring 4-hr work followed by 4-hr rest (4:4 schedule) addressed this question (Adams and Chiles, 1961, 1963a, 1963b). These investigators found that subjects in the 4:4 condition consistently maintained better performance than either of the 4:2 groups. From these studies Chiles et al. (1968) conclude that subjects working 12 hr/day on a 4:4 schedule can maintain a generally high level of performance when compared with subjects working 16 hr/day on a 4:2 schedule, and can do so for a longer period. However, they did find some interesting exceptions. Arithmetic tasks were performed about equally well by all groups, and some highly motivated subjects could work 16 hr/day on the 4:2 schedule with essentially no decrements over a period of at least 15 days, and possibly for as long as 30 days.

Stress— Even with nominally acceptable work schedules, performance reserves may be diminished and crew safety endangered in emergencies or during periods of stress. Performance degradation has been demonstrated in studies of the combined effects of sleep loss and work-rest schedules (Alluisi et al., 1964; Chiles et al., 1968). Subjects working on the 4:2 and 4:4 schedules were given additional tasks to complete during their regularly scheduled rest periods during days six and seven of a 12-day study. These additional tasks prevented their sleeping for periods of 40 to 44 hr. The results show a general trend toward greater decrements in those tasks which required either sustained concentration or systematic shifts in attention or mental set. The results further indicate that subjects in the 4:2 condition showed greater performance decrements resulting from sleep loss compared with subjects working on the 4:4 schedule, and did not recover as well following the sleep-loss period. These results strongly suggest that the more demanding the schedule, the more severe the performance impairment both during and after the depletion of performance reserves by the stresses of lost sleep and additional workload.

It seems clear that a high work/rest ratio is not likely to be practical for long-duration missions. We need to determine what specific work-rest schedules would be best over extended durations. Factors to be considered include the type and amount of the work to be accomplished, the types and extent of environmental stressors present, the length of the mission, the level of task proficiency required by mission standards, etc. Several of these variables are considered in the following sections.

Factors Affecting Workload

Klein (1970) refers to workload as the requirements of task performance that can be specified without reference to any operator response or effort actually applied to satisfy those requirements. A clear distinction is made between the demands for action implied in the tasks and the capabilities necessary for effective response (see also Chiles and Alluisi, 1979). A second conceptualization of workload focuses on how much the operator has to do and/or how hard he or she must work to satisfy a set of demands (Cooper and Harper, 1969; Jenney, Older, and Cameron, 1972). A related emphasis is provided by Welford (1968) in characterizing effort as the intensity of work. A final conceptualization of workload emphasizes effort and accomplishment. Here the focus is on the total activity of the operator in completing the required tasks (Cantrell and Hartman, 1967).

For purposes of the present discussion, workload will be defined broadly as the total work demand placed on the operator, including the amount and intensity of operator effort required to accomplish the task. If we think of a continuum of workload, we can assume that some optimal work load exists for a given individual at a given level of skill. Thus, the optimal load varies within certain limits according to individual differences in capacity, skill, and reactions to situational factors. The extremes of this continuum have adverse implications for performance and they are of primary concern to us here.

Fatigue— If we assume that for a given individual there is an optimum level of task demand, then we may define "overload" as an increase in the task demand beyond the individual's optimum level. Similarly, when the demands of the task are lowered below some critical level, task underload results. Many investigators have attempted to unify research on the effects of these two extreme loading conditions through the hypothetical construct "fatigue." As with workload, there is no single definition of fatigue that is universally accepted. As Gartner and Murphy (1976) point out, at least five different conceptualizations of fatigue are present in the performance literature. We will employ one of these conceptualizations:

Fatigue represents a reversible impairment of performance as a consequence of over- or underloading. Here, fatigue takes the form of lowered sensitivity, responsiveness, or capacity, which may show itself as a reduction in the amount of information that can be handled at any one instant, or in the amount that can be handled over a given period.

In structuring the work regime of the astronaut, the goal is to maintain an optimal workload so that fatigue does not occur. In the event that the reserve capacities of the astronaut are depleted because of either overloading during emergencies (e.g., continuous performance of highly complex tasks) or underloading during mission cruise phases (when few task demands are placed on the operator), we can anticipate that fatigue may be a problem. Gartner and Murphy (1976) suggest four broad categories of behavioral deterioration associated with such over- or underloading: motivation decrement, skill or proficiency decrement, psychological stress, and performance decrement. Fatigue can both slow performance and result in irregular or disordered performance (see review of the Effect of Loading, in Welford, 1968).

One review (Beljan, 1972) of complex task performances of fatigued subjects revealed the following degradations: (1) range of attention is restricted; (2) responses to signals are made independently of the occurrence of the signals; (3) mood is changed, and impatience occurs or increases; and (4) orientation becomes less acute.

It is clear that fatigue is a potentially serious problem for astronaut crew performance and well-being. Management of this condition can be approached in several ways. The traditional approach used with airline crews has emphasized adequate rest, physical fitness, weight control, nutrition and diet, and moderation in the use of alcohol and tobacco in order to offset the negative effects of over- or underloading (Schreuder, 1966). Pharmacological agents, especially various amphetamine-like drugs, have also been used to reduce astronaut fatigue (Parin, Kosmolinskiy, and Dushkov, 1970). However, as discussed in chapter II, the long-term effects of drugs in space are not well understood.

There is an obvious need to explore how mission activities might be integrated to minimize fatigue, both with respect to the absolute amount of work required and to the sequential processing of tasks involving various levels of complexity and effort. The delicacy of balancing workload was demonstrated in certain problems encountered in Skylab, where mission programmers may have planned too busy a schedule in their attempt to prevent the adverse effects of underloading (Cooper, 1976). Such a situation invites danger if a continuous misloading results in significant deterioration in performance levels.

It is also important to understand the proper sequencing of tasks. Simple tasks can be performed effectively at much higher levels of fatigue than more complex tasks (Welford, 1968). In designing a daily schedule, it might prove useful to assign complex tasks only at the beginning of the work day or following substantial rest periods. Simple tasks could then be assigned to follow complex tasks or could be completed at the end of the day when accumulated fatigue might be expected to be greatest.

From our review thus far, we can see that the type of work-rest schedule most beneficial for space missions must depend a great deal on the degree of loading involved in the tasks. Chiles and Adams (1961) posed the following guidelines regarding the duration of work periods as a function of loading:

- A maximum duty period of 4 hr is the most efficient when a passive task is combined with one or more active tasks, the workload is not too great, and a high level of performance must be maintained.

- When a passive task occurs by itself, attention cannot be adequately sustained to ensure satisfactory performance for periods longer than a maximum of 2 hr.

- When there is a considerable variety and active participation in the major tasks, or passive tasks have readily detectable signals to which the operator must respond, the duty period may be extended to 10 hr.

Each of these recommendations will no doubt have relevance to various phases of extended missions. It would be desirable to develop a taxonomy of tasks and duty hours as they apply to each crew person at different points during the flight.

Thus far, we have assumed that over- and underloading result in the same state — "fatigue." However, various researchers take exception to this position. For example, Welford (1968) defines fatigue as resulting from conditions of overloading, but prefers the concept of "monotony" as related to conditions of underloading. Monotony can occur when tasks require little cognitive processing or attention, are highly repetitious, are not complex, and/or have been extensively overlearned. Under these conditions, Welford predicts that performance may be generally poor as with overloading, but that the nature of performance degradation is fundamentally different. When the operator is overloaded, action is confused and judgments or graded responses tend to be undifferentiated. In contrast, when the operator is underloaded, attention tends to drift, signals are missed, and performance is lethargic. Whether or not fatigue and monotony are fundamentally different states, underloading related both to the task and the environment can be a potentially serious problem during extended missions.

The most extensively studied condition of task underloading involves the requirement for visual or auditory vigilance. One of the earliest of the laboratory studies of this phenomenon is the test employed by Mackworth (1950), where subjects attend to the pointer movements of a clock and respond appropriately to randomly presented anomolous movements. This and similar tasks, using a wide variety of signals such as a faint spot of light (Wilkinson, 1961), small changes in sound (Mackworth, 1950), or occasional features interspersed in a regularly presented series of digits (Bakan, 1953) have indicated significant performance decrement over relatively short periods of time. The decrements are shown not only by missed signals, but by slower responses to those that are detected (Wallis and Samuel, 1961; McCormack, 1962; Surwillo and Quilter, 1964; Buck, 1966). Detection failures do not appear to be the result of looking away from the signals, since Mackworth, Kaplan, and Metlay (1964) have shown that signals are often missed even when the subject's attention is fixed on them. Nor is the performance decline the result of a simple lack of activity. Whittenburg, Ross, and Andrews (1956) repeated Mackworth's clock test, but required subjects to respond to every jump (making a different response for small and large jumps), and found that performance continued to show declines even when warnings of signals were given.

Performance decrements in response to routine tasks occur in a variety of situations. For example, Adams and Chiles (1961) found that performance of monitoring and vigilance tasks indicated decrements over a 15-day confinement period, and Altman and Haythorn (1967) found poor performance on monitoring tasks among two-man groups confined for 10 days. Burns and Gifford (1961) found that decrements occurred when subjects had to operate routine tasks without apparent consequence. However, when the vigilance tasks were embedded in a multiple-task performance battery, and occurred with active tasks like arithmetic computations, performance appeared to be no different than for the active tasks. If anything, the vigilance performances deteriorated less and recovered more quickly than did active task performance (Alluisi, Coates, and Morgan, 1977).

It is clear that we must guard against the problem of task underload if optimal performance is to be attained and sustained in space. There appear to be several ways of minimizing the decrement in vigilance. Duty periods should be no more than 30 min if monitoring or vigilance comprises the sole task or a major part of the job to be accomplished within that time frame. Making the display and monitoring signal more intense appears to be a useful aid (Mackworth, 1950; Broadbent, 1958). Periodic changes of activity (Bevan, Avant, and Lankford, 1967), brief rest periods (Bergum and Lehr, 1962), or the presence of others in the room (Fraser, 1953; Bergum and Lehr, 1963; Williams, Kearney, and Lubin, 1965) all help to sustain satisfactory vigilance levels.

In considering the design and scheduling of tasks to be accomplished by spacecrews, it will be important to determine whether task underload may become a problem during particular phases of the mission. Although certain tasks, such as those requiring monitoring or vigilance, could give rise to underloading problems during flight, it is more likely that underloading difficulties will be related to the sheer repetitiveness of the tasks demanded of spacecrews, and the degree to which they are so overlearned as to no longer be interesting.

Uniform environmental conditions— Although the term "loading" generally refers to the demands of the task, a broader definition encompassing environmental stimulation can also be useful. Research strongly indicates that performance can be expected to deteriorate when the environmental input is at a reduced level. Studies of extreme sensory or perceptual deprivation conducted primarily in the 1950s and 1960s (see Zubek, 1974 for overview and synthesis) demonstrated significant impairment in certain types of performance.

Bexton, Heron, and Scott (1954) and Scott, Bexton, Heron, and Doane (1959) found that performance on measures of verbal fluency, anagrams, and various numerical tasks was significantly impaired during several days of perceptual deprivation. No deficits were found for measures of digit span, analogies, or associative learning. Zubek, Aftanas, Hasek, Sansom, Schludermann, Wilgosh, and Winocur (1962) report that, during 7 days of perceptual isolation, impairment occurred in arithmetic problem-solving, numerical reasoning, and recognition, but no significant changes were found in digit span, rote-learning, recall, and verbal reasoning. Impairment of visually cued reaction time following 2 days of perceptual deprivation has been demonstrated in studies by Nagatsuka and Suzuki (1964). Further, auditory (Zubek, Pushkar, Sansom, and Gowing, 1961) and visual (Zubek et al., 1961, 1962) vigilance have deteriorated during experiments involving low levels of sensory input. These results confirm that a reduced or nonvarying work environment can produce dramatic decrements in various aspects of performance.

Less extreme deprivation experiments have also shown that isolation in a nonvarying environment can sometimes adversely affect performance. For example, space-cabin simulator studies conducted at the USAF School of Aviation Medicine (Flinn, Monroe, Cramer, and Hagen, 1961; McKenzie, Hartman, and Welch, 1961; Morgan, Ulvedal, and Welch, 1961) found that two-man groups isolated for 17 days or for 30 days showed a gradual increase in average response time to a 14-task multiple-test battery. Similar findings demonstrating minor changes in performance have been reported from sealed-submarine experiments (Fawcett and Newman, 1953), fallout-shelter studies (Hammes, 1964), and in laboratory-isolation investigations (Altman and Haythorn, 1967).

Subjective reports of crews stationed in the polar regions frequently include complaints that long wintering-over periods affect memory and concentration (Mullin, 1960), but no systematically obtained data involving actual mission tasks have been presented to verify or deny that significant performance impairments have occurred. However, there are indications that long-term environmental underloading can disrupt performance. For example, crews in the polar regions have demonstrated markedly low involvement with intellectual activities, and many persons slated for isolated duty indicate an intent to occupy themselves with intellectual pursuits, but few people actually follow through (Rohrer, 1961; Gunderson and Nelson, 1963). It seems probable that some degree of performance attenuation will occur among crews isolated for long periods, perhaps owing to decrements in morale and motivation rather than to any cognitive or psychomotor deterioration. Research involving limited environments is certainly warranted in order to explore more fully just how an unchanging environment affects performance under long-term conditions.

Variety of work available— As Kubis and McLaughlin (1967) summarize, overlearning of the mission duties; extensive training in the diagnosis of a means to deal with simulated system failures; and comprehensive training in egress, escape, and survival techniques all have been emphasized as means of building confidence, reducing and controlling anxiety, and minimizing the potential problems that might occur in flight. Without question, this approach has been a valuable facet of astrononaut training, given the emergencies that have occurred and the skill with which the astronauts have responded. However, in anticipation of future longer-duration missions, some attention should be devoted to the potential problems that such a comprehensive training approach may generate.

Given the extreme importance that isolated individuals place on stimulating, challenging, and meaningful work, we must ensure that the training program, with its rigorous overtraining goals, does not reduce the interest value of mission responsibilities, adding to the potential for monotony and boredom in a long-term mission. Although crews need to be fully prepared, we must remember that being well trained exacerbates the dangers of boredom and makes the designing of jobs to be interesting and intrinsically motivating even more important. Apparently we cannot count on the anticipation or excitement of spaceflight to offset monotony; in the Mercury project, Gordon Cooper fell asleep inside the capsule during a long hold before liftoff (Wolfe, 1979).

One strategy for addressing monotony and boredom in spaceflight might be to overtrain crewmen on those tasks that are crucial for mission safety and success, while leaving noncritical procedures to be scheduled for in-flight learning. If it were possible to schedule some training exercises and academic learning during the flight itself, the additional challenge and stimulation of meaningful work would be available. This possibility would seem particularly appropriate to the cruise phases of long-duration missions, where only routine duty tasks are required, where the environment may be relatively unstimulating, and where few important mission goals are involved.

Crewmembers might also be given greater flexibility in determining their own work schedules. Both American and Russian space crews have repeatedly asked to have some options in the day-to-day organization of their work and leisure schedules — options allowing them more control over the effectiveness of their work (Cooper, 1976; Gazenko, Myasnikov, Ioseliani, Kozerenko, and Uskov, 1979). American crews have argued that rigidly structured schedules lead to confusion and as a result, experiments may not be conducted as carefully as desired. A varied approach to work scheduling seems worthy of trial; however, there appear to be no plans to institute this type of arrangement in the Shuttle Program (Morris, 1975; Watters and Steadman, 1976).

One further possibility for varying crew responsibilities is the rotation of crew duties. If each crewmember is held responsible for the same duties during an extended flight, the chances of boredom will increase; rotating the assignment of noncrictial duties could be used to minimize the chances of performance decline. Research is needed to investigate this and the other possible performance-enhancement techniques.

The Effects of Desynchronosis

It is well documented that many functions of the human body, including temperature, metabolic rate, and endocrine output among others, are regulated on a near 24-hr basis (Conroy and Mills, 1970; Colquhoun, 1971). Alterations in these cycles occur on Earth in unusual environments or through displaced or disrupted sleep or work, with potentially detrimental effects. Disruption in the body's circadian activity, known as desynchronosis, can produce physical symptoms such as malaise, insomnia, appetite loss, and nervous stress (Hauty and Adams, 1966a, 1966b, 1966c).

Desynchronosis problems associated with sleep or work schedules have been observed in space. For example, up to and including the Apollo 9 mission, it was common practice to employ staggered sleep schedules for crewmembers. This scheduling resulted in shifts of as much as 6 to 10 hr from the Earth-based time during which sleep normally occurred. Such disruption of the established cycle was described as producing "a most unsatisfactory situation in flight" (Berry, 1969). Even during later flights when crewmembers slept simultaneously, hectic schedules required displacement of the sleep phase by several hours, leading to complaints of fatigue and sleep problems. For future extended missions, the problems of adjusting work and sleep schedules to the demands of mission requirements will be an issue of importance. We need to understand how alterations in circadian rhythm can affect performance and what potential areas of research should be favored.

Studies of desynchronosis on Earth have included a number of different situations and techniques. Experiments with constant illumination (Webb and Agnew, 1974), day lengths extending beyond 24 hr (Webb and Agnew, 1975), industrial shift workers whose daily wake-sleep cycle is often displaced from the norm (Aschoff, 1978; Reinberg, Vieux, Ghata, Chaumont, and La Porte, 1978; Monk, Knauth, Folkard, and Rutenranz, 1978), bedrest patients (Winget, Vernikos-Danellis, De Roshia, and Cronin, 1974; Winget, De Roshia, and Sandler, 1979), etc., have all contributed to our understanding of desynchronosis. Alluisi and Morgan (1981) recently reviewed the effects of temporal factors on human performance and productivity. Although a discussion of each of these topics is beyond the scope of this book, one clearly relevant area, transmeridian flights, will be reviewed. The transmeridian flight situation is similar to that experienced by astronauts, in that phase shifts accompany translocation. Also, the published studies provide performance data that permit some generalization to the spaceflight situation.

Transmeridian flight— In 1966, Hauty and Adams used reaction time, decision time, critical flicker fusion, and numerical ability to assess the effects of translocation of subjects to a new time zone. Measurements taken following long westward flights showed that reaction time and decision time were both significantly impaired during the first day. Although similar measures taken following eastward flights tended to show decrements, the observed impairment did not prove to be statistically significant.

More extensive studies have been conducted by Klein. In one study, Klein, Bruner, Holtman, Rehme, Stolze, Steinhoff, and Wegmann (1970) measured pilot performance in a flight simulator. Pilots were tested in Germany, then in the United States, and again in Germany. Measurements of the magnitude of performance alteration showed that the change in performance level was more adversely affected and of higher significance following eastward flight from the U.S. to Germany than following westward flight from Germany to the U.S. In a second study (Klein, Wegmann, and Hunt, 1972), eight subjects were measured, again following both westward and eastward flights. In this experiment, the performance measures used included a reaction time test, a symbol cancellation test, a digit summation test, and a complex psychomotor performance test known as the Kugel test. Klein et al. (1972) found that after eastbound flight, it took 12 days for performance on the Kugel test to recover, whereas it took 9 days for the simpler performance tasks to readapt. Recovery time for the Kugel test following westward travel was 10 days, whereas the simpler tasks required 6 days.

Klein and his colleagues do not believe that there is any difference in psychophysiological response due specifically to the direction of flight. Rather, they ascribe the greater performance decline found in their studies in eastward travel to factors which, for a variety of reasons, lead to greater fatigue. Although some suggestions have been offered for distinguishing fatigue effects from circadian effects, it remains extremely difficult to separate the two when abnormal sleep is involved (Dodge, 1982), and in most studies to date, fatigue and circadian effects are compounded. (For a discussion of this and related issues, see Alluisi, 1977.)

From these and other transmeridian flight studies (e.g., Fort, 1969; Conroy and Mills, 1970; Berkhout, 1970; Wright, Voget, Sampson, Knapik, Patton, and Daniels, 1983) several inferences can be drawn regarding the relationship between desynchronosis and performance. First, decrements in performance are found for a range of behaviors from simple reaction time to cumulative exercise capacity. Second, the more complex the task, the more detrimental is desynchronization upon task-completion efficiency. Whereas simple tasks in both the Hauty and Adams (1966a, 1966b) and the Klein et al. (1972) studies were affected by shifts in circadian rhythm, the greater deterioration was clearly associated with complex tasks. Third, wide individual differences in adaptability occur. Some individuals may adapt to a phase shift within 1 day, and others appear to adapt only after extended periods of time (Strughold, 1952). Finally, the majority of research indicates that although performance is affected by desynchronosis, the time course for adaptation is less than that for the body rhythms themselves.

Some important research questions can be drawn from these studies. For example, with few exceptions, there have been essentially no studies of the combined effects of monotony, desynchronosis, and other temporal factors (Alluisi and Morgan, 1981), conditions that are likely to prevail in spaceflight. Another issue involves the possible differential effects associated with flights headed in different directions, and the underlying mechanisms of circadian-rhythm phase shifting. Controlling for fatigue would help to isolate potential direction effects. Also important is the question of individual differences in phase-shift adaptability. If, as it appears, wide differences exist in the degree to which individuals can adjust to changes in circadian rhythms, it is important to understand the behavioral and physiological correlates of these differences.

Zeitgebers— A factor related to circadian rhythmicity concerns those variables that regulate the 24-hr clock. Such variables are referred to as zeitgebers and consist of a wide range of physical, temporal, and social "cues" that serve to entrain sleep and wakefulness to a particular rhythm. Very few physical cues to diurnal cycles exist in space. Story Musgrave, Mission Specialist on STS-6, reports that time has little meaning in spaceflight except as it relates to elapsed time or work schedules (personal communication). When zeitgebers are not present, certain rhythms become "free running" and as a result may vary from a 24-hr schedule. Perhaps the most significant diurnal cue is the day/night cycle. In space, this cue is at best lacking, and at worst it can introduce distortions by driving rhythms in deleterious ways. For instance, the path of the orbiting Space Shuttle exposes crews to multiple sunrises and sunsets in a single Earth day. Before we attempt to reproduce or replace Earth zeitgebers under spaceflight conditions, we must determine which cues are involved with what specific rhythms. We can assume that artificial lighting will serve as a strong cue to periodicity. Also, social cues revolving around daily meals, work-rest schedules, and evening leisure activites can be important. Indeed, some research indicates that social cues may be even more important than lighting conditions for regulating work performance (Aschoff, 1978). Given that circadian rhythms and the issues that surround them have important implications for performance in space, methods of entraining rhythms to the most advantageous wake-sleep cycles constitute an important research area.

Performance as a circadian function— The performance levels of different tasks have been found to fluctuate during the day. It appears that performance may be subject to cyclic regulation of circadian periodicity in much the same way as many physiological functions. This finding has important implications for operators in space.

For example, performance as represented by the average number of simple sums computed in a 48-min period follows a cyclic pattern coincident with diurnal changes in body temperature (Colquhoun, Blake, and Edwards, 1968). Similar patterns have been observed for auditory vigilance tasks measuring the average number of signals detected and the average response time. Also, performance rhythms have been found within the workday portion of the 24-hr cycle.

Kleitman (1963), regarded as a leading authority in this area, has concluded that there is a recognizable general pattern in the fluctuations of performance levels throughout the work day which in many cases can be related to diurnal variations in body temperature. Results indicate that performance on many tasks shows a morning rise with a peak somewhere in the afternoon, followed by a fall. Although the exact timing of these points is disputed by various investigators (see Colquhoun, 1971, for example), all appear to agree that there may be recognizable patterns in the fluctuations of performance levels during the work day. In their review of temporal factors in human performance and productivity, Alluisi and Morgan (1981) conclude that whether or not performances show the circadian fluctuations, the underlying physiological rhythm is a bodily factor that must be taken into account. This relationship poses some important research questions for space mission planners seeking to optimize work output and accuracy.

For example, from a rhymicity perspective, it may prove advantageous to schedule delicate and sensitive tasks after the "morning rise," and to relegate less demanding or less critical tasks for the period following the "afternoon fall." This recommendation fits in well with our discussion of task overloading/underloading and the need to balance work load across the day. Such a schedule would require constructing a taxonomy of tasks most appropriate for different phases during the daily work schedule, which in turn requires a greater understanding than is currently available of what types of requirements and tasks are most affected by circadian periodicity.

Sleep Disturbances

Implicit in the study of effective work-rest conditions for spaceflight is the need to identify relevant factors associated with sleep, and to determine how these factors may affect performance during waking hours.

In confined, isolated environments such as habitation in the polar regions, sleep difficulties have been a repeated problem. For example, Natani, Shurley, Pierce, and Brooks (1970), commenting on the rather severe alterations in sleep observed among the early Antarctic exploring parties, noted that conditions of insomnia and changes in the quality and pattern of sleep occurred, although the actual quantity of sleep did not vary. Such findings have been confirmed in investigations of personnel stationed for many months in Arctic and Antarctic outposts. Here, sleep quality deteriorates and the men suffer from a dramatic form of insomnia in which the dream phase of sleep declines significantly, the deeper stages of sleep decrease or disappear altogether, and only the lightest stages of sleep are actually reached. The restorative benefits of sleep may be severely limited under such conditions. The subjective reports that are associated with such disturbances indicate that persons are physically uncomfortable, withdrawn, easily annoyed, and preoccupied with vague physical complaints, and have varying degrees of depression.

Findings of quantitative and qualitative sleep loss in other isolated and confined conditions suggest that it is important to determine how sleep may be affected in the confined, isolated quarters of space. A basic question is whether "normal" sleep patterns can ever be achieved in space. The astronauts' often choppy schedules may decrease sleep duration or displace the time when sleep occurs. For example, during the Gemini VII mission, both crewmen felt fatigued, yet averaged only 5.3 hr of sleep a night throughout the flight. They slept less than 5 hr a night during the final 4 days (Berry, Coons, Catterson, and Kelly, 1966), and sleep was particularly poor on the first night (a fact reported by numerous astronauts). These problems appear to be related to high noise and vibration levels, general tension during the mission, staggered sleep schedules, and bed designs that were inadequate to ensure comfortable sleeping positions. Similar difficulties have been reported by the cosmonauts (Oberg, 1981).

During the later Apollo missions, the amount and quality of sleep were still unresolved issues. Certainly there were occasions when crews experienced sleep difficulties, but these difficulties may be explained by the prevailing physical conditions. For example, the crew of Apollo 14 experienced a large displacement in the normal terrestrial sleep cycle, a phase shift of some 7 to 11.5 hr later than they were used to. The crew had difficulty sleeping, probably owing to the lack of kinesthetic sensations and to muscle soreness in the legs and lower back. Throughout the mission sleep was intermittent and never exceeded 2-3 hr of deep and continuous sleep. In the lunar module, crewmen slept little. Difficulty in finding a place to rest their heads, discomfort of the pressure suit, and a 7° tilt of the module, all combined to make sleep difficult (Strughold and Hale, 1975).

Some encouraging results can be noted when the results of the Skylab missions (Frost, Shumate, Salamy, and Booher, 1977) are examined. During the 59- and 84-day missions, there was no significant decrease in the average amount of sleep time. A 1-hr decrease in sleep time was observed on the 28-day mission, but this was strictly voluntary and not due to insomnia. Sleep latency, or the length of time needed to fall asleep, was relatively long in the early portions of the 84-day mission, but returned to values typical of pre- and postflight conditions by the latter half of the mission. No significant change in sleep latency was seen in the 59-day mission, whereas the in-flight latencies were actually significantly lower than preflight latencies during the 28-day mission.

Significantly, the most marked changes in sleep occurred not in flight but upon return to Earth (with alterations more of quality than quantity). This suggests that readaptation to a 1-g environment may be more disruptive to sleep than adaptation to a 0-g environment.

Performance correlates of sleep loss— Against this background of sleep disturbances in space, let us consider how sleep loss on Earth has affected performance. The two-man Space Cabin Simulator Studies (Cramer and Flinn, 1963) showed that sleep loss directly affects the performance of tasks requiring sustained concentration or vigilance. Performance also can be affected indirectly, since there are strong relations among mood, attitude, motivation, and performance efficiency. Other studies have confirmed these results. For example, Wilkinson (1969) kept a group of men awake at night for various lengths of time over a 6-wk period. After each experimental night, the men were kept busy for most of the next 15 hr while auditory vigilance was tested at five different intervals lasting 1-hr each. The results indicated that following a full night's sleep (7.5 hr), about 65% of the signals were correctly detected. When a sleep debt of 5.5 hr occurred (i.e., sleep lasted only 2 hr), performance during the following day was significantly reduced. Interestingly, when sleep was reduced to 5 hr/night for two consecutive nights, a similar deterioration in performance was noted. These findings indicate that even relatively minor sleep loss, when accumulated over several evenings, can adversely affect performance in a manner similar to acute sleep loss on a single night. Wilkinson also found that sleep loss had a significant impact on decisionmaking as measured by a five-choice display panel task. A single night without sleep reduced the rate of response and increased the latency of response. Similar results have been found for tasks involving psychomotor performance on a rotary pursuit task and on a reversed-digit writing task (Chambers, 1964).

In another study, Wilkinson (1963) found that a night of sleep did not entirely compensate for a night without sleep. Twenty-four men were deprived of sleep and required to keep busy for 24 hr, the sleep deprivation portion of the experiment ending in early evening. Most of the men then slept for about 2 hr and later went to bed at the usual time, resulting in about 9.5 hr of sleep by the second day. The subjects were then tested on a vigilance monitoring task or on the five-choice decision task. With both tasks there was a significant deterioration in performance compared with the control condition after normal sleep. The effect was more pronounced during morning trials than during afternoon trials, indicating that by the end of the second day the effects of sleep loss from the previous evening had begun to dissipate. In addition, Alluisi, Coates, and Morgan (1977) present data clearly showing that the effect of sleep loss interacts with the circadian rhythm, as does the recovery from sleep loss.

Sudden awakening— One final issue regarding the relationship between sleep and performance entails the ability of subjects to function optimally immediately upon awakening. During some of the past space missions, events have occurred that required the crew to arise from sleep and effect repairs immediately. How good is performance immediately after sudden awakening? If less than optimum, what is the time required to reach normal waking levels of performance?

Langdon and Hartman (1961) investigated this problem by assessing the performance task of turning off lights at a desk console. Subjects were awakened either at midnight or between 3 and 4 a.m. In this study, the subjects required not less than 7 min after being awakened to achieve normal efficiency in performing the task. Similar findings are reported by Webb and Agnew (1964), who awakened subjects from the deepest stage of sleep. These and other findings (Wilkinson and Stretton, 1971; Kleitman, 1949) suggest that performance upon first awakening is degraded, a point which should be taken into account in scheduling activities, when possible.

From the results reported above, we must conclude that sleep disturbances could pose a significant hazard to spacecrews. Efforts must be made to minimize such disturbances and to ensure that wake-sleep cycles do not adversely affect sleep. There are some unanswered questions surrounding the issue of sleep as affected by weightlessness. A fundamental question concerns how, over the long term, sleep requirements change in space. Although early mission crews experienced great difficulties in sleeping, the later crews of Skylab reported very few problems. In fact, some of the Skylab astronauts reported that they thought sleep was actually better in space than on Earth, since they required fewer hours of sleep and felt less fatigue during the workday. Perhaps the lack of gravity and consequent diminution in the forces acting on the body (particularly the muscles) may be responsible for this interesting effect.

Relevant and potentially important research can be conducted on sleep issues without going into space. For several years levitation via airflow (Scales, Winter, and Block, 1965) has been used for the treatment of hospitalized severe burn victims (Whitaker, Graham, Parsi, and Olson, 1976; Demling, Perea, Maly, Moylan, Jarrett, and Balish, 1978). Since the degree of pressure placed on the body using air is minimal, air levitation might provide a good analogy to the sleeping conditions of weightlessness. Less exotic, but also potentially useful, would be similar research employing water levitation techniques.

SUMMARY AND CONCLUSIONS

In this chapter we have examined some of the issues involved in defining and measuring performance on Earth and in space. The measurement of performance under space-operational conditions poses many problems. Because of the complexity of the tasks and the environment, it is seldom easy to investigate the many basic human abilities that must work together for a mission requirement to be completed successfully; nor is it easy to understand and predict the impact of the myriad of overlapping environmental stressors that may affect that performance.

For example, navigation of a spacecraft may require the execution of a number of human abilities including arm-hand steadiness, finger dexterity, hand-eye coordination, perceptual speed, and rapid reaction time, all happening with such fluidity and coordination that to assess the individual level of each factor at a given moment is not feasible. Furthermore, even if a change in the quality of vehicle control is observed, it may be difficult to determine whether this change is due to fatigue, space sickness, increased arousal, or loss of sleep, or to some combination of these and/or other factors.

Investigators have sought to develop means outside the operational setting to aid in the understanding of processes present within the operational setting. One approach is the discrete-task assessment technique. A task is selected which purports to measure some aspect of performance believed to be important in the operational setting. For example, rotary pursuit testing may be used to measure tracking ability, or visual monitoring may be used to measure the subject's vigilance. This approach is low cost, permits precise measurement of performance, and can be used to investigate the effects of individual environmental stressors. However, there remain significant limitations and important research questions regarding this approach. Frequently there is little similarity in the type of task employed in the laboratory to measure a certain ability and the task requiring that ability in the operational setting. The predictive validity of individual laboratory tests as they relate to space missions needs to be determined.

The discrete-task technique frequently fails to tap the more fluid, simultaneous demands of the operational setting. For this reason, some investigators have relied on multiple-task performance batteries or "synthetic" work. Here the time-sharing demands of simultaneously monitored and performed tasks approximate mission conditions. However, predictive validity is still an issue. Greater predictive validity can be obtained through the use of partial and full-scale simulation; but it is often difficult to isolate individual work units or to identify the particular environmental stressors of interest.

We cannot rely on any single approach to supply all of the information necessary to plan performance requirements for future missions. What appears to be most needed at this time is a more integrated approach using various techniques, along with a greater emphasis on data from actual missions. Using data from missions, researchers could begin to unravel the many complex factors involved in the definition and measurement of performance in space. A kind of performance laboratory in space could provide the opportunity to explore the complexities of performance unhindered by the requirements of mission operations. Findings from such studies could be validated by the selected measuring of performance on mission tasks. Recent interest in and support for the concept of such a Space Station has been encouraging. It is hoped that continued efforts can be made to bring this objective to fruition.

Beyond the basic issues of defining and measuring performance, we are faced with the task of developing a better understanding of the many factors of work in space that may influence performance levels. Weightlessness, space sickness, biomedical changes, and protective clothing, all may operate to reduce the work capacity of crews in space. Continuing research is needed to consider how these individual elements may affect work capacity, particularly under the extended exposures of a long-term mission. Mission goals, the amount and sequencing of daily work schedules, and the type and amount of preflight training must all be geared to the capabilities of the operator in space.

It may be necessary to balance work schedules in space to minimize the problems of overloading/underloading and the resulting condition of fatigue that can adversely affect performance. Such approaches as crew-selected work schedules or rotational work schedules should be examined. Also, the degree of overtraining required prior to flight for all tasks regardless of their relative importance must be questioned. If all work is overlearned, the problem of boredom may be exacerbated. The prospect of initiating a program of on-board learning should be considered. In this way, at least some aspects of mission-required work can continue to be novel and stimulating.

Desynchronosis can adversely affect performance. Precautions must be taken to ensure that controllable conditions of space and the requirements of work schedules do not jeopardize the synchronization of circadian rhythms. Continuing research is needed to determine the significance of various zeitgebers or cues for particular rhythms and their adapatability to space. It is important, too, that planners recognize that performance itself is subject to circadian cycles. Research could help identify patterns of performance rhythms and determine how work schedules could be arranged to maximize productivity.

Coincident with the study of work schedules and desynchronosis, the importance of sleep must be considered. Changes in either the quality or duration of sleep can have a profound impact on performance. Questions remain as to how sleep may be affected by long-term confinement and isolation, by the restrictions of the space vehicle, and by lack of gravity. One ground-based research avenue might involve the use of air- or water-body support systems to study the effects of reduced body pressure on sleep patterns.

Sustained, high-quality performance of crews in space is necessary for the safety of the crews and the successful completion of mission tasks. As we look toward future extended flights, we can anticipate a need to continuously refine our understanding of the complexities of work in the space environment.

Artist: Tom O'Hara

V
SMALL GROUPS

INTRODUCTION

The purpose of this chapter is to examine how people's relationships with one another may affect the psychological functioning and welfare of the individual astronaut and the performance and morale of the entire crew. The primary focus is on crews that are "small" in the sense that each crewmember has the opportunity to interact with each and every other crewmember on a person-to-person basis. The two-person groups (dyads) and three-person groups (triads) that have completed space missions thus far qualify as small groups, and so do the crews of six members or so that are considered appropriate for the Space Shuttle and for the first interplanetary missions (Stockton and Wilford, 1981).

During initial years of the space program, psychological concerns centered around the effects of weightlessness on astronaut performance, and upon man-machine engineering (Gerathewohl, 1959). By the mid 1960s, however, interests had expanded to include social psychological variables. Over the following decade, a number of theoretical papers and reviews appeared, the most salient including those by Berry (1973a); Haythorn, McGrath, Hollander, Latané, Helmreich, and Radloff (1972); Helmreich, Wilhelm, and Runge (1980); Kanas and Fedderson (1971); Kubis (1972); Rawls, McGaffey, Trego, and Sells (1968); Sells (1966); and Sells and Gunderson (1972). These reviews firmly established interpersonal and group variables as important determinants of crewmember performance and well-being in space.

INDIVIDUAL CHARACTERISTICS AND CREW COMPATIBILITY

A prevalent image of the astronaut is that of the rugged individualist who is prepared to set forth to open new frontiers. Kubis and McLaughlin (1967, p. 361) offer the following summary of early NASA selection programs:

> The search was for an individual with a high degree of intelligence, preferably characterized by mathematical and spatial aptitude. He was to be sufficiently creative to contribute not only to the development of test and space hardware, but also to the planning necessary for the success of the space program. With an ability to work closely with others, he was expected to tolerate extreme isolation without anxiety. Though reliable and consistent in his behavior, he was to possess the necessary flexibility and adaptability to meet any emergency without psychological disintegration. Deliberate rather than impulsive, and with outstanding capacity to tolerate stress, his motivation for volunteering in the space program was to be mission-oriented rather than based on personal need for achievement.

Given the nature of early pioneering efforts in space, the rugged individualists chosen as astronauts appear to have been very appropriate. However, there are important differences between the missions thus far accomplished and those envisioned for the future. Given that future missions will require increasing levels of cooperative functioning, selection and training procedures must not only yield effective individuals, they must yield effective groups.

Voluminous literature addresses the problem of selecting, as crewmembers, people whose behaviors are compatible with one another as well as with the environmental systems (Altman and Haythorn, 1965, 1967a, 1967b; Haythorn, 1968, 1970, 1973; Haythorn and Altman, 1967; Haythorn, Altman, and Myers, 1966; Haythorn et al., 1972; Kanas and Fedderson, 1971; Kubis, 1972; Natani, 1980; Sells and Gunderson, 1972). Crewmembers may be considered compatible to the extent that each member shows qualities and emits behaviors that the other crewmembers consider desirable and appropriate under spaceflight conditions. The problem is not merely finding people with the right kinds of behaviors, but finding people whose behaviors intermesh in a good or positive way.

The issue is exceedingly complex, because so many variables are suspected of entering in (Kubis, 1972).

According to Natani (1980), leaders who are experienced veterans of the isolation and confinement experience can assemble an effective crew if they have a large pool of applicants and the time to conduct thorough interviews. Selection based on interviews by psychiatrists, psychologists, and other mental health professionals has generally been less successful, and selection based on the results of psychological tests has been the least successful of all (Natani, 1980). Given that future space programs will require increasingly large numbers of astronauts, it seems worthwhile to perfect psychological tests that can be simultaneously administered to large groups of candidates and then scored by computer (Natani, 1980). Ideally, future testing programs will screen out candidates with undesirable characteristics, select in candidates with desirable characteristics, and take into account group as well as individual considerations.

The limited evidence available confirms the expectation that a lack of social compatibility has an adverse impact on morale and performance. Social compatibility emerged as the foremost factor in analyses of supervisory ratings and peer nominations at polar stations, and compatibility has been related to whether or not the Antarctic adventurers had a "good year" or a "bad year" (Gunderson, 1963, 1968; Gunderson and Mahan, 1966; Gunderson and Nelson, 1963, 1965, 1966; Nelson, 1965). In simulation research by Altman and Haythorn and their colleagues, members of isolated and confined groups who were incompatible showed increased stress, withdrawal, and territorial behaviors. In addition, they made more attempts to withdraw from the study (Altman and Haythorn, 1965, 1967a, 1967b; Altman and Taylor, 1973; Haythorn, 1968, 1970, 1973; Haythorn and Altman, 1967; Haythorn et al., 1966, 1972). Russian observers have reported associations between social compatibility on the one hand and morale and performance on the other (Leonov and Lebedev, 1975).

Gender

Space travel has been a male-dominated enterprise, but at least eight women astronauts are in training and it is recognized that in the long run some sort of sexual parity is likely to be achieved (Shurley, Natani, and Sengel, 1977). Extremely little is known about women in space. Women have visited polar stations, lived in underwater habitats, and participated in fallout-shelter studies, but the vast bulk of the data come from all-male preserves. Those few studies which have involved women have not focused sharply on sex or gender variables.

Several issues are involved when we consider women entering space. Perhaps the least of these is whether or not women are equipped for the rigors of life in space. Early doubts are giving way to a conviction that women can do the job and have the right to be there. This conviction reflects, in part, an increasing recognition of women's capabilities outside of the traditionally feminine sphere, and a growing recognition that technical systems are as easily engineered to meet women's as men's requirements.

There is evidence of attitudinal barriers against attaining sexual parity in space. Some space veterans, for example, have expressed the view that space tasks are men's tasks, and that women are unlikely to be able to meet the challenges (e.g., see Cunningham and Herskowitz, 1977; Oberg, 1981). Cosmonaut Valentina Tereshekova's orbital flight in 1963 has been dismissed as a political stunt, and her contributions to space exploration discounted, ostensibly because unlike the U.S. spacecraft of the era, the Russian craft were almost entirely controlled from the ground (Cunningham and Herskowitz, 1977; Oberg, 1981). The same critique has not been applied with equal force to Yuri Gargarin and other of Tereshekova's male contemporaries. In recent years, secondary "feminine" roles such as physician or stewardess have been deemed appropriate for female cosmonauts (Oberg, 1981); the second female cosmonaut, who joined a Salyut space station in 1982, was in fact a physician.

There are many complex issues regarding mixed-sex crews. Inclusion of crewmembers of both sexes can increase social diversity and provide the opportunity to exercise role options which might otherwise be forfeited for the duration of the mission. On the other hand, there may be a certain lack of ease in dealing with members of the opposite sex under conditions of isolation and confinement, and the attitudes that astronauts of each sex hold about the other could prove problematical. For example, difficulties could arise if crewmembers of either sex have strong prejudices about the capacities and/or appropriate roles of the other. A review by Adams (1980) suggests that sexually mixed groups within the military are fraught with conflicts, often centering around male officers' expectations concerning female personnel. Other spectres include the failure of male subordinates to respond appropriately to female leaders' commands, stress-induced strengthening of sexual stereotypes (leading, for example, to men taking unnecessary risks to impress women or women faking helplessness in the presence of men), and clique formation along sexual lines. Additional research is required to identify ways to select crewmembers who are flexible and tolerant in their attitudes and responses toward members of the opposite sex.

Intimate sexual relationships could form within same-sex or mixed-sex crews, but concerns about such relationships are heightened in the case of mixed-sex crews. The reason for this is that two powerful factors are likely to prevent the formation of intimate sexual bonds within same-sex crews: the clear preponderance within the general population of heterosexual preferences, and strong cultural taboos that prevent the expression of homosexual impulses. Sexual bonds of any kind are potentially disruptive because jealousies may arise as the result of other crewmembers "pairing off." In addition, a terminated intimate relationship which proves merely painful under normal conditions could prove devastating under conditions of isolation and confinement.

There is little basis for predicting sexual behavior in space. On the one hand, the periods of deprivation are likely to be so long that otherwise effective internal restraints may lose their effectiveness. Also, space travel can be very exciting, and a high state of excitement or arousal can fan sexual passions (Berscheid and Walster, 1978). On the other hand, weightlessness and a rigorous program of technical activities may detract from sexual interest. In addition, there is at least some evidence that, within some small social systems, there develops social norms (i.e., shared expectations regarding appropriate attitudes and behaviors) which discourage members from choosing one another for sexual liaisons. Members of these systems seem to recognize that endogamous choices can fan jealousies and reduce privacy to a dangerously low level. The findings are tentative, however, and come from kibbutzim (Talmon, 1964) and residential colleges (DeLamater, 1974), which maintain relatively permeable boundaries and thereby make exogamous choices possible. Such expectations do not develop within all small social systems, for within the Navy, shipboard pregnancies have become a problem (Adams, 1980).

In summary, there are many unanswered questions regarding interaction within mixed-sex crews. Intolerance toward members of the opposite sex; counterproductive sexual stereotypes, attitudes, and activities; the formation of cliques on the basis of sex; and the generation of rivalries and jealousies are among the problems that need to be understood.

Age

Although polar camps, subaquatic dwellings, and space simulators have tended to be male preserves, inhabitants have ranged in age from their late teens to their mid-forties or beyond. Group members who have deviated noticeably from the group's mean age have, like other members, been physically and mentally fit to stand the environmental rigors and make positive contributions to the group. Within the ranges studied, age has not emerged as an appreciable source of friction. Indeed, there are special advantages to including mature individuals. Research in subaquatic environments suggests that a mature individual may serve as a parent-surrogate who satisfies important needs of younger crewmembers (Radloff and Helmreich, 1968). It is significant that both members of the highly successful pioneering Space Shuttle crew were over 40.

Both relatively young and relatively old candidates have something to offer a spacecrew (e.g., peak energy and physical fitness in the former instance, and experience and perspective in the latter instance), but it will remain necessary to identify people who are biologically too young or too old to satisfy mission requirements. Computations of upper bounds, of course, must take mission duration into account; a person with sufficient youth and vitality to complete a 6-month mission may not be able to satisfactorily complete a 2-yr mission. It is also necessary to identify calendar ages at which people are likely to be *perceived* as too young or too old by other crewmembers. It is hypothesized that the inclusion of a person who is (justly or unjustly) defined as unfit by other crewmembers will generate negative attitudes which undermine the person's self-esteem and disrupt group functioning.

Another area of concern is the changes that astronauts may undergo during truly extended missions. Thus far, even the longest studies of isolated and confined individuals have involved but a very small segment of the participants' lifespans. But developmental changes which are undetectable on short missions may become prominent during missions measured in years. Work in the emerging field of adult developmental psychology suggests that people undergo fairly pronounced changes at several points during their adulthood (Kimmel, 1980). The major changes of interests and goals which sometimes accompany these transitions could reduce the astronauts' fitness for the technical side of the mission, and also their social compatibility. At present, missions are not measured in years, our knowledge of adult development is modest, and it may well be that commitment to a mission may prevent major changes of interests and identity. However, potentially significant developmental changes require consideration when planning a truly extended mission.

In summary, the evidence available suggests that tomorrow's astronauts will not have to be selected from a narrow range of ages. Additional research is required to identify means for establishing an appropriate age range and mix for a mission of given specifications. Eventually, some attention will have to be directed toward understanding adult development in space, and how such age-related changes are likely to affect fitness for and commitment to a mission.

Culture

The Americans who have visited space thus far have primarily been exemplars of the nation's dominant caucasian ethnic group; this is rapidly changing. Future missions will include a larger proportion of Black Americans, Latinos, and representatives of other minority groups. There is a possibility that, despite careful selection and training, prolonged isolation and confinement will bring long-standing prejudices to the fore. Although Kanas and Fedderson (1971) have discussed some of the implications of ethnically mixed missions, racial and cultural backgrounds have *not* been major variables in studies of isolated and confined groups. The literature, in fact, provides some basis for optimism. Specifically, three conditions associated with life in space may minimize certain ethnic prejudices.

First, some prejudice appears to be the result of an assumption that people from other ethnic groups maintain attitudes that are different from one's own (Stein, Hardyck, and Smith, 1965). In fact, astronauts are likely to discover that they have many interests and values in common (e.g., those centering around the mission). Such similarities should militate against prejudice.

Second, some prejudice flows from the perception of low social status rather than the perception of race or ethnicity *per se* (Allport, 1954; Amir, 1969). Since space voyagers are likely to come from a highly select population in terms of ability, education, and health, premission social status is unlikely to contribute to prejudice. However, care has to be taken not to somehow create invidious status distinctions along ethnic or cultural lines during premission activities or during the mission itself. Such distinctions could arise, for example, if minority astronauts were not fully trained, or if they were limited to relatively menial roles during the flight. In the

Russian program, guest cosmonauts from Eastern bloc nations received less training than their Russian counterparts, a discrepancy which seems to have occasionally led to some frictions (Oberg, 1981).

Third, under certain conditions, interaction is likely to lead to a reduction of prejudice (Allport, 1954; Amir, 1969). Two of the most important conditions — cooperation and the pursuit of common goals— are likely to be found in space missions.

Prejudicial attitudes are but one factor potentially affecting the compatibility of ethnically heterogeneous crews. First, as noted in our chapter on communication, ethnically mixed crews, especially international crews, may experience language problems. Second, it may prove difficult to identify equipment, provisions, and schedules of activities that are acceptable to representatives of different cultural groups. For example, individuals from different nations may have different dietary preferences and aversions, and seek different types of leisure activities. Such within-group similarities and between-group differences may encourage the formation of cliques along national lines. Once such cliques have been formed, it may be but a small step to international disputes. Yet another problem is that representatives of different cultures may have conflicting views regarding appropriate astronaut behaviors. For example, Americans have relied upon the image of the assertive, competitive, achieving pioneer (Wolfe, 1979). American astronauts may thus find it difficult to accept an astronaut from a culture that stresses mild manners, decorum, and the tight control of hostility.

Thus, there are many important questions concerning the effects of cultural variability on the compatibility of tomorrow's spacecrews. Although little data are presently available, there are some interesting prospects for future research. The mixed-nationality crews of the large oil supertankers (Mostert, 1975) may provide a useful analog of international spacecrews. Increased interaction between the European Space Agency and NASA, and the inclusion of international crews aboard the Space Shuttle may also prove to be revealing.

Personal Attractiveness

Rawls, Hopper, and Rawls (1969) instructed college students to "List as many things as you can possibly think of that would determine how closely you would be willing to interact with another individual." The other person's attractiveness in terms of such considerations as cleanliness, appearance, dress, and general demeanor emerged as a major consideration.

The search for complex bases for social compatibility should not cause personal attractiveness to be overlooked. Given that the need to staff large missions or to simultaneously staff a number of different missions will necessarily result in decreased selectivity and increased crew heterogeneity, it is necessary to learn more about the attributes which make an individual personally appealing in light of spacecrew norms. It might be useful, in this regard, to devise and validate an instrument for identifying personal characteristics which crewmembers are likely to find distasteful or annoying. This might involve a listing of physical characteristics, personality traits, and mannerisms (unkempt hair, dirty fingernails, stubbornness, and so forth) to be rated in terms of irritation value. Once perfected, such an "annoyance questionnaire" could be used in two ways. First, norms could be established to provide a basis for eliminating "unattractive" spacecrew candidates. Second, the instrument could be used for weeding out "finicky" individuals who find too many human frailties aversive.

Emotional Stability

A highly emotional or uncontrolled individual poses an unacceptable threat in any hazardous environment. Accordingly, it has been noted that Antarctic personnel place high premium on having calm, even-tempered, emotionally mature companions (Doll and Gunderson, 1971; Law, 1960; Lugg, 1973).

Many of the research questions surrounding emotional stability are questions of selection. Much more is known about how to exclude people who are liable to react badly than how to choose people of exceptional psychological health (Perry, 1965, 1967; Natani, 1980). Whatever the ultimate screening procedures, there is no getting around the fact that as more and more people are chosen for space missions, a few "high risk" individuals will inadvertently be included. Mission planners and managers need to know more about the kinds of supports or props that can be used to help people preserve or restore their emotional stability under conditions of isolation and confinement.

In some cases, crew perceptions of emotional stability may be more important than the facts. Behaviors which result in the inference that the astronaut is emotionally unstable (whether or not that inference is correct) may demoralize that astronaut and undermine the confidence of the remainder of the crew. Of particular interest is identifying those conditions under which undue significance is read into an outburst or other act, with the result than an effectively functioning crewmember is considered no longer a member of the team.

Competence

Spacecrews are likely to have a strong work orientation. Like inhabitants of other dangerous environments, they will no doubt recognize that poor or incompetent performance on one person's part can jeopardize everyone's welfare. Thus, they are likely to exhibit a strong preference for colleagues who are able and willing to get the job done. Gunderson and Nelson (1965) found that "task motivation" related to "good years" and "bad years" in the Antarctic, and Shears and Gunderson (1966) reported that both personal motivation and perceptions of the group's achievements were related to satisfaction with the Antarctic assignment. Studies undertaken by the Alaskan Air Command also suggest that marginal performance is correlated with poor adjustment and dissatisfaction (George Washington University, 1974), and Day (1969) has offered a fascinating account of adverse reactions generated by crewmembers who failed to fulfill their performance requirements in the days of sailing ships.

Generally, attention has focused on technical or task competence at the expense of interpersonal or social competence. Because selection procedures are such that people who lack essential physical and intellectual skills are likely to be eliminated at the outset, the success of a group in an exotic environment may very well depend upon the group members' abilities to cope with one another. Along with adding interpersonal performance measures to research on competence, we need additional data regarding peer perceptions and ratings of competence. For example, as in the case of emotional stability, a competent crewmember who is not seen as such may have more of an adverse effect on performance and morale than an incompetent crewmember whose inadequacies are not correctly identified. Alternatively, anyone who is able to convey an impression of knowledge and skill may have a calming effect on the rest of the crew. A high degree of perceived competence is particularly crucial for those who lead isolated and confined groups (Leonov and Lebedev, 1975).

Cooperativeness

Space voyagers embark on highly interdependent ventures which require the utmost cooperation for success. According to McClintock (1972, 1978), people vary in terms of their interests in coordinating their efforts for mutual gain. Individuals who are governed by own-gain motivation tend to be interested in achieving what they can for themselves, and to be relatively uninfluenced by the anticipated effects of their actions on other people. Individuals who are governed by relative-gain motivation tend to be interested in doing better than other people; they view the absolute level of reward that they get as unimportant relative to whether or not they "come out on top" in the competition. Individuals who are governed by joint-gain motivation tend to prefer courses of action which yield benefits for other people as well as for themselves; thus, they show a sensitivity to other people's needs and a concern for their welfare.

McClintock and his associates hypothesize that each person is consistently governed by one of these three motives (Maki, Thorngate, and McClintock, 1979; McClintock, 1972, 1978). Each motive is believed to stem from early childhood socialization and reflects both familial and cultural values. A better understanding of these motives may prove useful in the flight-personnel selection process, or in establishing the most effective reward structures in the space-capsule microsociety.

Relevant to both competence and cooperativeness is Helmreich's work on the achievement orientation, or need achievement (Helmreich, et al., 1980). Classically, need achievement has been defined as a persistent preference for engaging in success-related activities (Atkinson, 1958). People with high need achievement have many admirable qualities, but problems may arise aboard a space vehicle if attaining standards of excellence involves "prima donna" behaviors or a put-down of other members of the crew. According to Helmreich, need achievement can be divided into three independent factors. Work orientation refers to the motivation to work hard because work is a valuable activity in and of itself. Mastery orientation refers to a desire to continually improve one's own performance. Competition refers to an attempt to do better than other people. Helmreich et al. hypothesize that the combined interests of task accomplishment and social compatibility will be best served if crewmembers show a strong work and mastery orientation but relatively little competitiveness. The hypothesis is provocative, given the competitive orientation of the early astronauts (Cunningham and

Herskowitz, 1977; Wolfe, 1979). The rationale is that competitive individuals are likely to generate interpersonal stress and hostility, adversely affecting collaborative performance and further undermining the quality of a social environment whose quality has already been heavily sapped by the conditions of life in space (Helmreich et al., 1980). The available evidence suggests that work orientation and mastery orientation positively correlate with performance, and competition negatively correlates with performance (Helmreich et al., 1980). Additional research is required to test this hypothesis under conditions analogous to extended-duration spaceflight. In addition, it would be of interest to explore the possibility that the extent of the frictions caused by one crewmember's competitiveness may depend on the orientations of the other crewmembers.

Social Versatility

As noted in our introductory chapter, the conditions of spaceflight will restrict the opportunity to engage in varied roles. Crewmembers who can easily perform a wide range of role-related behaviors in flight should prove valuable, for they should help reinstate lost social opportunities. We hypothesize that the value of selecting crewmembers on the basis of an ability to engage in a wide variety of social behaviors should be inversely proportional to crew size and directly proportional to mission duration.

In Western society, men are expected to engage in task-oriented instrumental activities, and women are expected to engage in socioemotional or expressive activities. Men are expected to be autonomous, independent, somewhat dominating and aggressive, and emotionally inhibited. Women are expected to be warm and nurturant and to openly display their feelings (Bem and Bem, 1970; Broverman, Vogel, Broverman, Clarkson, and Rosenkrantz, 1972). There are, of course, great individual differences in the extent to which members of each sex adopt and enjoy such traditional social roles (Bem, 1974, 1975).

Research by Helmreich and Spence (Helmreich, Spence, and Holahan, 1979b; Helmreich et al., 1980; Spence and Helmreich, 1978; Spence, Helmreich, and Holahan, 1979) suggests that some people have the capacity for both goal-oriented instrumental behavior and interpersonal sensitivity. The extent to which these capacities are exercised depends largely upon situational demands. Such people, who are referred to as androgynous, appear to have positive self-concepts and build rewarding interpersonal relationships. Helmreich et al. (1980) note that androgyny appears highly desirable for astronauts, for a strong instrumentality combined with interpersonal sensitivity should be associated with both task accomplishment and social harmony.

In effect, androgynous individuals are individuals of either sex who can perform multiple roles, or at least central components of traditional masculine and feminine roles. Androgynous crewmembers, then, may have the value of increasing social variety within a crew. Future research should explore additional dual traits. As an example, consider the Jungian introversion-extroversion dichotomy. Who might be more valued under conditions of isolation and confinement: the shy and self-contemplative introvert, or the sociable and outgoing extrovert? On the one hand, introverts should be self-sufficient and accommodating of others' privacy needs. They may be well-equipped to weather extended communication breaks with Earth. On the other hand, extroverts should provide social stimulation and help satisfy others' affiliative needs. They may also thrive on attention from Earth when communication with Earth is possible. The hypothesis that each type of person offers mutually exclusive qualifications for extended-duration spaceflight poses a dilemma for those involved in the selection process. This dilemma would be resolved if future research revealed a third type of person, a person who exhibited behaviors commonly associated with both extroversion and introversion. In theory, this relatively flexible "ambivert" might be better able to deal with the rigors of space than the relatively inflexible introvert or extrovert. There are many other dimensions in which one might find that the best crew candidate is the one who possesses a duality of traits.

Similarities and Complementarities

There are numerous personal qualities and attributes whose effects on group compatibility are determined by the qualities and attributes of the other people in the group. In some cases, it is people's similarities that make for social compatibility; in other cases, people's differences intermesh. Such factors include attitude and value homogeneity, skill complementarity, and need compatibility.

First, conflicts of social, moral, and ethical values have proven to be a problem in some of the fallout-shelter studies (George Washington University, 1974), and almost all reviewers have tended to accept the position that homogeneous attitudes, values, and interests will militate against interpersonal conflict. The expectation that crews composed of individuals with shared attitudes and values will tend to be compatible is certainly supported by studies in other contexts. Results from the field and from the laboratory have been spectacularly consistent: attitudinal similarity is a powerful determinant of mutual attraction. It has been repeatedly found that the *proportion* of shared attitudes determines the extent to which people find each other attractive (Byrne, Gouaux, Griffitt, Lamberth, Murakawa, Prasad, Prasad, and Ramirez, 1971). Group members do not have to be completely "like-minded," for attitudes vary in terms of their relevance to the group. A group that reacts in a highly spirited fashion to dissimilarities on issues that it feels are significant may allow considerable latitude for differences of opinion in areas unrelated to its purposes and tasks (Schachter, 1951).

Thus, attitude and value similarity on mission-related issues should be a powerful determinant of crew compatibility. Still, one would hope to find, within a given crew, sufficient attitudinal variability to generate interaction and provide new ideas during problem-solving sessions. Additional research is necessary to identify the appropriate balance between similarities and differences in attitudes, opinions, and beliefs. We hypothesize that it is essential for crewmembers to share certain general values and desirable for them to exhibit variability in the ways that these values are expressed. Value homogeneity is likely to decline as crew size increases, but the development of new selection and indoctrination procedures should help to offset this decline.

Second, as noted by Haythorn and his associates, interlocking or complementary abilities should enhance group compatibility (Haythorn et al., 1972). One type is skill complementarity, which exists when one person is skilled in an area in which the other person is unskilled. Another is cognitive complementarity, which exists when people have nonoverlapping knowledge and must learn from or rely upon each other. Complementary abilities should allow each crewmember to contribute to the crew's welfare, sensitize each to the importance of the others' contributions, and in consequence promote solidarity and high morale. However, there is little or no research characterized by systematic efforts to relate complementary and overlapping abilities to compatibility within isolated and confined groups.

Third, people's needs may fit together in such a way as to affect group compatibility. Particularly important for present purposes is Haythorn's version (1968), which has been tested under conditions of isolation and confinement. This theory involves three patterns of needs. Congruent needs are similar-appearing needs such that the satisfaction of one person's need results in the satisfaction of the other person's need. For example, two people who have needs to affiliate could find mutual satisfaction by affiliating with one another. Complementary needs are different-appearing needs such that the satisfaction of one person's need also satisfies the other person's need. For example, a person with a need to teach might establish a satisfying relationship with a person with a need to learn. Competitive needs are such that the satisfaction of one person's need results in the frustration or aggravation of the other person's need. This might occur, for example, when two people vie for dominance within a group.

Important tests of this theory have been performed by Altman and Haythorn (Altman and Haythorn, 1965, 1967a, 1967b; Altman and Taylor, 1973; Haythorn and Altman, 1967; Haythorn et al., 1966). In these studies, isolated and confined subjects were paired to form dyads varying in homogeneity or heterogeneity along dimensions of dogmatism, dominance, need for achievement, and need for affiliation. Adaptability in isolation was a direct function of need-compatibility. All of the isolated dyads that experienced serious difficulties had been composed in such a way as to be incompatible. None of the nonisolated control dyads composed of individuals with incompatible needs showed comparable levels of difficulty. These results suggest that need compatibility gains importance under conditions of isolation and confinement.

Research to date thus suggests that it would be both useful and desirable to mount a comprehensive effort aimed at a better understanding of need compatibility. Such a program should attempt to identify relevant needs, show how they fit together, and spell out the consequences of compatibility and incompatibility. Ultimately, screening procedures may be devised for weeding out crew candidates whose needs are too likely to conflict, or ways may be found for keeping competitive needs under control. Researchers involved in such a program should remain sensitive to the possibility that incompatibility may not be a problem if conditions conspire to prevent crewmembers from detecting their differences, and that incompatibilities that disrupt one group may not affect another.

Group Homeostasis

According to Russian researchers, people who can synchronize their activities in very minimal kinds of social situations tend to form highly compatible and effective groups (Leonov and Lebedev, 1975). This research involves sitting groups of three or more candidates in front of a "homeostat," which consists of individual panels mounted with dials, pointers, and other controls. Each candidate is instructed to adjust his or her controls according to specifications; unknown to each candidate is the fact that as they proceed with the task they alter the settings of their colleagues' controls. The difficulty level of the task can be varied to study such variables as differences in group functioning under various levels of frustration. The ability to achieve the correct setting of one's own controls without preventing others from achieving their correct settings defines group homeostasis. As group members learn to coordinate their activities to operate the homeostat, their psychophysiological responses (heartbeat rate, respiration, certain brain waves, and so forth) also become synchronized. The groups that achieve homeostasis are seen as particularly suitable for isolated and confined settings and as highly capable of dealing with threatening or demanding conditions. The research on group homeostasis is intriguing, and needs to be replicated and extended in the West.

Crew Size and Social Compatibility

Increasing crew size increases the number of possible dyadic relationships within the crew according to the formula $(n^2 - n)/2$, where n is the number of people in the crew. Thus whereas a three-person crew could generate 3 dyadic relationships, a six-person crew could generate 15 dyadic relationships and a twelve-person crew could generate 66 dyadic relationships. Increasing crew size increases the number of possible social relationships and, among other things, options for social stimulation, options for developing friendships, and options for exercising varied role behaviors.

The evidence from studies of isolated and confined groups is a bit sketchy, and is complicated by the problem that relatively large groups may be stationed at a relatively comfortable main base, whereas relatively small groups may be located in primitive quarters that offer few of the main base's amenities. However, S. Smith's (1969) review indicates fewer emotional and interpersonal problems in relatively large isolated and confined groups. In one study, Doll and Gunderson (1971) found that Antarctic parties varying in size from 8 to 10 reported less in the way of compatibility and accomplishment than those varying in size from 20 to 30. They also report that military personnel stationed at small bases were more hostile than their counterparts at more heavily populated bases. Although cross-study comparisons are difficult, it is interesting to note that Georgia fallout-shelter studies (Hammes, Ahearn, and Keith, 1965; Hammes and Osborne, 1965; Hammes and Watson, 1965), which imposed very Spartan conditions on unselected but unusually large groups, had very low defection rates. S. Smith and Haythorn (1972) found triads more harmonious than dyads in a simulation study.

Because increasing crew size should increase the level of social stimulation, the number of friendship options available, and the opportunity to exercise role-related behaviors, it is expected that, socially, membership in a large crew should be more easily endured than membership in a small crew. However, the relationship between group size and interpersonal compatibility is not well understood, and (because few studies have involved varying group size while holding other variables constant) we lack the necessary bearings for making confident predictions. A basic research question is in identifying the functional form of the relationship between crew size (particularly over the range from about 2 to 30 crewmembers) and social compatibility.

Assembling Groups

Much of the literature on selection is based on the assumption that there are certain qualities that will affect a person's overall suitability for membership in a spacecrew. To some extent, this assumption is reasonable; for example, it is difficult to imagine conditions under which an incompetent, emotionally unstable astronaut would make a positive contribution to crew performance and morale. However, to some extent, people's strengths and weaknesses are to be found in the eyes of the beholder. For example, the quality of the contributions of female or minority astronauts might depend in part on the prevalence of racist or sexist attitudes among the rest of the crew. Furthermore, as we noted in our discussion of research on attitudinal similarities, complementary needs, and group homeostasis, compatibility often depends on the way that different people's qualities fit together. Consequently, it is necessary not only to consider each crewmember's personal qualities, but to relate these to the personal qualities of the rest of the crew.

The end result is that future selection research may require a shift of focus from characteristics of isolated individuals to combined characteristics of different members of the crew. Associated with this shift are many complex issues. First, there are theoretical and empirical issues regarding the interplay of different people's characteristics. Second, there are technical issues regarding the assessment of these characteristics. Third, there are extremely knotty questions regarding appropriate statistical-analysis techniques.

INTERPERSONAL DYNAMICS

In the present section, our focus turns to some social processes that are likely to occur within small crews. We shall consider the specific topics of leadership, cohesiveness, conformity, group performance, and group change over time.

Leadership

Leadership is a social influence process. According to Hollander (1978), in the course of this process, leaders organize, direct, and coordinate followers. They also exert influence to (1) help the group maintain harmony and stability, (2) interpret the conditions that confront it, (3) set goals, and (4) meet challenges posed from without. The leader's right to exert this influence may be conferred through appointment by a higher authority, or by the group itself.

Although one group member may be appointed or elected leader and assigned distinguishing tokens of status and rank, the leader/follower distinction is oftentimes blurred (Hollander, 1978). First, leadership is a relational concept, with the result that the person who is leader from one perspective is a follower from another perspective. In any multilevel organizational hierarchy, most people will fill both leader and follower roles. Second, leadership involves a two-way influence process. Although the focus of discussions of leadership is likely to be on the leader's influence over his or her followers, the followers influence leaders in return. Leaders, for example, need followers' inputs to reach informed decisions. The appointed leader who fails to respond to followers' influence attempts may find it increasingly difficult to lead, and the nonresponsive elected leader may be booted out of office. Finally, despite the fact that there is an identifiable leader, certain leadership functions may be dispersed among various group members. As we shall see, one person may organize and direct the group while another attempts to satisfy the group members' human needs.

The course and outcome of the leadership process depends on three general factors (Hollander, 1978). The first is the characteristics of the leaders themselves, including legitimacy, ability, experience, motivation, interests, expectations, and so forth. The second is the characteristics of the followers, again including such considerations as ability, motivation, interests, and expectations. The third is the characteristics of the situation, including the nature of the physical and social setting, the available physical and social resources, and the structure of the group's task. Prescriptions for good leadership often dwell upon the selection and training of leaders. However, such prescriptions could also involve the selection and training of followers, and the structuring of situations and tasks.

Heavy demands will be placed upon people performing leadership functions in space-capsule microsocieties. These demands are expected to become increasingly burdensome as the mission continues.

First there will be the stringent technical requirements associated with operating safely in a hostile environment. Although there will be advance preparation and some degree of communication with resource people at mission control, supplies will steadily decrease and as distance increases it will become increasingly difficult to maintain good communications with Earth. To a great extent, problems will have to be solved using the highly limited resources available in the closed environment of the space capsule.

Second, the demands on leaders' interpersonal skills are likely to be formidable. As noted earlier, it has been hypothesized that isolation, confinement, risk, and other conditions associated with spaceflight are conducive to deteriorating interpersonal relationships (George Washington University, 1974). Because of this possibility, people in leadership roles will have to be shrewd judges of human nature and display superior interpersonal skills.

A failure to fulfill the requirements of leadership can lead to severe penalties for an isolated and confined group. In the 1959-1960 fallout-shelter studies, a deliberately passive role on the part of the shelter commander was credited with a general lowering of standards of behavior and a loss of interest in matters of civil defense (Strope et al., 1960, 1961). The Georgia fallout-shelter studies also revealed that a failure in management led to increased friction and decreased morale (Hammes et al., 1965; Hammes and Osborne, 1965; Hammes and Watson, 1965). Weak or incompetent leadership has spelled disaster for a number of polar expeditions (Leonov and Lebedev, 1975). Competent leaders, on the other hand, may serve as models whose enthusiasm and even temper are emulated by the crew. Good leaders can prevent factionalism and ease group members through troubled relationships.

Task and socioemotional leadership activities— Repeatedly, distinctions have been made between task activities (also known as initiation of structure and concern for production) that help the group get the job done or move toward its goals, and socioemotional activities (also known as showing consideration and concern for people) that promote harmonious relations within the group. Group functioning requires people who take the initiative in each of these areas. Socioemotional leadership is at least as important as task leadership, and perhaps more so, judging by some of the research.

It is not clear how often the same individual can satisfactorily fill both task and socioemotional leadership roles. The pioneering research by Bales and his associates (Bales, 1950, 1953, 1958, 1970) found that some group members engaged in more task and socioemotional activities than others, and as a result were offered leadership status. But it was also found that the person who engaged in the most task activities was not the same person who performed the most socioemotional activities. There were, in effect, two leaders: the task leader, who was rated as having the best ideas, offering the most guidance, and being most influential in forming the group's opinions; and the socioemotional leader, who was the best liked. The usual explanation for the emergence of the second leader is that a task leader's sense of purpose gives rise to activities (unpopular orders, sharp criticism, etc.) that hurt group members' feelings. The second leader emerges to smooth things over and restore harmony to the group.

However, it should be noted that the initial studies involved emergent leadership. That is, unacquainted individuals joined in a discussion, and social structure emerged as interaction progressed. The task leader took a role of power and influence, and it may have been his presumptuousness that caused the internal conflicts. According to Burke (1972), when a leader is designated by a higher authority and is hence perceived as "legitimate," group members are more accepting of heavy-handed task acts and the need for the second leader diminishes. Certainly, Thor Heyerdahl and many of the other expeditionary leaders discussed by Leonov and Lebedev (1975) appear to have had heroic capacities to perform both task and socioemotional leadership roles. Yet, the evidence reviewed by Katz and Kahn (1978) suggests that only under rare conditions are task and socioemotional leadership roles best filled by the same individual.

Researchers, planners, and managers are thus confronted with the problem of understanding the optimal distribution of leadership behaviors within the crew. Specifically, to what extent should various task and socioemotional leadership behaviors be concentrated in the hands of a specific leader, rather than distributed across two or more people within the crew? Of particular interest in light of Burke's arguments is determining the extent to which a given crew-member should attempt to manage both task and socioemotional leadership roles. To begin with, Leonov and Lebedev's and Katz and Kahn's conclusions need to be reconciled. One possibility, suggested largely by Burke's research, is that structural factors, or possibly certain kinds of leader-follower compatibilities, make it possible for people such as Thor Heyerdahl to perform both roles. Another, definitely counterintuitive but intriguing possibility is that isolation, confinement, and risk make it relatively easy for one individual to perform both task and socioemotional leadership roles. Yet another alternative is that under conditions of isolation, confinement, and risk, followers have a strong defensive need to perceive their leaders as highly competent on both task and socioemotional dimensions. Evaluation of these alternatives awaits the results of further research.

Personality, situation, and leadership— A prevalent theme is that certain personality traits can be identified with effective leadership. Summarizing the results of scores of studies, Mann (1959) reported that intelligence, adjustment, and extroversion are moderately related to leadership, whereas dominance, masculinity, and interpersonal sensitivity are somewhat less closely related. Observations of Sealab II led Radloff and Helmreich (1968) to suggest that people under stress in isolation and confinement may not need a young, action-oriented leader as much as a mature individual who inspires identification and provides reassurance. Citing work by Misumi and Shirakashi (1966) and Cooper (1966), Kubis (1972, p. 55) derived the following composite picture of the effective space-crew leader:

... he elicits the best from his men ... is himself personally competent ... is interested primarily in results and achievement ... but is always aware of the normal human needs of the group and attempts to provide opportunity for their satisfaction ....

A person who is a highly competent leader under one set of conditions may prove to be a marginal or incompetent leader under another set of conditions; thus, both personality variables and situational variables influence quality of leadership (Fiedler, 1967, 1971, 1978; Hollander, 1978; Katz and Kahn, 1978; Mann, 1959). Perhaps the most promising theory which simultaneously considers personality and situational factors is Fiedler's (1967, 1971, 1978) contingency theory of leadership. Concerned with predicting performance rather than satisfaction or morale, the theory has been tested successfully in many military and civilian settings, and deserves close attention from space mission planners. The independent variables are situational favorableness and leadership style, and the dependent variable is leadership effectiveness.

Situational favorableness refers to the structural and social-climate variables which make a group easy (high favorableness) or difficult (low favorableness) to lead. These include (1) the extent to which the leader is accepted and respected by the group, (2) the extent to which the group's goals are clear and structured, and (3) the extent to which the leader has been invested with the power to reward and punish group members.

Leadership style refers to the leader's orientation toward tasks and people. This can be measured by asking the leader to evaluate the least preferred co-worker (LPC) with whom he or she has ever worked. High scorers, who tend to give favorable ratings to the least preferred co-worker, are relatively socioemotional in outlook. Low scorers, that is, people who assign harsh ratings to their least preferred co-workers, have more of a no-nonsense, task orientation. In Fiedler's theory, high scorers and low scorers are referred to as high LPC leaders and low LPC leaders, respectively. Leadership effectiveness, the dependent variable, can be measured by any objective measure of task accomplishment.

According to contingency theory, different degrees of situational favorableness require different leadership styles. Under conditions of very high or very low situational favorableness, the task-oriented, low LPC leader is likely to prove most effective. As Jacobs (1971) so aptly puts it, the leader can afford to be firm when accepted by the group, pursuing clear goals, and invested with power to reward and punish. He or she must be firm when rejected by the group, grappling with ambiguous goals, and lacking the power to reward or punish. Under conditions of intermediate situational favorableness, the interpersonal sensitivity of the high LPC leader is likely to be of use in working through the moderately troubled relations within the group, thereby freeing the group to continue toward its goal.

Careful planning may create and maintain a high degree of situational favorableness on short-term missions, but such conditions may be difficult to sustain on prolonged flights. For example, it may be relatively easy to link the on-board leader's evaluation of crewmembers to the latter's continuation and advancement within the space program. However, as noted in chapter VIII, as the link with Earth becomes tenuous, traditional bribes and threats may lose force. Thus, we hypothesize that whereas task oriented, low LPC leaders may do best on carefully planned short flights, socioemotionally oriented, high LPC leaders may have an edge on extended flights.

Contingency theory posits that leadership style is fairly firmly ingrained. This implies that whereas leaders may be selected for missions of varying degrees of situational favorableness, training programs intended to change their styles could be ineffective. However, Fiedler and his associates have developed a self-instructional program called LEADER MATCH, which helps leaders self-select and gain control over such variables as those which determine situational favorableness (Fiedler, Chemers, and Mahar, 1976; Fiedler, 1978). This program, as it stands, or with some modifications, may prove useful for space-mission leader preparation.

From Fiedler's contingency theory of leadership flow many lines for future research. Efforts must extend beyond assessing the situational favorableness of a given mission and then choosing the leader with the most promising style. First, it is necessary to understand fluctuations in situational favorableness over time. Second, procedures must be developed to help spacecrew leaders learn to identify shifts in situational favorableness, and change their behavior accordingly.

Autocratic and participative procedures— Leaders who make decisions without soliciting subordinates' inputs are said to use autocratic procedures. Leaders who solicit subordinates' inputs are said to use participative procedures. These latter procedures include (1) consultive decision making in which the leader seeks opinions of informal leaders and of rank and file, (2) representative democracies, and (3) full democracies.

Early "leadership climate" research undertaken on the eve of World War II suggested many advantages to the democratic approach (Lewin, Lippitt, and White, 1939). Most reviewers seem to believe that modal group members can offer very useful information, and conclude that, more often than not, the quality of a decision will be enhanced by membership participation (Kleinhans and Taylor, 1976; Steiner, 1972, 1976). In addition, it has been found that organizational members are likely to feel a more personal commitment to decisions they have helped make than to decisions imposed from above (Coch and French, 1948; Hollander, 1978). However, the overall picture contains many complexities, and the optimum point along the autocratic-democratic continuum depends upon such variables as the personalities of group members, the distribution of knowledge and skills within the group, the group's size and organization, and the degree of structure of the problem (Hollander, 1978; Vroom, 1976; Vroom and Yetton, 1976).

Mission managers may find it neither necessary nor desirable to establish procedures such that all decisions can be traced to one point on the autocratic-democratic continuum. For example, the Russians have reported success with procedures whereby mission decisions were made by the commander and crew issues were decided by democratic votes (Leonov and Lebedev, 1975).

An analysis of situations, personnel, and likely scenarios would seem essential for maximizing the advantages and minimizing the disadvantages associated with autocratic and democratic decision making procedures. Perhaps a useful starting point for such research is the Vroom-Yetton normative model of participative decision making (Vroom, 1976, Vroom and Yetton, 1976). In essence, this model is applied by first answering seven questions (regarding such issues as the availability of information, the degree of conflict among subordinates, and the need for subordinate acceptance of decisions). On the basis of the pattern of answers, one of five decision-making procedures is prescribed.

In summary, future research on small-crew leadership should include an expanded range of leadership alternatives and options.

Particularly pressing problems include identifying the optimal distribution of task and socioemotional activities, and achieving a good fit between structural characteristics and leader characteristics. In addition, we need to know more about the consequences of various autocratic and participative decision-making procedures under conditions of isolation, confinement, and risk.

Cohesiveness

Cohesiveness refers to the strength, solidarity, or unity of a group. It is roughly synonymous to elan or morale. Although originally intended to be a unitary construct, cohesiveness sometimes designates a group with energy and a strong sense of purpose, and other times a group whose members have a strong sense of liking for and involvement with one another and who express positive attitudes about the group. Whereas drive and amiability often covary, it is possible for a group to be characterized by one of these attributes but not by the other (Lott and Lott, 1965).

In some cases, adverse conditions and suffering seem to increase cohesiveness (Aronson and Mills, 1959; Gerard and Mathewson, 1966). In effect, undergoing a trying initiation encourages people to rationalize the discomfort by telling themselves that membership in the group is extremely desirable. However, most discussions focus on the rewards or satisfactions of group membership as the major cause of cohesiveness. Cartwright (1968), for example, has defined cohesiveness as the sum of the satisfactions which membership accords all of the members of the group. Satisfaction is likely to be high to the extent that the group (1) engages in activities that the members find intrinsically satisfying, (2) pursues goals of importance to the members, (3) provides social support and emotional gratifications, and (4) serves ulterior motives. Thus, a crew might be expected to be cohesive when the crewmembers (1) enjoy flight and adventure, (2) subscribe to the mission's overall goals, (3) encourage each other, and (4) help fulfill each others' needs in areas tangential to or unrelated to the mission.

The establishment and manipulation of group goals is one commonly prescribed way to promote cohesiveness (Sherif and Sherif, 1969). Goals refer to objectives or end states which have motivational properties in that they instigate, strengthen, and impart direction to behavior. Group goals refer to objectives or end states which are endorsed or found compelling by the different people who comprise the group. Group goals encourage people to coordinate their activities for mutual gain, and hence are likely to affect the tone of interpersonal relations within the group. The isolation and confinement literature, for example, suggests that individuals may be able to suppress their differences in the interest of group goals. In Sealab II, some aquanauts commented that teammates who didn't always see eye to eye were able to get along for the period of the mission (Radloff and Helmreich, 1968). In a field experiment involving preadolescents who were camping in the wilderness, Sherif, Harvey, White, Hood, and Sherif (1961) found that incompatible or separatist goals promoted prejudice and open conflict, whereas goals that were of importance to all of the members of the group promoted amiability and unity.

First, mission planners and managers should take steps to help spacecrews identify group goals. Such goals should be superordinate in the sense that they override individual or subgroup goals which, if pursued, might encourage behavior that is detrimental to the overall mission. Crewmembers must feel personally committed to these goals; it will not suffice to simply impose them from above. Additionally, these superordinate goals must require a high degree of cooperative activity (Sherif and Sherif, 1969).

Second, steps should be taken to ensure that group goals are clear and well understood. Discussing Antarctic groups, Natani and Shurley (1974) have noted that scientists are given a brief introduction to the "big picture" at an orientation conference, but that their goals remain basically individualistic. Navy personnel are given only a minimal understanding of their science support role, with the result that they find it difficult to become firmly committed to the overall mission. Under such conditions, unwelcome forms of socioemotional behavior are likely to take precedence over welcome forms of task behavior.

Third, a means must be found to maintain astronauts' interest in distant goals over prolonged periods of time. It may thus be desirable to establish a number of interim goals which can be pursued and savored. Perhaps this has been best expressed by Sells and Gunderson (1972, p. 82):

> . . . To maintain group integrity and motivation of group members, the void between initiation of a mission and final attainment of its goals must be filled with richly detailed programs of activities that permit achievement of meaningful interim goals. It is also important that both the

ultimate and intermediate goals be expressed in a manner that permits assessment of success in such a way that it is compatible with supervisory controls, available rewards, and individual career growth . . . .

Another way to promote cohesiveness would be to help minimize interpersonal conflicts. Presumably, initial social compatibility (as discussed earlier in this chapter) will be a major factor. Both task and socioemotional training can also be expected to help minimize or contain interpersonal conflicts. People who do not know what to expect and do not know how to do their jobs are likely to frustrate and annoy one another. In addition, people who are unskilled may respond to a poor overall level of performance by acting toward one another in negative ways (Shurley et al., 1977). Both Kubis (1972) and Berry (1973a) have advocated direct training in human relations. Such training may involve the entire crew, or, if this is impossible, crewmembers who are in managerial roles or whose assignment is to work with other people. Training in interpersonal relations was considered valuable by subjects in the Douglas simulation study (George Washington University, 1974).

To devise adequate training programs, more must be learned about the on-board diagnosis and management of interpersonal frictions and conflicts. Such research would be aimed at uncovering techniques that astronauts might use to identify and combat the underlying sources of interpersonal stress. It might address ways of recognizing and managing one's own rising tensions and tensions in other people. Satisfactory experiential training in interpersonal relations may require an authentic setting characterized by isolation, confinement, and stress.

Another technique for increasing spacecrew solidarity is derived from manning theory as developed by Barker (1968) and Wicker (1979). An implication of this theory is that the size of a spacecrew relative to the amount of work that needs to be done will affect the level of energy that the crew will apply to the task, the favorability of the attitudes that crewmembers maintain toward one another, and the overall level of involvement of the crewmembers in the group.

Adequate staffing occurs when there is a good match between the number of people available, on the one hand, and the situation's technical and social demands, on the other. Under conditions of adequate staffing, people are neither rushed nor are their abilities allowed to languish. The group's interests and energies cover the immediate demands, but do not extend much beyond those demands.

Understaffing occurs when the number of people available is less than typically hoped for or expected, given the task demands. Wicker (1979) hypothesized that understaffing encourages people to (1) work close to the limits of their abilities and skills; (2) assume difficult, important, and varied tasks; (3) feel important, responsible, and versatile; (4) monitor and improve each others' performance; (5) display tolerance of the other people's occasional lapses; and (6) ignore personal characteristics (such as race, sex, and personality) that are irrelevant to performance. Thus, members of a "short" or understaffed spacecrew may feel a bit harried and pressured, but they may also work hard, feel good about themselves, and be supportive of one another.

Overstaffing occurs when the number of people available is greater than typically hoped for or expected, given the situations' demands. Wicker (1979) hypothesized that overstaffing (1) encourages perfunctory, lackadaisical performances; (2) fosters a high degree of specialization accompanied by a jealous guarding of all activities which fall within the associated narrow domain; (3) undermines one's sense of self-esteem; (4) encourages little concern of the crew for each others' progress; (5) develops cynical attitudes of the crew toward one another; and (6) encourages the crew to dwell upon race, sex, personality characteristics, and other task-irrelevant individual differences. Working in an overstaffed setting may be "easy time," but it may also be psychologically and socially unhealthy.

The theory of manning grew from studies of people engaging in voluntary activities (e.g., extracurricular school activities and church affairs) and in service professions, and most of its important tenets are yet to be tested in varied organizational settings. Nonetheless, this theory raises the interesting possibility that, subject to the limitations imposed by safety requirements, spreading a spacecrew just a little bit "thin" may enhance the crew's functioning. It must be emphasized, however, that manning theory needs to be evaluated in isolated and confined settings, that optimal understaffing is a question of degree, and that understaffing might have disastrous consequences when a group must continue its activities uninterrupted over a prolonged period of time.

Some ambiguity surrounds the relationship between drive and amiability, on the one hand, and actual performance, on the other. Cohesive groups often are efficient and effective (Cartwright, 1968; Lott and Lott, 1965; Shaw, 1976). However, this does not necessarily mean that enhancing cohesiveness improves performance. First, successful performance can be a cause, rather than an effect, of cohesiveness. This is most clearly shown in Bakeman and Helmreich's (1975) study of divers aboard Tektite. It was found that whereas measures of cohesiveness obtained early in the mission were unrelated to measures of performance obtained later in the mission, measures of performance early in the mission were predictive of subsequent crew cohesiveness. Second, group standards or norms mediate the relationship between cohesiveness and performance. If norms support performance-related activities, then cohesiveness is likely to improve performance. If, on the other hand, norms support limited output or engagement in irrelevant tasks, cohesiveness may undermine performance (Berkowitz, 1954).

In summary, cohesiveness is typically viewed as a function of the rewards and satisfactions of group membership. Major research and planning issues are associated with promoting a high degree of cohesiveness within tomorrow's spacecrews.

Compliance, Conformity, and Independence

A certain amount of social activity aimed at ensuring adherence to group norms is generally regarded as beneficial, because it promotes coordination of efforts and a sharing of values within the group. However, such influence processes have certain potentially adverse effects which may become pronounced under conditions of isolation and confinement. Strong social pressures can inhibit the flow of creative ideas, particularly in a cohesive group. Individuals may fear that unorthodox suggestions will incur the leader's displeasure, undermine morale, or lead to rejection. This fear is justified, for it has long been known that a person who violates group norms is likely to trigger a specific series of events (Schachter, 1951). The initial reaction is an increase in communications intended to bring that person back into line. If these attempts are unsuccessful, communication ceases and the deviant is ignored. Under normal conditions, such ostracism may simply result in the deviant leaving the group, but under conditions of isolation and confinement, the deviant cannot leave the group.

Prolonged rejection may lead to the pathological "long eye" syndrome (Haggard, 1964; Rohrer, 1961). Noted primarily in polar camps, this syndrome may involve hallucinations, tears, loss of appetite, silence, suspiciousness, and sloth. This is not only extremely punishing to the rejected individual, but it penalizes the group by robbing it of the services of one of its members. This can be a major problem in small crews that begin the mission only minimally staffed. Rohrer (1961) documented at least 10 cases of "long eye" in the Antarctic. Fortunately, the effects of "long eye" are temporary and vanish quickly when the individual is reaccepted by the group. Further research is required to better understand the causes and consequences of ostracism under conditions of prolonged isolation and confinement. It is essential to find ways to soften the blow of social rejection without entirely eliminating it as a mechanism of social control. Finally, it is important to find ways to reintegrate the deviant into the crew once the episode is over.

Strong conformity pressures may reflect attempts to maintain group harmony. Such harmony may be purchased at the expense of performance. Janis has coined the term "groupthink" to refer to conditions under which efforts to maintain group harmony undermine critical thought and lead to poor decisions (Janis, 1971, 1974; Janis and Mann, 1977). Spacecrews may be quite vulnerable, since groupthink becomes likely when the group is concerned with maintaining amiability, when there is little or no communication with people outside of the group, and when the group is confronted with a threatening situation. Among the most important characteristics of groupthink are (1) false optimism and lack of caution, (2) direct pressures on nonconformers, (3) a fear of disapproval for expressing new alternatives, (4) an illusion of unanimity, (5) the emergence of "mind guards" who protect the leader from criticism, and (6) efforts to deny or rationalize all ill-omens.

Safeguards against groupthink include soliciting external inputs during the decision-making process, appointing a devil's advocate to challenge majority views, and reconsidering decisions before action is taken. It is not clear that all of these safeguards are effective; for example, Bennis (1976) argues that a devil's advocate is ignored because the group recognizes that he or she is merely playing a social role. Those safeguards which are effective under "normal" conditions may or may not be workable under conditions of isolation and confinement. Specifically, some of the remedies proposed for groupthink presuppose a social system with a relatively permeable boundary. More research is needed on the emergence and control of groupthink under conditions of isolation and confinement.

Special problems arise in emergency situations in which leaders are likely to exert authority to gain prompt acceptance of a course of action. Ideally, leaders would invariably make appropriate decisions which would then be gracefully accepted and implemented by the crew. Actually, leaders may make faulty decisions because they may lack certain information or fail to process information correctly. There are cases on record when aviation accidents or near accidents could have been averted if crewmembers had forcefully drawn the pilot's attention to unnoticed conditions or had openly questioned the pilot's instructions (Yanowitch, 1977; Murphy, 1980). That is, correctable pilot errors have gone uncorrected because of unquestioning attitudes, a lack of assertiveness, or deficient communication skills.

Certainly, we may expect selection and training procedures to yield leaders who have a higher degree of competence than the average member of the crew. Furthermore, we do not expect a leader's performance to be enhanced if he or she is constantly bombarded with wrong-headed suggestions. However, even as a leader's inability to induce compliance can be regarded as a failure in leadership, so can an inability to attend to relevant subordinate input. It is necessary to find ways to help crewmembers identify the conditions under which they should speak up, and to devise training procedures to ensure that they have sufficient assertiveness to do so. Moreover, it is necessary to find ways to help leaders respond appropriately to unsolicited follower inputs. In summary, it is necessary to identify and achieve a healthy balance between crewmembers' acceptance of social influence and a willingness to engage in appropriate independent action.

Group Performance

The determinants of group performance are many and complex. Some degree of order and coherence is imposed by the Hackman and Morris (1975, 1978) process model of group performance. The model's name is derived from its emphasis on the processes that occur as group members interact with one another.

According to the model, three factors or sets of variables affect group performance. The first factor is the knowledge and skills which group members are able to bring to bear. The second factor is the amount of energy or effort that the group members apply. The third factor is the performance strategies or procedures that the group members follow while addressing their tasks. To the extent that conditions foster the application of skill and effort, and to the extent that groups choose appropriate performance strategies, the group will prove to be efficient and effective. To the extent that the application of skill or effort is dissuaded, and to the extent that groups follow inappropriate performance strategies, poor performance will result.

Hackman and Morris identify three levels of inputs which directly affect group processes and indirectly affect performance outcomes. Individual-level inputs include such variables as the group members' skills, attitudes, and personality characteristics. Group-level inputs include such variables as group size, morale or cohesiveness, communication patterns, and social norms. Environmental-level inputs include such variables as the nature and structure of the group's tasks, and the level of environmental stress. Through the manipulation of inputs, group processes can be altered and performance outcomes improved.

First, by manipulating group composition, changes can be effected in the knowledge and skills which group members bring to bear on the group's tasks and problems. For example, it might be possible for space crewmembers to be chosen, in part, on the basis of complementary skills and interests (Haythorn et al., 1972). A careful analysis of mission requirements and of the people who might satisfy them could result in a large crew with a good spread of abilities and skills.

Second, the model suggests that the manipulation of task characteristics can increase group members' commitment to work hard on the task. Drawing on the work of Hackman and Lawler (1971) and Hackman and Oldham (1974), Hackman and Morris (1975, 1978) suggest that a high degree of individual-effort expenditure becomes likely when individual assignments (1) provide the opportunity to use a variety of personal skills and abilities, (2) involve "whole and visible" pieces of work, (3) are related to the physical or psychological well-being of other people, (4) make some allowance for individual initiative and discretion, and (5) are structured in such a way as to allow the person to monitor his or her level of performance.

Finally, through manipulating social norms it should prove possible to influence task performance strategies. For example, as already noted, groups whose norms foster a high degree of conformity may have difficulty solving problems that require innovative solutions. Pressures "not to be different" discourage group members from sharing new or unusual ideas. If new norms could be developed that encourage independence of thought and expression, the group could rapidly uncover a wide array of creative solutions to its problems. Hackman and Morris urge group members to continually ask themselves if their usual performance strategies are the most appropriate ways of proceeding under emerging conditions. They suggest that an open preperformance discussion of strategies can pay big performance dividends, particularly if the group finds itself confronted with a nonroutine problem or task.

This promising model of group performance has received only limited testing, and these tests have not involved isolated and confined groups. Certain of the issues raised by Hackman and Morris are prime research issues. These revolve around constituting crews in such a way as to ensure a desirable range of knowledge and skills, structuring tasks in such a way as to maximize motivation and commitment, and identifying and encouraging social norms which foster appropriate performance strategies under conditions of isolation, confinement, and risk. Some of these issues will be further addressed in our discussion of motivation in chapter VIII.

Temporal Dynamics

Interpersonal dynamics within a crew can be expected to change over time. Building on work by Bales (1950), a number of researchers including Bennis and Shepard (1956), Mann, Gibbard, and Hartman (1967), and Tuckman (1965) have identified developmental stages within small groups. Although these formulations differ in terms of their specifics, there is some agreement that early in the group's life, activities center around the feelings that members have about each other and their relationship to the leader. Only after these issues are resolved will the group become highly cohesive and gain the ability to effectively focus its energies on the job at hand. Illustrative is Tuckman's (1965) formulation based on studies of therapy groups. According to Tuckman, groups pass through four sequential stages: (1) testing and dependence or "forming," (2) intragroup conflict or "storming," (3) development of group cohesiveness or "norming," and (4) task functioning or "performing."

The research on group development implies that using, as a crew, a well-established or mature group that has thrashed out norms, tested limits, come to grips with authority relations, and reconciled individual differences would help minimize interpersonal conflicts within the spacecraft. A relatively mature group may have another advantage: It can be studied as a group prior to departure and this study could provide a backup to the initial selection process (Perry, 1965, 1967). The disadvantage of a well-established group is that the members may find that they have little new to offer each other socially. That is, as the natural processes of social penetration continue (Altman and Taylor, 1973), crewmembers will become increasingly familiar to one another, find increasingly less to discuss, and perhaps lose interest in one another. Future research is required for selecting and training crews that are highly coordinated but retain the capacity to provide appropriate levels of mutual stimulation.

As missions increase in duration, crews will lose the characteristics associated with short-term groups and gain the characteristics associated with long-term groups. Ziller (1977) has identified four important ways in which short-term groups and long-term groups differ. First, whereas short-term groups are locked into the here-and-now, long-term groups have a sense of continuity and view themselves in relationship to the past, the present, and the future. Second, whereas short-term groups tend to be dominated by immediate outcomes, long-term groups have images of the future and may be willing to forego short-term gains in order to achieve long-term goals. Third, relationships within short-term groups tend to be impersonal and reflective of social roles, whereas relationships within long-term groups tend to be personal and based on strong attachments among members. Finally, whereas short-term groups tend to be relatively static, long-term groups tend to be dynamic, or, in Ziller's words, always "in process." Most group dynamics research involves short-term groups, yet it is long-term groups that we must understand to plan extended-duration space missions.

Not all microsocieties in space will be closed systems. There is likely to be some turnover in orbiting laboratories or settlements. This raises the problem of introducing and assimilating newcomers into preformed groups. According to a review by Crandall (1978), because newcomers do not share the continuing members' knowledge and attitudes, they may unintentionally act in disruptive ways and hence come to be seen as disloyal to the group. Aware of this problem, newcomers themselves are likely to be anxious and prone to conform.

Crandall reviews several methods for easing the integration of newcomers into ongoing groups. Each of these procedures may reduce conflict and attrition, and hence deserves careful consideration for incorporation into mission plans. First, there is preentry therapy, which is expected to encourage anxiety control and reduce the need to conform. Second, newcomers can become acquainted with current or former group members prior to their entry. Third, newcomers can be given candid and realistic (as compared with guarded and idealistic) information about the group. Finally, newcomers can be sponsored; that is, an established group member can introduce and tutor each newcomer.

In the area of personnel rotation, there are many topics for future research. One hypothesis is that there is an advantage to using, as newcomers, people selected and trained along with those who have already entered space. Another possibility is that there is an advantage to letting the crew help select its own new members. Still another possibility is that assimilation is enhanced as a result of telecommunication with the newcomer prior to the newcomer's departure from Earth. Finally, it would be useful to know more about the kinds of conditions that will result in newcomers being given more time to gain acceptance by the group.

Yet another issue is fixing the number or proportion of crewmembers that should be rotated or replaced at any one time. In the military, piecemeal replacement has not been particularly successful. On the other hand, introducing large groups of newcomers means that many people have to be socialized simultaneously and that oldtimers will feel particularly threatened. It is also necessary to find means for selecting individuals who are to be placed during a given personnel exchange.

In the future, it should become increasingly likely to actually train crewmembers or entire crews in space. This might be done, for example, inside a well-established satellite or lunar station. Such bases could provide a relatively safe environment where new astronauts could learn to function under conditions of weightlessness and come to grips with the psychological and social realities of isolation, confinement, and risk. In addition, the promise of being in outer space at a relatively early point in one's career might help generate and sustain a high level of motivation. On-board training should ease a new crew's transition from training to operational status. Seasoned operational crews might be more willing to accept new crewmembers who have already gained experience in space than those whose prior experiences have been limited to Earth.

In summary, crew dynamics will change over time. Future research is required to better understand these changes. In addition, future research is needed to ease the integration of newcomers into crews that will have rotating memberships. On-board training may serve a number of useful purposes, including the integration of newcomers into seasoned operational crews.

SUMMARY AND CONCLUSIONS

This chapter reviewed some of the ways in which interpersonal and small-group variables may affect crew performance and the quality of life in space. A central theme in the literature has been choosing as crewmembers people who are compatible in the sense that their abilities, interests, and motives prompt them to act in ways which other crewmembers consider desirable and appropriate. Almost nothing is known about all-female or mixed-sex groups operating under conditions of isolation, confinement, and risk. On the one hand, mixed-sex crews offer social diversity; on the other hand, sex-role stereotypes and negative attitudes toward astronauts of the opposite sex could prove disruptive. Further research is necessary to discover patterns of interaction within all-female and mixed-sex crews, and to ensure that crewmembers are flexible and tolerant in their dealings with members of the opposite sex. Moreover, means must be sought to minimize the potentially disruptive effects of pairing-off and to ensure that sexual expression assumes acceptable forms.

The available evidence suggests that there may be certain advantages to including both relatively young and old members among a crew. Desirable age mixes need to be better understood. In addition, we need to know more about adult development under spaceflight conditions.

It is expected that social change coupled with a growing likelihood of international missions will increase cultural heterogeneity within spacecrews. The intergroup relations literature specifies certain conditions which are likely to reduce prejudice, and many of these conditions can be established in space. In addition to prejudice, problems include language or communication difficulties and differing, culturally related preferences regarding food, facilities, and activities. Further research is a prerequisite for achieving coordination within ethnically mixed crews.

A number of personal qualities have been found to be of high value to most members of isolated and confined groups. One such quality is personal attractiveness. It should prove useful to select-out individuals with characteristics that most other crewmembers will find aversive and individuals who find too many human qualities aversive. Second, emotional stability is very important. Future screening procedures must extend beyond selecting-out emotionally unstable candidates, and select-in candidates of exceptional psychological health. Perceived emotional stability may be as important as actual emotional stability, so it is necessary to study the psychological processes upon which judgments of emotional stability rest. Third, a high level of technical competence is crucial, especially for leaders. Competence research should be extended to include interpersonal as well as technical competence, and, as in the case of emotional stability, research should be expanded to include perceptual and judgmental processes.

Two other important qualities are cooperativeness and social versatility. Research regarding own-gain, relative-gain, and joint-gain motivation may be of use, as people with joint-gain motivation may prove to be of value in the highly socially interdependent setting of space. Perhaps even more important is further work to bridge achievement motivation and cooperation. Research in this area suggests that need achievement subsumes three variables known as work orientation, mastery orientation, and competition. It is hypothesized that individuals who are strong on work orientation and mastery orientation but weak on competition will best serve the joint task and socioemotional requirements of space. This hypothesis requires additional study under conditions analogous to those found in space.

The conditions of spaceflight will restrict the opportunity to enter into varied social relationships. Thus, social versatility may also be an asset in space. Relevant here are studies of individual differences in the ability to perform both the task-oriented masculine role and the socioemotionally oriented feminine role. Additional research is required to better understand the consequences of androgyny under spaceflight conditions, and to identify additional traits which might be related to social versatility.

Our discussion of social compatibility then turned to an examination of similarities and differences among crewmembers. It has repeatedly been found that attitudinal similarity is a powerful determinant of interpersonal attraction. However, extreme attitudinal homogeneity may cause boredom and undermine a crew's problem-solving potential. A major issue is identifying conditions under which consensus is preferable to diversity, and conditions under which diversity is preferable to consensus. Also, interlocking abilities and needs are expected to contribute to compatibility. Relevant to the former are skill complementarity and cognitive complementarity; although these variables have appeared in theoretical discussions, they are not well researched. Relevant to the latter are congruent, complementary, and competitive needs. Although some progress has been made studying such needs, it is time to consider a comprehensive program to identify relevant needs, discover how they fit together, and demonstrate the consequences of compatibility and incompatibility.

Russian researchers report that people differ in the extent to which they can effectively coordinate their activities in certain kinds of experimental social situations. Accompanying a high degree of interpersonal coordination in voluntary responses is the synchronization across group members of certain nonvoluntary psychophysiological responses. The Russians suggest that groups whose members develop such synchronicity tend to do well under spaceflight conditions and to be particularly well suited for dealing with danger. This intriguing line of research needs to be replicated and extended in the West.

Because increasing crew size should be associated with an expanding array of social options, crew size is expected to correlate positively with crewmember satisfaction. However, the data on this point tend to be anecdotal and are all to often confounded by variables other than group size. A thorough understanding of the effect of crew size on crewmember satisfaction awaits the results of studies of different sized groups operating under spaceflight-like conditions.

In concluding our discussion of social compatibility, we proposed a shift in emphasis from selecting crewmembers on the basis of their individual characteristics to selecting entire crews on the basis of their combined characteristics. This is desirable because com-compatibility depends in part on the intermeshing of different people's qualities and upon different people's perceptions of and attitudes toward one another. Accompanying this shift are many complex theoretical and methodological issues. To compose large crews, and to staff multiple missions, we will have to devise reliable and valid means for assessing large numbers of people in relatively brief periods of time.

Strong demands will be placed on crewmembers who occupy leadership roles, and the penalties for incompetent or weak leadership will be high. Our discussion of leadership began with the common distinction between task and socioemotional leadership activities. Case histories suggest that under conditions of isolation, confinement, and risk, a leader can perform both task and socioemotional leadership roles, but influential reviews of leaders in everyday environments show that normally the two roles are differentiated. Future research must address different divisions of task and socioemotional activities within spacecrews. One possibility for reconciling conflicting claims is that under conditions of isolation, confinement, and risk, people have a strong defensive need to perceive their leaders as competent on all dimensions.

Contemporary leadership theory focuses upon the interaction of structural and personality variables rather than upon either variable alone. The most influential theory is contingency theory, which suggests that situational favorableness and leadership style combine to determine leadership effectiveness. This theory posits that task-oriented leaders are maximally effective under conditions of extremely high or extremely low situational favorableness, and that socioemotionally oriented leaders are maximally effective under conditions of intermediate favorableness. Tomorrow's missions are likely to vary in terms of their situational favorableness, and on any given mission, situational favorableness may decline over time. Additional research is required to achieve a good match between situational favorableness and leadership style throughout the course of future flights. Researchers should consider an expanded range of leadership alternatives, and should remain sensitive to the possibility that no single decision-making structure will prove "best" for all purposes.

Most analyses describe group cohesiveness or unity as resulting from the rewards of group membership. We expect morale to be high when crewmembers are working toward well-defined superordinate goals. It is hypothesized that such distant goals should be supplemented by interim goals. Through reducing the costs of group membership, task and socioemotional training should also raise cohesiveness. Investigators must also examine the procedures that astronauts can use to diagnose and contain on-board frictions. Also relevant to understanding spacecrew cohesiveness is manning theory, which explores the consequences of involving few people or many people relative to the job to be performed. Early findings suggest that there may be both psychological and social benefits when the number of participants is slightly less than normally expected, given the task demands. Manning theory requires particularly careful study prior to possible application in space-capsule microsocieties.

Many complex issues surround compliance and conformity. Social pressure is essential for interpersonal coordination. However, rejection or ostracism may produce unacceptable personal consequences in space. Further research is necessary to find ways of softening the potentially adverse effects of conformity pressures and to find ways to reintegrate deviants into the crew. Conformity pressures can also restrict the flow of innovative solutions to problems. Additional research is required to understand the development and control of "groupthink" under spaceflight conditions. A similar problem to combat is an unquestioning attitude in a leader's presence. Training procedures must be found to help crewmembers separate useful from distracting decision-making input, and to ensure that leaders are receptive to the useful input.

Group performance reflects a complex interplay of many different factors. According to the Hackman and Morris (1975, 1978) process model of group performance, there are essentially three techniques for developing efficient and effective crews. These are (1) composing crews in such a way as to ensure a good representation of the desired technical and social skills, (2) structuring tasks in such a way as to maximize motivation and commitment, and (3) manipulating social norms to encourage the development and use of appropriate performance strategies. These techniques need to be evaluated as applied to isolated and confined groups that are operating under conditions of risk.

Both recently formed (young) crews and well-established (mature) crews offer advantages and disadvantages. Ideally, a crew entering space would be young in the sense that the members have not yet become bored with one another but mature in the sense that they have achieved a high degree of interpersonal coordination. Additional research is required to identify ways to simultaneously achieve a high degree of social variety and interpersonal coordination. Finally, certain types of crews will have rotating membership. A central issue here is identifying ways to ease the integration of newcomers into crews. It is hypothesized that the process of assimilation may be eased if newcomers undergo preentry therapy, are sponsored by specific crewmembers, and/or are in frequent telecommunication with the crew prior to their own visit to space. Training crewmembers in space itself may help novices adapt to the conditions of space and facilitate the transition to operational status.

Artist: Paul Arlt

VI
COMMUNICATION

INTRODUCTION

As commonly used, the term "communication" refers to the exchange of information between one person and another or between one social system and another, for the purpose of reaching a shared understanding. Conrath (1972) and Katz and Kahn (1978), among others, place the management of the communication process at the essence of organization. Through the control of communication, managers seek to provide individuals with access to the information they need for the competent performance of their jobs, while limiting their access to inessential communications in order to reduce the overall level of distraction.

Planning a space mission involves all the complex communication issues found in any large-scale endeavor. It also includes those issues that are special to isolated and confined individuals and to groups separated by large distances. Future flights, involving crews away for long periods of time, will further complicate communication questions. Because of the importance of such issues to the success of future space missions, this chapter focuses specifically on the process and effects of communication in extended spaceflight.

Space missions involve communication at several levels, the precise number of which depends on such factors as the size and complexity of the mission and the availability of communication media. The three simplest levels include (1) communication within the confines of the space capsule; (2) communication between a space capsule and another capsule, satellite, or space habitat; and (3) communication between the mission and Earth. Communication which takes place at these different levels may differ in the rate and quality of information flow; for instance, discontinuities may be introduced when communication must depend on electronic mediating systems.

Communication in space as elsewhere serves both task and socioemotional functions. In terms of task functions, both hierarchical and lateral communication are necessary to organize and coordinate crewmembers in pursuit of mission goals. Information must flow to managers to ensure that informed decisions are reached; information must flow from managers to subordinates to ensure that decisions are understood and implemented. Lateral communication is necessary for the general coordination and implementation of activity. Communication is essential also for maintaining the psychological and social well-being of the crewmember; it provides updated knowledge of other people's attitudes and views, which, as noted elsewhere in this book, is necessary for social comparison processes and for conflict management. In addition to its use for casual enjoyment, communication is a prerequisite for social recognition and censure, both of which have been shown to be potent controllers of human behavior. Finally, miscommunication (i.e., a failure in the communication process) can contribute to interpersonal friction and conflict within the crew or between the crew and those on the ground.

DIRECT INTERPERSONAL COMMUNICATION

Even direct, person-to-person communication can be altered by the conditions of the space environment. The space atmosphere and weightlessness, as well as the presence of life-sustaining equipment, can interfere with both verbal and nonverbal exchanges.

Verbal Communication

Verbal or linguistic communication refers to communication through grammar and syntax — that is, "what is said." High levels of ambient noise, along with atmospheric attenuation within the space vehicle, can interfere with spoken communication. In the future, increases in the spatial volume of the vehicle's interior, along with larger and more diverse crews, are likely to increase the likelihood that the essentials of the message will be missed or misunderstood.

Poor signal-to-noise ratios— As noted in our discussion of habitability (chapter III), space habitats are acoustically noisy environments, with noise intense during launch and reentry. Although training may yield the high degree of interpersonal coordination required to function effectively during these phases, there is some risk that a need for deviations from prearranged procedures will arise and that queries and instructions will be difficult to communicate orally. During the cruise phase, the operation of life support systems makes the space environment at least as noisy as a busy office (Berry, 1973a). Thus, major propulsion-system noise and aerodynamic noise may make oral communication virtually impossible during some periods, and the ambient noise level may make oral communication difficult under normal cruise conditions.

Additional complications are related to the channel of communication within a spacecraft. Artificial atmospheres, such as those used in space and in undersea environments, can hamper direct communication. First, if low air pressure is used, sound transmission is impaired. Under this condition, astronauts would have to speak louder than usual to be heard at a given distance. Also, some inert gases used to prevent excessive oxygen richness have voice-altering qualities. In artificial atmospheres, then, there may be a tendency to speak too softly to be heard, and voice quality may be altered in such a way as to make it difficult to be understood. To avoid problems in direct verbal communication, ambient noise must be limited and the atmosphere configured to include considerations of speech-transmission requirements.

Crew size and message complexity— As long as spacecrews remain small and select, each member can be expected to be well-versed in the same technical language. Astronauts, like other specialists, have developed a special language in which acronyms and other speech habits are used to improve efficiency by condensing words. Myasnikov, Panchenkova, and Uskov (1977) term such specialized language "normalized." As crews become increasingly large and are selected on a broad range of criteria, the use of such language could lead to misunderstanding and conflict with other crewmembers who are using the commonly employed language or language "normalized" to another field of interest.

Procedures must be identified to ensure that important messages are received and understood. These procedures should emphasize characteristics of clarity, redundancy, and feedback. To be clear, a message must be put in "the right words." That is, messages must be encoded or framed to take into account the receivers' abilities, training, and attitudes, remembering that successive generations of astronauts are likely to differ appreciably from each other. Redundancy or repetition should be employed to increase the opportunity for a message to be received and understood. Redundancy may involve repeating a message at two points in time, or transmitting a message to different people who have similar responsibilities or who perform similar functions. Beyond a certain level, redundancy is counterproductive because once a message has been received and acted upon, redundancy serves solely to block channels which are needed to transmit "new" or additional information. Feedback assures the source or sender that a message has been received. Certainly the simplest and probably most effective feedback procedure, when bidirectional verbal communication is possible, is to ask the receiver to restate the communication in his or her own words. However, even the practice of simple acknowledgment has been shown to result in reduced errors among aircrews (Foushee, 1982).

Once the message has been correctly framed, the task of mission planners and managers is to identify the optimal level of redundancy needed and the type of feedback required. Presumably, for a given message, the need for redundancy is inversely related to the availability of feedback.

Ethnic diversity and language barriers— Ethnically mixed crews, especially international crews, may experience serious language difficulties. The most extreme example to date of language diversity in space has been the use of "Ruston" (Russian-Houston) on the Apollo-Soyuz flight. After spending 1000 hr studying the language of the other, astronauts attempted to address the cosmonauts in Russian, and cosmonauts did their best to respond in English. However, these attempts were mainly symbolic, since the level of proficiency involved would be insufficient for communication on a typical flight. The flights which have brought together Soviet and other Eastern European crewmembers are more significant to the question of ethnic diversity. Here, Russian is the official flight language and non-Russian crewmembers must demonstrate an acceptable level of fluency. This requirement has not been overly restrictive, since most cosmonauts have shared a common Slavic background. It has, however, introduced new variables into the process of communication in space. Remek, a Czechoslovakian cosmonaut, recounts an interesting problem that he encountered in attempting to broadcast to his countrymen from space (Remek, 1977). As he was speaking, he realized that he had never before translated the specialized language of space from the Russian in which he had learned it. Struggling to find Czechoslovakian equivalents of Russian "space-eze" resulted in a very uncomfortable broadcast, which was later analyzed to contain three times as many "ers" and "ahs" as his usual broadcast.

The ability of humans to process information is reduced when the environment is complex and demanding (Schroder, Driver, and Streufert, 1967) and linguistically heterogeneous crews add to the complexity of the environment. In the related area of flight control, solutions have been proposed for dealing with problems of linguistic differences. The Canadian government has experimented with the use of bilingual controllers (Stager, Proulx, Walsh, and Fudakowski, 1980), and the International Civil Aviation Organization has recommended that one language (English) be available on request at designated airports and on international service routes. A significant factor in this latter approach is that English may be the first language of neither the pilot nor the air traffic controller. Although both of these suggested approaches work reasonably well, neither provides a total solution. Billings and Cheaney (1981) have identified several incidents in which language difficulty played a significant role in aggravating problems of an aircraft in distress.

Although to date no serious incidents have resulted from linguistically heterogeneous spacecrews, cross-cultural activity in space is in its infancy. In fact, Spacelab offered the first opportunity for a non-American to fly a NASA mission. Therefore, it is conceivable that in times of emergency or overload there would be a tendency for international crews to miscommunicate, for instance, by reverting to their native tongues. With linguistically heterogeneous crews, it might be necessary to rely more heavily on the slower, but more precise, written mode. It may also prove helpful to code much of the instrumentation and equipment with international symbols, or at least with symbols whose meaning has been previously agreed upon (Remek, 1977).

Even in a monolingual crew, there could be problems related to accents and regional dialects. Within the English language there also are subtleties of intonation, inflection, context, meaning, and interpretation which can influence the degree of communication and understanding among users. In addition, it can be anticipated that dissimilarity in background will make message interpretation difficult. Rogers and Agarwala-Rogers (1976) report that heterophilous communication (i.e., communication between senders and receivers differing in education, status, beliefs, etc.) often leads to message distortion, long reaction times, restricted channels, etc. In terms of cultural differences, much remains to be understood about how miscommunication could develop in the space environment, and how such miscommunication could be prevented or corrected.

Nonverbal Communication

Communication through the linguistic mode (grammar and syntax) is supplemented by communication through three other nonverbal modes. These are (1) the paralinguistic mode (communication through the amplitude, rate, and tenor of speech); (2) the kinesic mode (communication through facial expressions and gestures); and (3) the proxemic mode (communication through distancing or placement). Research suggests that a large proportion of affective information transmitted under face-to-face conditions is conveyed through nonverbal channels (see Mehrabian, 1971). Nonverbal messages can substitute for verbal messages, modify verbal messages, and contradict verbal messages. For this reason, nonverbal factors must be considered an integral part of communication in space.

The space environment— High ambient noise and low air pressure will interfere with paralinguistic and with linguistic communication. Background noise of certain frequencies and timbres may mask informative intonations. Artificial atmospheres which transform or alter voice quality could prove particularly troublesome. The extent to which such limitations on paralinguistic channels could impair or distort task and socioemotional communication flow needs to be examined.

Weightlessness has the potential of interfering with both kinesic and proxemic aspects of communication. Weightlessness is associated with a certain puffiness or immobility of the face. Communicators may lack emotional expressiveness or else convey misleading or inappropriate emotions. Myasnikov et al. (1977, p. 2) note:

> According to the experience of Soviet cosmonauts and American astronauts, strongly expressed puffiness distorts usual facial expressions, [and] impoverishes the set of mimics required by a given situation. Under such conditions the relationship "speaker-listener" is changed to one unusual for partners in intercourse ....

Under normal circumstances, people express attraction to one another by decreasing interpersonal distance (Altman and Taylor, 1973). In space, floating and the use of safety lines, clamps, and other anchoring devices, and operating in cramped quarters tend to reduce the expressive value of interpersonal distancing. Additional limitations are imposed when communicators are encased in protective garments.

Weightlessness can change the physical and anatomical cues on which people depend for information about others, leading to false or prejudicial interpretations. Space travelers must be sensitized to these changes related to weightlessness, and compensatory mechanisms must be devised.

Crew heterogeneity— We can expect that increasing crew heterogeneity in space will be associated with an increase in the likelihood of nonverbal and verbal miscommunication. Pace of speaking, intonation, and the like vary from one group to another, with or without a common language. Problems in proxemic and kinesic expression are also likely to arise with increasing crew heterogeneity. For example, interpersonal distancing is a culturally conditioned response (Hall, 1959). Since appropriate interpersonal distance zones are specific to a society or subgroup, one participant in an intercultural exchange may view the other as too close, whereas the other is likely to view the one as too distant. Such discrepancies could be exacerbated by the cramped and isolated conditions associated with spaceflight.

Ekman and his associates (Ekman and Friesen, 1972, 1975; Ekman, Friesen, and Ellsworth, 1972) have found that a given emotion will yield similar facial expressions across cultures. Although this reduces the possibility of kinesic miscommunication, there remains intercultural variability in the stimuli which give rise to specific emotions among representatives of different cultures, and in the display rules or strategies used for managing or controlling expressions within different cultures. Such variability could cause significant problems under conditions of isolation, confinement, and risk. For instance, a verbal report of a potentially dangerous situation might not be given sufficient weight if the facial expression of the communicator suggests nonchalance to a recipient from a more expressive culture. On the other hand, a rebuke that was intended to be mild might be perceived as severe if the communicator's face seemed "overemotional" to a recipient from a less expressive culture. Our limited experience to date provides only hints concerning intercultural, nonverbal aspects of communications in space. We need to determine which cues lead to misinterpretation and what problems are likely to follow.

MEDIATED COMMUNICATION

Many communications that are essential for the space traveler will not occur under face-to-face conditions. Interorganizational communications between physically remote units, such as a satellite and a surface explorer or a space vehicle and the Earth, must be transmitted electronically. Although there are numerous similarities between direct and mediated communication, these forms of communication differ in many obvious and not-so-obvious ways. In this section we will consider the special characteristics of mediated communication.

Planning Considerations

We are in a period of rapid growth in computer applications. At the same time, we are witnessing the long-predicted revolution in the field of communication. But more important than the growth of either of these industries is the fact that these two giants are now merging to provide capabilities far in excess of the added possibilities of both. Distinctions that formerly existed between communication and computers have blurred as computers control and direct communication operations while voices and dynamic video images are digitized, compressed, stored, and even sent over telephone lines. New and exciting products that promise to change the way we conduct our businesses and run our homes and our lives appear with a frequency that bedazzles even the most committed electronic communications buff. It is in this rapidly changing milieu that planners must attempt to address the communication needs and possibilities of extended spaceflight. In technological development, production trails knowledge, leaving innovations outmoded even as they are disseminated; when the vehicle using the technology is in space, the planning problem is truly challenging.

From the design perspective, one of the most pressing requirements for communication systems for space is that they be capable of adapting to industry changes. In the early evolution of computer technology, emphasis was placed almost entirely on hardware design. A shift in attitude occurred with the realization that software could introduce enormous flexibility to a system. A third phase evolved when it became obvious that the software portion of the system, being skilled-labor-intensive, was the most costly component of the system. The present and presumably final stage of the process acknowledges the high costs of software by placing even greater emphasis on this component. The emerging view is that hardware may last only a short period, but that software must be preserved. With this view, the emphasis shifts to the coordination and integration of programming packages and interfaces that will allow developed techniques to be used with each other and with any equipment. Although the hardware to be used in space cannot follow the throwaway approach, the current direction of systems' development toward greater flexibility and compatibility should benefit the space application by providing a wide range of opportunities to update the system without changing the equipment.

Systems Requirements

A distinction that is often drawn in the communication literature is that between "data" (emphasizing the capacity of the carrier or the rate of the transmission) and "information" (emphasizing the intelligence in the message).[1] Great quantities of data can contain little information, whereas important information can be conveyed in a single (yes/no) bit.

Following STS-1, astronauts commented on the excessive amount of useless data they had received during the flight. Similarly, principle investigators reported receiving so much data that were irrelevant to them during the Spacelab simulation (Helmreich, et al., 1979a) that important information may have been overlooked. A significant consideration for any system, but especially for one in space, is that the communication contain a high percentage of useful information. Although the goal of providing useful information is easy to enunciate, it can be difficult to put into operation.

---

[1] Information Theory (see Shannon, C. E.; and W. Weaver: The Mathematical Theory of Communication. U. of Illionis Press, Urbana, Ill., 1963) is a significant exception to this lexical distinction.

What is "useful" information depends both on the particular circumstances surrounding the event and on other considerations. Dervin (1981) draws an interesting distinction between two contrasting assumptions concerning the utility of information. The first view of information is that it has some intrinsic or objective value which the user must be persuaded to adopt. The second view is that information gains its value only through interpretation and use by the individual. The importance of Dervin's distinction lies primarily in the standards it suggests for communication systems. For instance, the objective view could be satisfied by a system that is predominantly one-dimensional; the subjective or interpretive view would require a highly interactive system. As Dervin points out, the objective or persuasional approach is the dominant model of most delivery systems in use today. It is perhaps overly ambitious to expect that spacecrew systems will be among the first to deviate from this pattern. Yet neither should we feel bound by limiting or inappropriate models. In designing systems for future spacecrews, we must attempt to eliminate unnecessary data and to strike some balance between providing necessary information (as judged by others) and meeting the perceived needs of involved individuals.

Even the best-designed communication systems have finite information-processing capabilities. Information overload occurs when the system is confronted with more information than it can effectively or efficiently process. Miller (1960, 1978) has identified several possible responses to input overload which are common to all information systems. These include queueing, selective processing, omission, and error. Two general procedures are available to ensure that important information is not missed. One is a system of prioritization such that critical information is given preferential treatment in the course of queueing and filtering. The other involves the establishment of reserve or emergency communication channels. Both of these procedures are probably needed to ensure that the communications systems used in each mission will make provision for identifying and promptly processing crucial information, even under overload conditions.

Systems Effects

During the last two decades there has been increased interest in understanding the effects resulting from the use of mediated communication systems, such as those which will provide the link between the spacecraft and the ground. This area of investigation is directly applicable to the requirements of spacecrews and is presented here in some detail.

An operating communication system involves a complex mix of hardware and people variables. Several authors have developed descriptive taxonomies in attempting to put into order the many variables which influence systems use and impact (Bailey, Nordic, and Sistrunk, 1963; Casey-Stahmer and Havron, 1973; Johansen, Miller, and Vallee, 1975). Broadly represented, these taxonomies divide the various influencing factors into antecedent variables (conditions which precede the interaction, such as the physical environment and the hardware/software system), and intervening variables (conditions which operate during a particular interaction, such as "gatekeeping" activities or the availability of feedback). Criteria variables are the measures by which the system is assessed. Criteria variables include both effectiveness and satisfaction measures.

Mediated systems influence the dynamics of the communication process in two related but distinct ways. First, there are the effects common to all mediated modes. Second, there are the effects specific to particular classes of systems.

Overview of mediated systems effects— Numerous studies (see Chapanis, Ochsman, Parrish, and Weeks, 1972; Westrum, unpublished Ph.D. Thesis, Purdue University, 1972) have shown that mediated communication differs most from direct face-to-face communication when the activities engaged in require a high level of interaction among participants, and when the expression of emotion and its accurate perception are related to the outcome. Compared with direct communication, mediated communication contains reduced socioemotional content. As a result, mediated communication tends to be less effective for tasks such as getting acquainted with another or tasks involving bargaining or negotiation, than for tasks such as giving and receiving information, asking questions, and exchanging opinions (see reviews by Short, Williams, and Christie, 1976; Hough and Panko, 1977; Johansen, Vallee, and Spangler, 1979). Studies by the Communications Studies Group (Champness, 1971) also indicate that people are more willing to compromise or "go along" in a direct exchange and are less easily moved from their established positions in a mediated exchange. If follows from the various findings that mediated communication is better equipped to handle routine, businesslike, and formal exchanges than significant, social, and informal exchanges.

Various hypotheses have been offered to explain the differences observed between face-to-face communication and mediated communication. These hypotheses can be subsumed under three general categories. The social presence hypothesis (see Short et al., 1976) emphasizes how real or immediate other individuals seem to the communicating individual. The channel hypothesis assumes that changes in the nature of the exchange are a result of changes in the amount of information received. Direct contact brings into play numerous communication channels such as visual, auditory, olfactory, and tactile. With information flowing over so many pathways, the system is more than adequate, it is naturally redundant. As the number of channels is reduced, redundancy is lost; as the number of channels is further reduced, information is lost. The task/socioemotional ratio hypothesis states that, as the medium of communication changes, the relative importance of interpersonal information and task-related information changes. Here the emphasis is not on the net gain or loss of total information, but rather on the relative emphasis given to different kinds of information.

Some data are better explained by one hypothesis than another. For instance, it has been demonstrated that channel impairment can adversely affect the assessment of another's opinion and the reaching of an agreement in bargaining situations (Communications Studies Group, 1972). In support of the task/socioemotional ratio hypothesis, it has been found that real opinion change (as opposed to willingness to compromise) is greater after an audio-only conversation than after a face-to-face meeting (Communications Studies Group, 1972; Short, 1972a, 1972b, 1973). However, all three explanations describe many of the differences observed between direct and mediated communications. They are distinguished primarily by the value they assign to particular modes of communication. The social presence and channel hypotheses suggest an overall superiority of direct communication; the task/socioemotional ratio hypothesis does not suggest a superiority of one system over the other, except in relationship to a particular task or goal.

Differences among media— As discussed above, face-to-face communication can be contrasted with mediated communication. It can also be conceptualized as one point on a communication continuum. Other major points are full-motion video, audio, and computer or telegraphic communication. Full-motion video is often referred to as broadband communication because its transmission requires approximately one thousand times the bandwidth of an audio link. Audio, computer, or slow-scan video can be transmitted over physical links which have only limited capacity, and thus are called narrowband communication.

Differences among media parallel those of face-to-face versus mediated communication. Interactions involving tasks with little affective component, such as exchanging information, are not as influenced by the particular medium of communication; interactions with high affect, such as bargaining sessions, are more influenced by the choice of medium. Knowing the other person well can offset some of the problems of, for instance, remote negotiations; however, tasks with an emotional component are generally handled better with direct communication or via a video link than with audio-only or computer communication. Some interactive tasks such as problem solving may or may not be influenced by the medium, depending on the particular exchange.

Differences among media can also be explained in terms of social presence, channels, and the task/socioemotional ratio. For instance, extending the channel hypothesis, we can consider the various media as differing in the number of modes of verbal and nonverbal channels that they employ. As shown in table 1, when compared with direct communication, video, audio, and computer or telegraphic systems employ successively fewer modes.

Video systems— As noted above, direct communication, and to some extent systems with a video component, convey affective information. The visual dimension also serves another important function — it helps regulate the conduct of an interactive meeting. Eye movement and body position announce who wishes to speak and with what urgency, aiding the smooth flow of an interaction (Argyle, 1969). In a group situation, visual information also provides a continuous status update, keeping the remote participants aware of who has joined and who has left. Visual information also serves to "refresh" impressions of participants.

When more than two people are communicating, the presence of nonverbal cues makes a direct or a video-supported exchange far easier to control than an audio or a computer/telegraphic exchange. However, there is no required relationship between the presence of these cues and the management of the interaction. Management without visual cues can be accomplished by providing a system that can identify speakers and give participants a way of indicating their wish to talk. (For a description of such a system, see Stapley, 1973.)

TABLE 1.— COMMUNICATION MODES AS A FUNCTION OF MEDIA

Mode	Face-to-face	Video	Audio	Computer/teletype
Proxemic (distancing or placement)	X			
Kinesic (facial expression and gestures)	X	X		
Paralinguistic (amplitude, rate, and tenor of speech)	X	X	X	
Linguistic (written or spoken word)	X	X	X	X

Among the media, video most closely approximates the effects of direct communication (Goldmark, 1973); however, significant differences remain. For instance, in a video exchange someone must decide who or what to focus on, whether to show close-ups or distant views, etc. This preplanned selection and focusing requirement considerably limits the choice and ease of viewing that is available in a face-to-face interaction. Also, the direction of gaze in a video exchange can introduce anomalies. When meeting face to face, people alternately establish and break eye contact. When meeting over video, one tends to fix on the monitor; this can result in the perception of being ignored (if the camera is displaced from the monitor) or of being stared down. One highly advanced (and costly) video design attempts to deal with some of the problems of direction-of-gaze. Decisions and Designs Inc., under contract to the Defense Advanced Research Projects Agency, have developed a concept which they call "virtual space." Through the use of multiple cameras and monitors, participants at each location appear to turn toward the particular individual they are addressing. Evaluators report that the system has been well received (Sticha, Hunter, and Randall, 1981), suggesting that the verisimilitude of direction-of-gaze may be a significant factor in the smooth flow of an exchange.

Audio systems— Audio systems can provide a great deal of information, not only of the subject being discussed, but also of the

It can be concluded, then, that although there is a basic preference for direct communication, mediated communication is not a poor substitute for face-to-face exchanges. Rather, it is another form (or several other forms) of communication. Although all forms of mediated communication seem to be aided by the previous acquaintance of the participants, preference for one system over another is highly dependent on the specific conditions.

System acceptance— Any communication system, whatever its capability, will fail if it is not used by its intended beneficiaries. Many systems have been resisted because (1) the designers did not understand fully the users' needs, (2) the potential benefits were never made clear to users, or (3) the users were never properly trained in the system (Elton and Carey, 1980). Systems have also failed because of unanticipated user fears or because users felt that the new medium would somehow disrupt the normal pattern of their lives. Elton and Carey have offered a number of useful suggestions on planning and implementing communication systems.

First, potential users should be included in the planning and implementation process as early as possible. There are three reasons for this: User involvement helps designers understand user attitudes and needs, users become committed to choices to which they themselves have contributed, and there develops a cadre of users who have sufficient familiarity with the system to be able to explain it to their peers.

Second, planning should take the psychological aspects of the physical arrangement into account. The physical arrangement refers to such variables as equipment location, physical setup, and operator-machine interface. A location should be sought which is at once accessible and private. Although it may be necessary to locate the equipment in a multipurpose area, steps should be taken to minimize the extent to which the system's use and other area activities interfere with one another. The equipment should be as simple and gimmick-free as possible in order to provide minimal distraction from the messages themselves. In general, users resist systems that are remote or inaccessible, that are difficult to use, or that raise issues of territoriality or privacy.

Third, system planning should take the social context into account. The social context includes such variables as group norms, expected degree of formality during the communication, user social status, and the like.

Fourth, hardware development should be accompanied by the development of a protocol that specifies who uses the system, how it is used, and under what conditions it is used. The intent of such protocols is to facilitate interaction, reduce the potential for embarrassment, and eliminate uncertainties which can contribute to negative attitudes. For example, an appropriate protocol would overcome such problems as the uncertainties generated by not knowing that others have entered or left the assemblage.

Fifth, once systems have been designed and installed, efforts must be taken to ensure that all of the system's potential users understand its strengths, limitations, and procedures for operation. The expectations of users have been found to play a significant role in system acceptance or rejection. Training should precede the system's first use, as users' initial experiences are likely to have a disproportionate impact on their attitudes about the system. Users will often balk if a system fails to perform to their expectations on demand. However, these same users will frequently persevere with a lesser system, provided its limitations are understood in advance. Similarly, the perceived reliability of a system may be at least as important in user acceptance as its actual reliability, with operational/environmental factors playing a significant role in the shaping of perceived reliability (Bair, 1978).

Finally, ongoing evaluations should continue throughout the life of the system. Such assessment should address the system's performance, its impact on relevant aspects of the users' lives, and users' attitudes. Maintaining a high level of rapport with users is essential to effective systems evaluation.

In summary, the abruptness with which communication systems can be dismissed or devalued suggests that the planning and implementation of communication systems is a major task and requires a high degree of effort and attention.

Application to Space

Careful planning can help avoid some of the problems associated with matching the requirements of users and the offerings of communication systems for space. However, the findings reported above suggest that some difficulties could persist, whatever the system or implementation plan. Mediated communication systems handle simple tasks better than complicated tasks, information exchange better than persuasion or influence, and work-related activity better than social activity. Yet it is anticipated that extended missions will increase the need to deal with complex tasks, to reach agreement on issues, and to engage in affective exchanges with those at home. Among mediated systems, broadband systems are more effective than narrowband systems in handling tasks that are complicated or that contain high emotional content. Yet it is anticipated that for extended missions, narrowband computer-based systems will have certain technical advantages that may counter other considerations. For missions involving very large distances, transmission delays may render simultaneous or quasi-simultaneous communication impossible. Here, computer communication would have an advantage, since it is easily used in a sequential mode.

Although suggestive, studies reported to date on the impact of mediated communication systems are not definitive for space application. First, the subjects of these studies have been neither isolated nor confined. Second, these studies have cast electronic exchanges as an option rather than a necessity for interaction, and users have not been required to rely on these systems exclusively for prolonged periods of time. Third, studies to date have focused on goal-oriented groups functioning specifically in their work roles, rather than on socioemotional groups functioning in their family or friendship roles — one of the requirements for space. Many of the questions concerning the relationship of media variables and social interaction will have to be reexamined to take into consideration the unusual conditions of space.

COMMUNICATION NETWORKS

In most organizations there is neither the opportunity, nor the need, nor the desire to interact with all other members. Functional communication networks describe the patterns of communication which evolve among individuals or nodes (i.e., who talks with whom). The unit of analysis in communication networks is not the individual, but rather the relationship between or among individuals (Rogers and Agarwala-Rogers, 1976). The methodological techniques used to study communication networks include traffic monitoring, studies of task requirements, participant-observation of interactions, cross-sectional surveys, small-world procedures (taking one part as representative of the whole), and studies of message diffusion. The uses to be examined by these methods are production (getting the job done), innovation (exploring new options), and maintenance (keeping the system and its components operational). In this section we consider how studies of communications networks can help us to understand information flow within a space organization and between the space organization and external systems.

Internal Communication

Information may flow in several directions within an organizational structure. Communication flowing in different directions serves different functions and is likely to encounter different obstacles or barriers.

Formal communication networks— Formal communication networks reflect management-imposed structures which channel interaction up and down the organizational line and, to a lesser extent, laterally among peers. Formal communication networks are highly important to the management function, and managers spend an average of 75% of their time in communication activities (Bair, 1980). Communication from managers to subordinates typically assumes one of five types (Katz and Kahn, 1978): (1) inculcation of organizational or group goals, (2) explanation of organizational policies and procedures, (3) job rationale, (4) job instruction, and (5) feedback. As Katz and Kahn note, although a manager's task may be simplified by offering subordinates only job instruction and feedback, more general instruction gives workers a framework for coping with nonroutine conditions and promotes identification with organizational goals. Therefore, the first task of a space manager is to identify and impart the goals of the organization or mission.

Once a communication framework has been established, attention can be turned to using that framework to meet organizational needs. Communication up the line helps managers monitor performance and provides information that is useful for making decisions. Communication down the line provides direction, course changes, and feedback. Although downward communication can be direct, as when top management issues an edict to subordinates at all levels, certain conditions serve as barriers to the upward flow of communication. First, role expectations cast subordinates as listeners rather than as communicators. Second, since managers control their subordinates' fates, subordinates tend to censor communications which might displease the manager or cast aspersions upon their own performance. Third, upward communication tends to proceed in a step-by-step fashion. At each step, someone is in a position to act as a gatekeeper. Although this individual may perform the positive function of preventing information overload, he or she may also filter, distort, or block needed information. In a steep, multilevel hierarchy, it is unlikely that messages from the lowest levels will ever reach the top. Space planners and managers must be aware of these barriers and establish mechanisms to smooth the upward flow of important information.

Horizontal or lateral communication involves the exchange of information among workers at the same hierarchical level. As Katz and Kahn (1978) note, horizontal communication eases some of the burdens of management because peers can indoctrinate and train one another, and can coordinate activities of groups of workers. Moreover, horizontal communication provides important and necessary socioemotional rewards. However, peer indoctrination and training can be faulty or even at cross-purposes to management goals. Some organizations adopt heavy-handed managerial techniques in order to discourage horizontal communication flow. However, this form of cure may be worse than the ailment, particularly under conditions of isolation and confinement. In space as elsewhere, the problem is not one of preventing horizontal communication, but of preventing such communication from assuming destructive forms.

Informal communication networks— In addition to the imposed organizational arrangement, group members construct their own communication networks. These informal networks have two basic purposes. The first function of an informal network is to compensate for inadequacies in the formal network by establishing new links to bypass obstacles that emerge in the formal structure. In many groups there is someone who can expedite the matter at hand; frequently, the expeditor is not the person charged with the responsibility. By locating the right person (who may be anyone in the organization), the informal network is used to serve organizational goals.

A second function of an informal communication network is to serve the needs of the individuals involved, whether or not these needs coincide with the goals of the organization. These informal networks link those with similar interests and experiences (Rogers and Kincaid, 1981) and those who simply like each other.

Informal networks serve a valuable organizational function: they protect the individual from becoming isolated from colleagues of his or her profession by stimulating him or her to more creative thinking and by supplying a psychological support group. It is not unusual for an ongoing, informal work relationship to obscure the formal organizational structure until, finally, the formal is forgotten. Within large spacecrews, we can expect that informal networks will arise. In examining the advantages or problems of an organizational unit, informal communication networks must be considered as carefully as formal communication networks.

Group membership— Different organizational subsystems can constitute themselves into cliques or factions which do not communicate effectively with one another. Members of particular groups may have difficulty identifying or appreciating each others' specific perspectives and goals. The attitudes and behaviors of most subgroups or cliques are strongly influenced by one or a few individuals called opinion leaders. It is important to know who is likely to emerge as an opinion leader in space, and how this individual's influence on other clique members can be directed toward positive and productive interactions with other subgroups. Likert (1961) has suggested that good communication and cordial relations can be promoted among different groups by having selected individuals maintain simultaneous membership in them. A person serving this "linking pin" function is likely to understand the needs and views of each group and to be able to represent each to the other. From previous confinement activity, we know that participation in two or more subgroups improves intergroup relations. In the Tektite underwater research project, benefits accrued from having engineers and scientist-aquanauts play active roles in each others' fields of expertise (George Washington U., 1974). The linking-pin function differs from the boundary-role or bridging function described in chapter VIII in that the former requires membership in several groups, whereas the latter involves holding a membership in only one group while acting as a point of contact for the other. A third role is that of the liaison, where the individual connects two or more cliques within a system without belonging to any (Rogers and Agarwala-Rogers, 1976). These roles share many of the same requirements. It would be interesting to determine which role accomplishes certain tasks with the greatest effectiveness and with the least burden on the role occupant. It would also be useful to understand more fully the desired characteristics of individuals occupying such roles.

Performance and satisfaction— The structural arrangement of a communication network can influence its effectiveness. Reviews of experimental nets (see, e.g., Shaw, 1976) suggest that when the task is simple and requires people in central positions to collect and distribute messages, centralized nets tend to be an effective arrangement; however, when the task is complex and requires people in central positions to perform numerous and varied operations, decentralized nets tend to be more effective. In terms of member satisfaction, decentralized nets have been found to offer participants greater feelings of self-determination and to provide more socioemotional gratification than do centralized nets. Conclusions such as these provide valuable insight but are far from conclusive. For instance, the experimental nets examined by Shaw consisted of stations staffed by single unsupported individuals. It is not clear that such findings can be extrapolated to other conditions. Centralized nets could be more effective than decentralized nets even when the task is complex if, for example, the centralized station has higher information-processing capability, as through multiple staffing or data-processing support. In addition, "centralized" and "decentralized" nets are two extreme arrangements. It may be that a preferred system would combine characteristics from each.

To date, spacecrews have been sufficiently small that internal networking considerations have been minimal or nonexistent. However, questions of structuring communication linkages can be expected to grow with the crew and to require more careful analysis in terms of both effectiveness and acceptability.

External Communication

A fundamental difference between space and other organizational arrangements is that informal contacts between the space vehicle and the ground will be severely curtailed. This limitation puts a large burden on the formal communication network.

For the crew of a space vehicle, external communication will assume one of two forms. These are (1) primarily task-oriented communication with ground control, and (2) primarily socioemotional communication with family and friends.

Communication with ground control— The most important requirements of a communication system relate to effectiveness and reliability. However, given an adequately operating system, communication-related problems can still plague the mission-ground relationship.

As we have seen (chapter I), isolated and confined individuals tend to direct hostility toward outsiders. A primary challenge, then, is to devise a communication arrangement which provides ground control with adequate access to the crew, but which does not interfere unduly in the crew's life space. Systems which allow ground control to "tap in" to the crew at will (and particularly systems which tap in unannounced) can be expected to arouse strong negative reactions. First, surveillance itself decreases autonomy and thereby decreases job satisfaction. Research in mundane organizations suggests that people who feel closely supervised tend to be dissatisfied with organizational rules and procedures, downrate their supervisors, and develop hostile attitudes toward management (see Day and Hamblin, 1964). Second, communication systems which allow unannounced eavesdropping or rude interruptions may be seen as an unwarranted infringement on the privacy of the group (see chapter III). Third, communication systems which are under the control of outsiders may raise conflicts concerning who is "in charge." Questions of control or authority can be particularly knotty when the ground believes an issue to be of concern to mission management and the crew believes the issue to be an internal matter.

There is also the problem of who on the ground should have access to the crew. Usually communication from ground personnel is relayed through a third party. (In the case of NASA missions, this responsibility lies with CAPCOM). Yet in a simulation of a Spacelab flight, researchers on the ground found that their ability to collaborate with crewmember experimenters was significantly impaired by this two-step process (Helmreich et al., 1979a). Alternately, if many individuals were to have access to the communication system, the aggregated messages could become overly burdensome and interfere with effective communication. Thus, a major problem confronting planners and managers is to devise a system that provides both ground control and spacecrews with an optimal degree of accessibility to each other. This requirement includes providing the crew with some opportunity for private, internal communication.

The Russian spaceflight experiences have led Gazenko, Myasnikov, Ioseliani, Kozetenko, and Uskov (1979) to offer two recommendations for aiding the spacecraft/ground-control interaction. First, these authors believe that two-way television communication initiated during the Salyut 6 flight was a definite improvement over earlier one-way video systems, in which only the cosmonauts were viewed. Two-way video provides balance between communicating nodes and also improves the ability of communicators to judge each others' emotional states. Second, Gazenko et al. recommend that radio and other communications operators receive basic training in interpersonal relations. When listening to recent exchanges between ground control and U.S. missions, it is obvious that ground-control personnel are keenly aware of the role they play in smoothing the interaction with the spacecrew. However, it is not realistic to expect this level of concern to be sustained forever. Training in interpersonal relations could offer an aid to the ground-control/spacecrew interaction when flights become long and routine.

Communication with family and friends— Personal communication that makes it possible to converse with family and friends has the potential of reducing anxiety concerning events at home and of reducing dependency upon fellow crewmembers for the satisfaction of all interpersonal needs. Extended spaceflight is expected to result in an increased desire to communicate with the folks back home. The Russian experiences have demonstrated that as mission length increases, so do the depth and duration of external communications (Gazenko et al., 1979). Berry (1973a), among others, argues that space voyagers should be given ample opportunity to communicate with people who are important to them personally, and that a scrambler system or comparable device should be used to ensure privacy. However, the value of ready access to people at home is not accepted by all. Some writers argue that it is to the advantage of an isolated and confined group that it remain out of contact with friends and relatives. In support of this position it should be noted that:

1. Certain individuals tend to monopolize communications systems and thereby give rise to conflicts within the group (McGuire and Tolchin, 1961).

2. The failure of someone at home to communicate on schedule provokes worry and fear (Pope and Rogers, 1968).

3. Worriers may remain "glued to the telephone" and neglect other duties (Radloff and Helmreich, 1968).

4. The messages received by isolated individuals tend to be bland and unsatisfying (Law, 1960; McGuire and Tolchin, 1961; Earls, 1969).

5. Even otherwise satisfying communications may be followed by a period of letdown or depression (McGuire and Tolchin, 1961).

Communication opportunities undoubtedly bring their share of problems as well as solutions. Observing this issue in space, Gazenko et al. (1979, p. 9) note:

> . . . it is still unclear which of the two alternatives suit (spacecrews) best: either facilitation or inhibition of normal ground-based ties . . . .

Some of the problems identified above may represent imperfections in the communication systems themselves. For example, the availability of private communication systems may allow for more spirited messages. Or, the failure of a family member or friend to communicate on time may be less worrisome if communication with home is so routine that it no longer is necessary to arrange each transmission well in advance. Post-communication letdown may be minimized if the next opportunity to communicate is in the near rather than in the distant future. Questions of the possible debilitating effects of bad news from home (or alternately of censorship) are also of issue here (see chapter VII).

As discussed earlier in this chapter, a question to be addressed concerns how mediated communication can be used in support of intimate and personal relationships with family and friends. A related question concerns how mediated communication with those at home affects direct relationships within the space vehicle. Mediated communication is generally employed in support of work-related tasks. One would not expect such exchanges to influence face-to-face relationships. In space, mediated communication with family and friends is likely to be emotionally charged and could influence the cohesiveness of the spacecrew. Research on how such communication opportunities could influence primary group relations is needed.

The questions surrounding contact with family and friends at home are among the more complicated interpersonal issues that space planners and space travelers must address. Although our emphasis here is on the spacecrew and their mission, whatever compromises are struck must consider the welfare of those at home as well as those in the space vehicle.

SUMMARY AND CONCLUSIONS

There are numerous unanswered questions surrounding communication of persons in space. Within the space capsule, the direct verbal communication among crewmembers may be influenced by attenuation related to transmission characteristics of the capsule environment and by high levels of ambient noise. Conditions of space (particularly weightlessness) can alter nonlinguistic communication cues such as voice tone, facial expression, and distancing. These variations, especially when complicated by the linguistic and cultural differences of multinational crewmembers, could result in miscommunication and interfere with the efficient work operations or social interactions of space travelers. Studies of international crews such as those proposed for exploring the ocean bottom (Glazer, 1974) would offer an opportunity to investigate the implications of linguistic diversity in space. In addition, procedures need to be developed to ensure that verbal messages are received and understood. More generally, greater attention needs to be given to understanding which cues survive and which fail the space environment, as well as the kinds of miscommunication that result from altered cues.

Equally demanding questions involve the establishment and maintenance of working and personal relationships with those on Earth or on other spacecraft. Space planners charged with providing for extended spaceflight face special problems in their attempts to implement communication systems in the face of rapid development in the technology. At the operations level, they must determine how such systems can be designed to avoid overload and to select and prioritize important information. At a more general level, they must determine how these systems can serve the information needs as defined by users and the information needs imposed by mission planners. Questions of protocol and of access must also be addressed.

It is predicted that space travelers, like other users of mediated systems, will find the medium better suited to simple rather than complex tasks and to work-related rather than social interactions. Unlike other users, space travelers will not have the option of substituting a direct exchange when they find the medium wanting. Space travel reverses the usual pattern of direct interaction with intimates and mediated interaction with secondary or work-related contacts. One of the more important questions for extended spaceflight concerns if, and how, a communication system can be used to deal with the kinds of complex and emotional interactions that are usually reserved for face-to-face communication. A related area requiring investigation is the effect of mediated/primary interactions with family and friends at home on direct/secondary relationships with crew members. Bedrest studies, such as those conducted at Ames Research Center (Chapman, Winget, Vernikos-Danellis, and Evans, 1975; Chapman, Winget, and Vernikos-Danellis, 1976; DeRosia and Sandler, 1979) could provide an effective vehicle for examining these issues, as well as explicating the biomedical and performance issues of weightlessness.

Space managers, like managers everywhere, must deal with problems of keeping needed communication channels open while discouraging those communication activities which tend to inhibit the goals of the organization. Cliques and factions evolve in all large undertakings; however, negative aspects associated with such subgroups could be particularly destructive in the closed environment of space. A possible solution is the involvement of individuals who share memberships in several groups and who can interpret each group to the other. What is not well known is whether the burden placed on these individuals by their intermediating responsibilities is manageable in the space environment.

The effectiveness and efficiency of operations and the satisfaction of participants have been related to the structure of the communication network. We need to determine whether centralized, decentralized, or some combination of these two arrangements are most appropriate to space and how the efficacy of these arrangements might change over very long periods of time.

A phenomenon which has emerged in many confinement experiences is the tendency of separated individuals to direct hostility to outside contacts. In space, this tendency could translate into problems between on-board crewmembers and ground control. Several suggestions, including the use of two-way video and the training of ground crews in interpersonal relations, have been advanced. The smooth interaction between the crew and the ground is a fundamental issue which needs to be addressed for extended spaceflight. Whatever solution or range of solutions are adopted must take into consideration both the access needs of the ground and the autonomy needs of the crewmembers.

Whether, overall, spacecrews will be aided or hindered by interaction with family members and friends at home is still open to question. However, the prevailing view is that opportunities for private exchanges with loved ones will be reassuring and sustaining to space travelers. Just what level of interaction will provide the most support remains to be determined.

Artist: Vincent Cavallaro

VII
CRISES

INTRODUCTION

Throughout this report we have been considering the "normal" or everyday problems of spaceflight. In this chapter we turn to the special problems arising from unusual and threatening events which we label "crises." For present purposes, a crisis may be defined as any situation which arises over a relatively short period of time and which could result in severe or even life-threatening consequences. Crises can arise either from physical events external to the crew, such as a meteor strike or an equipment malfunction, or from internal psychosocial causes, such as a severe psychological disturbance on-board. We will consider some of the situations which could place the crew under sudden and extreme stress, examine the research that relates to these issues, suggest ways in which these events might be avoided or dealt with, and specify those research questions which suggest themselves in coping with potential crises in future spaceflight.

EXTERNALLY PRECIPITATED CRISES

To date, most NASA missions have been "textbook" flights and all have ended with the successful recovery of the astronaut crew.[1] There is a danger that this success record could obscure the

---

[1] Three American astronauts were killed in a capsule fire on January 27, 1967, while training for an Apollo mission.

extraordinary, and sometimes heroic, efforts that were needed to complete some of these missions successfully. Space remains an extremely dangerous environment. A general history of emergencies which have arisen in space may serve to illustrate this point.

Experience in Space

The Mercury flights witnessed three emergencies, all related to reentry and splashdown. The hatch on MR-4 was blown prematurely after splashdown and astronaut Gus Grissom had to exit hurriedly and swim clear of his sinking craft. An emergency on MA-6 made it necessary for John Glenn to reenter the atmosphere uncertain if his heat shield was sufficiently secure to remain in place. On MA-7 Scott Carpenter overshot the designated splashdown area and a full hour intervened before ground control could be sure that he had landed and exited his spacecraft safely.

The Gemini and Apollo programs also had their share of difficulties. On Gemini 8, astronauts Armstrong and Scott experienced the first emergency to occur in space. After docking with an Agena rocket, the vehicle began to spin out of control. The crewmembers were able to escape by firing their retrorockets, returning to Earth 2 days ahead of schedule. On Apollo 11, the first moon landing, Commander Neil Armstrong was forced to take over control of the lunar module to avoid descending into a giant crater; a crater near-miss was also experienced by the crew of Apollo 16. On returning to Earth, the crew of Apollo 15 experienced a rough landing when one of their vehicular parachutes failed to deploy during final descent. The most critical U.S. emergency to date occurred on Apollo 13. With the spacecraft almost a quarter of a million miles from Earth, an oxygen tank exploded. The astronauts moved to the lunar lander for emergency return to Earth. Again, possible damage to the heat shield added to the concern during reentry.

There were also some dangerous situations during the Skylab series. Skylab 1 (unmanned) arrived in orbit with its meteorite/thermal shield torn away, with a solar wing broken off, and with the second solar wing jammed. The Skylab 2 crew had the unenviable job of trying to correct these problems so that the main habitat could be made operational. Astronaut Paul Weitz engaged in hazardous extravehicular activity in an unsuccessful attempt to release the jammed solar panel, after which Commander Charles Conrad attempted for 4 hr to dock with the damaged Skylab. The docking finally succeeded, but the crew never really knew until the end of the mission whether or not they would be able to undock for the return trip home. Working in extreme heat, the crew managed to deploy a parasol to shield the vehicle from the Sun, allowing the Skylab missions to proceed. On Skylab 3 an emergency flight home was contemplated for a time when a leak was detected in the command-module thruster.

Our information concerning Russian flights is incomplete, but we know that they have experienced emergencies, occasionally with severe consequences. Although the Russians had experienced difficulties with some of their earlier flights, they suffered their first known loss when Vladimir Komarov, flying Soyuz 1, was killed when the parachute controlling his landing craft became tangled. The second known Russian space tragedy occurred on Soyuz 11. Three Russian cosmonauts lost their lives when a valve failed to operate properly, causing a sudden decompression in their craft. The crews of Soyuz 13, Soyuz 15, Soyuz 21, and Soyuz 23 also apparently experienced some kind of space emergency, since they were returned to Earth hastily, under suboptimal landing conditions.

On Soyuz 26, cosmonaut Giorgi Grechko, returning from a space walk and closing the external hatch, found the monitoring instrument in the transfer compartment signaling a valve malfunction. Such a malfunction would have made repressurization of the compartment impossible. Without repressurization, neither he nor his space partner could enter Salyut, and therefore could not return to Soyuz for the trip back to Earth. The crisis was resolved when repressurization was attempted and it was discovered that the malfunction had been in the signal and not in the valve.

In the only case to date of an abort prior to orbital insertion, Soyuz 18 (renamed Soyuz 00) was returned to Earth after a failure occurred in an upper-stage rocket. During the abort the cosmonauts were said to have "indicated substantial concern," not only about the hazards of their return but also about the possibility of landing in China. After a search lasting 1 day, the cosmonauts were located in the foothills of the Altai mountain range and were returned home.

The Soyuz 33 flight, with its Soviet/Bulgarian crew, also ran into difficulty when failure in the propulsion system prevented docking with Salyut 6 and its resident Soyuz 32 crew. The crew of Soyuz 33 were forced to fly an unusual reentry profile for their emergency return to Earth, ending what Soviet Deputy Flight Director Viktor Blagov termed "the most complicated flight we ever had."

Nor was the Apollo-Soyuz effort, which rendezvoused an American crew and a Soviet crew in space, without incident. First, there was a rough redocking of Apollo with Soyuz, which caused the Russian gound control to voice concern over the integrity of their vehicle. Then, after separation and while returning to Earth, the American crew apparently made an error in executing the landing sequence. As a result of actions taken to correct the original error, nitrogen tetroxide oxidizer, a highly poisonous gas, was released into the space capsule. After a very hard landing, the crew was able to escape the gas by donning oxygen masks, but not before one of their members lost consciousness.

This description of problems in space does not suggest that manned spaceflight has been more dangerous or less successful than anticipated. On the contrary, the safety record has been remarkable considering the incipient nature of this endeavor. It does point out that space is, and will continue to be, an extremely hostile and unforgiving environment. We can expect that added experience will tend to reduce the risk involved, but we can also anticipate that emergencies will continue to arise with some regularity.

Individual Response to Threat

Although it is unlikely that all future space travelers will possess the technical skills to deal with structural emergencies, all must possess the psychological constitution to withstand such threatening events. In the following section we will consider how individuals respond to fear-inducing situations generally, and the possible implications of these findings to the space environment.

Conceptual view of fear— The psychological literature contains a myriad of meanings of the words "fear" and "anxiety," as well as such related words as "stress" and "arousal." Spielberger (1972) describes the traditional distinction that is drawn between fear and anxiety: fear is the emotional response to a real or objective danger, whereas anxiety is the fear-like reaction that occurs in the absence of specific, objectively threatening circumstances. Spielberger further suggests that in this traditional view, fear is proportionate to the magnitude of the danger, whereas anxiety exceeds or is unrelated to the danger.

Such distinctions are not universally applied. Some theorists view fear as following from anxiety (Epstein, 1967). For instance, Danesh (1977) holds that a threat leads to anxiety, which in turn leads to one of a number of secondary, learned responses. One such secondary response is fear, which prompts the individual to escape or withdraw; another is anger, which inclines the person to attack. Other theorists use the terms fear and anxiety to describe contrasting reactions (e.g., see Sarnoff and Zimbardo, 1961). Still other theorists take the view that anxiety is a derivative of fear (see Izard, 1972). To add to the confusion, a distinction is often drawn between fear and fright. In contrast with fright, fear describes an instrumental response which allows the threat to be perceived and problem-solving behavior to be initiated. Fright describes a purely emotional and noninstrumental response to an ambiguous situation, one which thwarts rather than initiates problem-solving activity. Several recent attempts to untangle the lexical confusion of this research area have focused attention on the feeling state of the individual. Epstein (1972) and Spielberger (1972) suggest that if the feeling state in fear and anxiety is the same, no further distinction is needed.

In spite of the difficulties in describing this research area, there remain some substantive agreements. It is generally acknowledged that an emotional reaction to a threat of danger can have positive adaptive advantages, in that this reaction rallies the individual to act. It is also agreed that beyond a certain point the response becomes so strong that it impairs or even paralyzes action. There also appears to be a consensus that anxiety (and/or fear-fright) follows from stress. Spielberger notes that this latter conclusion "represents an important point of theoretical convergence in an otherwise chaotic literature" (Spielberger, 1972, p. 483).

The measurement of fear– There are two basic approaches to the measurement of fear: autonomic responses and subjective assessments. The former approach has the disadvantage of reflecting responses other than fear; the latter has the disadvantage of being colored by the perceived acceptability of reporting fear. In spite of these limitations, a positive correlation has been found between physiological measures of stress and subjective reports of fear, and between fear (however measured) and performance (Hodges and Spielberger, 1966; Thackray and Pearson, 1968). For certain situations, self-report has been found to be a convenient and a reliable predictor of performance (Pearson and Thackray, 1970).

Experimental findings– In this section we will consider the observational and experimental literature related to fear, with an eye to understanding the conditions under which fear could develop in space and the consequences that could follow.

Noyes and Kletti (1976) report that, for some individuals involved in serious accidents, a complete depersonalization occurs in the split second or so prior to the accident. These individuals describe a perceived slowing of time just before the accident, a heightening of awareness, and most significantly, a seeming separation of the experiential self from the physical body. They report a feeling of having "left their bodies" and of witnessing the accident from some distance away as disinterested third parties. Depersonalization, out-of-body experiences (Mitchell, 1981), and similar phenomena have been reported over the years by many people and under a variety of circumstances. The significance of the Noyes and Kletti study is that these authors found a high degree of response consistency, suggesting a possible tendency to depersonalize in reaction to a severe threat. The accidents described in this study were all unavoidable. A significant question is whether such a tendency would prevail if individuals, by their actions, could ameliorate the situation. A further question concerns how widespread such a response is, and how individuals differ in this response.

A series of experiments involving less immediate but serious threats to life were conducted at Fort Ord, California, in the early 1960s (Berkun, 1964). One experimental group consisted of Army trainees who were led to believe that (1) an aircraft in which they were passengers was about to make an emergency landing, (2) the outpost at which they were stationed was an artillary impact area, or (3) they had caused serious injury to a buddy by the faulty wiring of an explosive charge. The soldiers involved in the tests were required to perform certain tasks related to their duties. In reality these tasks were carefully controlled performance measures. Trainees in each experimental condition performed significantly poorer than did subjects who performed similar tasks under nonthreatening control conditions.

The relationship between fear and performance found for Army trainees is confirmed in the work of Helmreich (1967) and Radloff and Helmreich (1969). These authors found that those divers of Sealab II who reported themselves as most frightened performed worse and spent less time in the water than did their less frightened colleagues.

Although there appears to be a close tie-in between fear and impaired performance, the relationship between the stressful situation itself and the fear response is less predictable. In the Fort Ord study series, an experienced group of soldiers reported themselves to be less distressed in the experimental than in the control condition, and their performance improved in the "life-threatening" situation.

Time of arousal— Not all threatening events occur instantaneously and without warning. Fenz and his colleagues (Fenz and Epstein, 1967, 1969; Fenz and Jones, 1972) have shown that when one has some time to consider the threat, the temporal pattern of fear response can vary significantly. Using the sport of parachuting, these authors have detailed the most complete work to date on time-dependent responses to life-threatening situations.

Employing three basic response measures (physiological responses to the stress situation, physiological reactions to words associated with parachuting, and self-reports of fear) Fenz and his collaborators found that experienced jumpers generally have a lower level of anxiety than inexperienced jumpers. However, the main difference between the two groups is not in the level of response but rather in the pattern of response. For novice jumpers, fear response rises steadily from the morning of the jump until the "ready" signal preceding bailout, and drops to normal after landing. For experienced jumpers, the fear response peaks early on the day of the jump and drops to below normal just before bailout. Measurements taken after landing show that the fear response of experienced jumpers rises to normal shortly after landing.

In summary, the preponderance of data indicates that, unlike novice jumpers, experienced jumpers become fearful well in advance of the event, becoming calm as the event approaches. Similar results are reported by Capel, Youngblood, and Stewart (1970) and Knapp and Capel (1976), who found anxiety levels of aquanauts to be low prior to a hazardous dive, rising to normal immediately after the dive.[2]

---

[2] Although data showing different patterns of responses for novices and experienced jumpers are quite convincing, it should be noted that not all studies show the same consistent pattern. Hammerton and Tickner (1968), using heart-rate measures, found no differences among three groups: very experienced jumpers, inexperienced jumpers with full ground training, and inexperienced jumpers with abbreviated ground training. They did, however, find performance differences favoring training. Basowitz et al. (Basowitz, Harold, Harold Peraky, Sheldon J. Korchin, and Roy R. Grinker. *Anxiety and Stress*, McGraw-Hill:NY (1955)) found that base levels of anxiety in airborne trainees are often low during the initial stages of training, and rise after training is completed. Capel et al. (1970), examining anxiety in divers, reported that anxiety control became more difficult with experience. Thus, although experience may be a highly significant factor in the regulation of fear response, it appears that other factors also must be considered.

Mechanism of arousal— From the data on time of arousal, Fenz and his collaborators conclude that fear is inhibited under certain conditions, and that a particular pattern of inhibition is related to optimal performance. The optimal pattern (shown by experienced jumpers) allows fear to occur, but at a time considerably in advance of the fear event. Such a pattern calls attention to the danger, but leaves the person emotionally capable of performing necessary acts. Too much inhibition results in lack of attention to relevant details and consequently in poor performance, whereas too little inhibition leads to emotional chaos. These authors describe the pattern of novice jumpers as frequently reflecting an "all or nothing" inhibition process, with both extremes ill-suited to the task. Good performance results, not from allowing anxiety to be shut off completely, but rather from regulating the pace at which anxiety is experienced.

The question remains: What triggers the fear response to occur at a particular time? Fenz and his colleagues suggest that response corresponds to, and shifts with, decision activity. For instance, for the experienced parachutist, the decision point occurs on the morning of the jump. At that time the potential jumper assesses all conditions and decides whether or not to go. In this way, experienced jumpers have made their mental preparation and done their "worry work" in advance of the actual event (Fenz and Epstein, 1967). For the novice jumper, the final decision to make the jump is probably not made until the moment he or she hears the signal for bailout.

Individual differences— From the preceding discussion we can conclude that, although the mechanisms are complex, fear generally inhibits performance; however, optimal performance is achieved not by totally eliminating fear but by controlling the time and intensity of the fear experience. In this section we will turn to the question of how fear control relates to the personal and psychological characteristics of the individual, that is, to characteristics that might impact astronaut selection.

Danger control and fear control— Levanthal (1967, 1968, 1970) underscores the importance of the direction of mental focus in determining how individuals will respond to a threatening situation. If the individuals focus their attention on the external forces creating the threat, they prepare themselves to avoid or reduce the danger. If they concentrate on their subjective reaction to the danger, they prepare themselves to control their fear. Levanthal has termed these alternate coping strategies danger control and fear control. The successful strategist must somehow do both, either by dealing with fear in advance, as suggested above, or by dealing with both the danger and the fear (possibly by alternating attention between the two). Fenz (1973) found experienced parachutists to have an external orientation; this suggests that, although both danger and fear must be controlled, there is a success bias favoring those whose focus is predominantly external.

Levanthal and Lindsley (1972) imply that knowing certain personal characteristics of an individual (such as the ability to analyze and separate features of the perceptual field, or the inclination toward fantasizing) may help to predict an individual's inherent tendency to focus externally or internally in response to a threatening situation. These and related measures could be useful in studying the significance of external/internal focus on managing crises, including the potential crises of space.

Gender effects— Levanthal, Jones, and Trembly (1966) found that women report higher levels of fear response than men. Since men are generally constrained in expressing fear, it is frequently assumed that men experience a higher level of distress than their fear reports indicate. To test this possibility, Katlin and Hoffman (1976) examined the hypothesis that, for a given level of fear reported, men would experience a higher level of autonomic activity than women. These authors failed to confirm this hypothesis. Accepted at face value, this finding leads to two conclusions: (1) subjective reports are accurate measurements of disturbance and are sensitive enough to overcome even cultural biases, and (2) women actually experience more fear than men. However, the conclusions of Katlin and Hoffman should not be considered definitive. Although these authors report no statistically significant differences in autonomic response between men and women, the autonomic responses of women were less than those of men for each level of self-reported fear. Since the question of whether women actually experience more fear than men is extremely important for future spaceflight, this area needs further examination.

Birth order effects— Radloff and Helmreich (1969) found that first-born and only children did not perform as well in the hazardous environment of Sealab as did latter-borns (a finding which they report to be true also of fighter pilots during the Korean War). Lester (1969) also notes that there were few first-born or only children among the men on the climb of Mt. Everest, suggesting again the greater willingness of latter-borns to expose themselves to danger. Similar findings have been reported for women (Schachter, 1959;

Gerard and Rabbie, 1961). However, results in this area have been far from consistent. Gerard and Rabbie (1961) found that, unlike the women in their study, first-born men showed lower fear-induced skin conductance than latter-born men, whereas Miller and Zimbardo (1966) found first-born women to be less fearful than their latter-born counterparts.

A variable that presents such shifting results would not seem to have much to contribute to the understanding of fear responses or spaceflight selection. However, one cannot avoid being impressed by the strength and consistency of effects within individual studies. It has frequently been observed that many of the astronauts that have flown to date are first-born children, and 20 of the original 31 Mercury astronaut candidates were only or eldest children (Ruff and Levy, 1959). Since these statistics continue to command attention, it would be helpful to understand more fully their relationship, if any, to performance under stress.

Training for danger— A practical application of the theoretical and experimental investigation of fear and its control lies in the possibility of training individuals to deal with dangerous situations. The objective or mechanical skills necessary to reduce or eliminate a particular danger obviously can and must be trained. But how does one train an individual to deal with danger itself? There are several approaches to this question.

The first involves training a person in a particular stress situation for the same or a similar stress situation. There is considerable evidence that a person can be trained to deal with the dangers of a particular situation. Babin and Loiko (1973) report success in increasing the efficiency of pilots in their reactions to on-board emergencies. Hammerton and Tickner (1968) have shown that the level of ground training makes a significant difference in the ability to perform a tracking task just before a first parachute jump. And, the "experience" predictive of good performance in the studies of sports parachuting (Fenz and Epstein, 1967, 1969; Fenz and Jones, 1972) might more accurately be called training.

Analyses based on opponents processes provide an explanation of how such danger training might work (see Epstein, 1967). As applied by Solomon and Corbit (1974), a particular stimulus arouses an effect (State A). When the stimulus terminates, a new state (State B) opposite to the first appears. This new state persists for a while and then dies out. States can change in their quality and intensity with repeated stimulation. When this occurs, State A becomes weaker and State B becomes both stronger and longer lasting. For example, a person in a threatening situation experiences fear (State A) which, when the threat has been eliminated, gives way not to a neutral state, but to a state of elation or relief (State B). With repeated exposure, the aversive state becomes less pronounced and the reinforcing state becomes more pronounced and longer lasting. To the extent that this model accurately describes reactions to certain stimuli, the threatening stimulus becomes less fearful, and the experience of meeting the threat more rewarding.

A second major approach to stress training involves training an individual in one kind of stress situation for another kind of stress situation; that is, training a person to deal with danger generally. How much value this "irrelevant" stress training has for an individual is not totally clear, although many organizations which must deal with emergencies behave as if the value of such training were assured. For instance, a great deal of the training given to members of underwater demolition teams is for the purpose of pushing the individuals to their limits (Rubin and Rahe, 1974). Similarly, in describing stress training at the military academy, P. Patterson (1975) notes that cadets are stressed "for the purpose of experience, for the purpose of converting a young man into a professional soldier, and perhaps to identify cadets who cannot function under stress" (p. 2133). This explanation points out a second rationale for generalized stress training — its use as a screening device.

The Russian space program assumes that benefits follow from generalized stress training. Lieutenant General Georgy Beregovoi, head of crew training, believes that dealing with various life-threatening events prepares cosmonauts psychologically for the rigors of space. Parachute jumping and remote survival tests form the bases of this training program. Each cosmonaut must make at least 100 parachute jumps. As training progresses, the cosmonauts are required to perform tasks of increasing difficulty before releasing their chutes. Survival tests are similarly structured. Crews are deposited in extremely inhospitable environments (including the Siberian forests, Cuban jungles, and the Black Sea) and survive only by their own wit and endurance. There are no rescue teams to help out if trainees get into trouble (Bluth, 1981b).

The astronaut training program has not explicitly adopted the value of generalized stress training. However, elements of generalized stress training such as scuba diving, survival testing, and flight training historically have been part of the astronaut indoctrination program. However, their inclusion has frequently been defended on the grounds that such skills could be required for spaceflight. Now, for some astronauts, flight training can no longer be linked to any obvious spaceflight need. As a result, flight training has been dropped as a requirement for certain candidates (e.g., Mission Specialists and Payload Specialists on Shuttle). This move has met with mixed reactions: concurrence on the part of those who believe that flight training has needlessly restricted NASA's pool of qualified applicants (see O'Leary, 1970) and concern on the part of those who believe that an important element in spaceflight preparation has been eliminated.[3]

Significant questions remain concerning whether generalized training to deal with danger is effective, and if so, what skills need to be developed. Research to help resolve the debate on generalized training for danger and research to further the understanding of preparing for specific danger could have direct relevance to future astronaut training.

Group Processes

Thus far, we have concentrated on those qualities which might affect an individual's ability to respond to a life-threatening situation. In this section we will consider responses which emerge, not at the level of the individual, but from the group as a whole, either during or subsequent to a crisis situation. Before considering responses that we believe are likely to occur in space, let us consider one that, although it has been raised as an issue, we believe is unlikely to occur.

Some have expressed the concern that, in the immediate aftermath of a spaceflight accident, panic[4] could develop within the space crew, especially if the group were large and diverse (Hartman and Flinn, 1964). Early work on the panic phenomenon contained assumptions that made a panic response seem likely in a variety of situations. It was assumed, for instance, that emotion (as opposed to

---

[3] A difference in training among classes of astronauts raises broad issues. For instance, a lack of shared experience may lead to the perception that some crew members do not have the appropriate "membership characteristics." This issue is considered in chapter V.

[4] Panic (also known as "panic flight response") usually refers to the frantic, nondirected activity sometimes observed in crowds in response to a severely threatening event.

rational thought) influenced one negatively, and that crowd membership had an essentially brutalizing effect on people in terms of facilitating emotion. In 1951 Mintz proposed a theory of panic based, not on emotional demands, but on the reward structure of the situation. Looking at panic from a rewards perspective makes such behavior seem less likely, since in many situations the needs of individuals are best served if everyone behaves in an orderly and cooperative manner.

Assuming the rewards structure model, Rosengren, Arvidson, and Sturesson (1974) identified three conditions which, taken together, could lead to a panic response: an individual

1. sees a threat toward his or her own existence,

2. sees a possibility of escape, and

3. believes this possibility is soon to disappear (e.g., because it is not sufficient for all who wish to use it).

From this description one can see how panic, or panic flight, might develop in, for instance, a crowded, burning building. However, to the extent that panic does not contribute to an individual's chance for survival, as it is unlikely to in space, the probability of its occurrence is low. Individuals exposed to an accident in space may experience intense fear, but the probability of panic on-board is remote.

Responses to threat— W. Smith (1966) examined the interactions within a group of men facing the hazards of crossing an Antarctic crevice field. He found that reaction to danger was not uniform, but changed over time, dividing into two major phases. The first or ineffective phase consisted of three parts and began with the group's displaying an inability to comprehend or to accept the seriousness of the situation (perhaps related to the reactions reported by Noyes and Kletti, 1976). Next came a period of energetic but incautious activity which was followed by a period of inactivity. The second or effective phase was marked by cautious action in which the group members recovered from the disorganization that had marked the earlier phase and began to work together toward resolving their problem. It would be of interest to know if the phases observed in this experience represent a general pattern of group response to crisis, and if so, how training might eliminate or shorten the ineffective period.

An area of particular importance in marshalling a group to deal with a crisis involves the coordination of team members. Insight into this area can be gained from examination of aircraft crews. There is growing evidence that inadequate management of cockpit personnel can result in crew error, and ultimately in accidents or near-accidents (Ruffell Smith, 1979; Lauber, 1980). The inability of aircraft crews to coordinate their activities during critical periods has been shown to relate to communication styles, social-psychological factors, and personality characteristics (Foushee, 1982). Significantly, problems of crew coordination similar to those found among flight crews have been identified by Russian cosmonauts. In noting these issues, Leonov and Lebedev (1972) stress the necessity for understanding, and even anticipating, the responses of other crewmembers, a condition which they call "homeostatic balance."[5] These authors suggest that lack of such balance can result in failure to coordinate activities because of "the inability to understand one another in critical situations, the asynchronous nature of psychomotor reactions, and differences in the degree of attention, thinking, and other inherent or acquired individual characteristics . . ." (p. 6). Although the goal of homeostatic balance among crew members is intuitively appealing, more work is needed to determine how such balance is established and maintained.

One reaction which can be predicted with some confidence is a tendency of crewmembers to come together in response to a threat (Gerard and Rabbie, 1961). From a planning perspective, it is important to understand why this happens. It is usually assumed that persons who find themselves in a fearful situation seek the company of others to determine just how they should feel or behave (Schachter, 1959; Gerard and Rabbie, 1961; Wrightsman, 1960; Zimbardo and Formica, 1963). However, Miller and Zimbardo (1966) and Sarnoff and Zimbardo (1961) have found that fear affiliation goes beyond information seeking, since individuals seek out others with certain compatible personal characteristics. It would be desirable to understand more fully what needs are addressed by affiliation in a crisis to

---

[5] The concept of homeostatic balance as used here follows a similar use of the term in the family therapy literature. Since Jackson's original use of "homeostasis" or "homeostatic equilibrium" to describe the emotional state which every family seeks to maintain, the term has been used widely in conjunction with events or behaviors that contribute to group equilibrium (Jackson, D. D.; The Question of Family Homeostasis. *Psychoanalytic Quarterly Supp., 31*, 79-99, 1957).

ensure that contacts which are available are supportive of both the practical and the emotional requirements of the situation.

A crisis is accompanied by a general state of instability within the group (Hartman and Flinn, 1964). Effective management of a crisis requires, then, that certain actions be taken prior to and following an incident, as well as during the crisis itself. For instance, planners and managers must identify in advance individuals whose specific skills could be used to achieve particular goals, and then set up the infrastructures to engage these skills. The quality of leadership displayed during a crisis will determine to a large extent how quickly stability can be restored (Llano, 1955).

Post-emergency concerns— The disaster literature provides evidence that threatening situations do not necessarily lead to antisocial behavior. On the contrary, it appears that in most crises altruistic behavior becomes the norm (Quarantelli, 1978). However, when a normal routine is reestablished, a different kind of behavior may emerge.

Scapegoating— Attribution of blame is a common occurrence during post-crisis recovery. Based on the work of Bucher (1957), of Veltfore and Lee (1943), and of Drabeck and Quarantelli (1967), Wenger (1978) concludes that beliefs about blame take a period of time to evolve. Accusations of blame can be expected to emerge, not in the immediate aftermath of a crisis, but at some later time.

Blaming or scapegoating has been found to occur in those circumstances in which human action is involved and in which there is the possibility of recurrence. Bucher opines that blame results from seeking an explanation to something which cannot be explained satisfactorily in conventional terms. In other words, blaming or scapegoating serves a psychological need. It is an attempt to control the future by creating a structure whereby inexplicable events become explicable. Apparently, blaming also has an instrumental effect. A person is more likely to offer help if the injured individual implies that the person is to blame (Schwartz and David, 1976).

People in certain positions are more likely to be blamed than others. The lowly ranked individual is usually not an acceptable scapegoat, and the tendency is to reject that person in favor of someone higher in authority (Bucher, 1957). However, the highest ranked individual may also be an unacceptable scapegoat, since condemning him or her tends to condemn the values of the organization.

Focusing blame on the second- or third-in-charge would seem to satisfy the psychological needs associated with scapegoating, while leaving the values of the organization unchallenged. Excluding the "person at the top" from blame would appear to be more functional for group members than for outsiders. In terms of space travel, one might expect that crewmembers will hold a second- or third-in-command responsible, while ground control might look to the commander.

Since blaming or scapegoating gives one some control over the action of others, and at least illusory control over future events, it is reasonable to expect that a crisis in space could result in scapegoating. Specific efforts need to be taken to deal with this post-crisis phenomenon.

Hysterical contagion— In hysteria, or hysterical conversion, a person demonstrates bodily symptoms which are without an ascribable physiological basis. Hysteria occurs when a person experiences extreme anxiety which is relieved by converting the anxiety to physical symptoms (Frazier and Carr, 1974). In hysterical contagion, the symptoms are passed from one individual to another, triggered by a precipitating event which heightens the sense of an immediate and tangible threat (Kerckhoff and Back, 1968).

Although the general nature and course of hysterical contagion are reasonably well understood, there is some controversy as to the dynamics of this rather strange phenomenon. Kerckhoff and Back (1968), Smelser (1962), and Klapp (1972) favor the explanation that hysterical contagion is part of a negative cycle of stress-illness-greater stress. Gehlen (1978), on the other hand, argues that in hysterical contagion the behavior adopted is positively rather than negatively reinforcing, allowing the individual to relieve stress by participating in the benefits associated with the sick role. The latter explanation seems to describe the data of hysterical contagion better than does the runaway-fear hypothesis. Hysterical contagion, at least under Earth conditions, has a brief and well-defined life span of a few days, after which the symptoms rapidly disappear. An interesting aspect of the hysterical contagion phenomenon is that it is probably related to a conscientious work ethic. Kerkhoff and Back found that hysterical contagion is unlikely to occur among individuals who are willing to take an occasional day off.

An episode of hysterical contagion in the aftermath of a fear-inducing event is unlikely to occur in space in the foreseeable future.

In fact, our present generation of astronauts has incurred some displeasure by trying to conceal, rather than disseminate, information concerning physical symptoms (Cooper, 1976). However, future spacecrews are expected to be large and to include individuals with backgrounds different from those of today's astronauts. Further, the conditions of space would seem to support the development of this phenomenon. These conditions include an elevated level of anxiety associated with living in space, a highly directed work-oriented subsociety, and a closed environment in which information would travel rapidly and in which social pressure could encourage the spread of symptoms. It seems that some cathartic measures will need to be devised for diffusing anxiety in space, both for routine prevention and to avoid the possibility of post-crisis maladaptive responses.

Implications for Space

The research reported here suggests many parallels to space. We can expect that space travelers will continue to be subjected to fearful, and even life-threatening situations. This realization could help shape our training programs and influence our selection criteria. It is imperative that space travelers learn to recognize and acknowledge threat at the earliest moment, so that valuable response time can be saved. And they must deal with both the threat and their emotional response to it in an effective manner. Since many of the potential challenges of space cannot be fully anticipated, it is important to determine if generalized stress training can be of benefit.

A crisis situation can threaten the integrity and stability of a group. Structures and procedures must be in place that allow space travelers to coordinate their activities with minimal loss of time. The role of the leader or leaders and the effectiveness of coordination among crewmembers will be crucial in formulating group responses to a life-threatening event.

Some mechanisms for dealing with stress can result in unacceptable consequences. For instance, in the aftermath of a threatening but poorly understood event, blaming or scapegoating is a common response. Such blaming in the confinement of space could place an intolerable burden on an individual; and if, as is likely, the individual were in a position of leadership, blaming could contribute to further group instability. Space travelers must be helped to control fear without resorting to mechanisms which could themselves threaten the welfare of the community.

INTERNALLY PRECIPITATED CRISES

To many, the threat posed by a disturbance in an individual or in the relationship among individuals does not seem as serious as a threat posed by a physical event. However, such a disturbance in space could have a devastating effect on the entire crew and its mission. In this section we will consider some of the more serious breakdowns in personal or interpersonal functioning as they relate to extended spaceflight.

Psychological Episodes

In speaking of a psychological episode in space, we are not making a clinical assessment, but rather are alluding to those kinds of personality disturbances which would be obvious to an involved layman. It is difficult to imagine a less desirable place for such a disturbance to occur than in a spacecraft thousands of miles from Earth. Although the likelihood of a full-blown psychiatric episode occurring in space is small, the costs associated with such an event could be extremely high and would present a serious challenge to the entire crew.

Experiences in confinement — Several episodes involving psychological disturbances have occurred on Antarctic wintering-over expeditions (Gunderson, 1968). One such episode took place at a two-man outpost located approximately 50 miles from the main camp. Duty at the outpost was rotated among the entire crew. Each man served 2 wk, overlapping for 1 wk with one crewmember, and for the second week with his replacement. On the occasion of interest, the supply plane arrived at the outpost, picked up the member who had served 2 wk, and dropped off the new man. During the night the new man became increasingly agitated. His companion tried to calm him, thought he had succeeded, and eventually fell asleep, only to awaken in the middle of the night to find his crewmate dressing for the outdoors. When asked what he was doing, the new member announced his intention to walk back to the base camp — a clearly impossible task. All of the partner's skills were needed to try to keep the disturbed man in the habitat while at the same time struggling to reach the base camp by radio. As soon as weather conditions permitted, the plane was dispatched and the disturbed individual was removed. In the case described, a failure to persuade the disturbed person would have resulted in his death. Under similar circumstances, another disturbed individual might have turned his anxiety against his partner, or have included his partner in his fantasy, with potentially serious consequences for both individuals.

Such an event has been reported by Ordiway and Raymon (1974). A commander and a navigator took off in a two-seater fighter craft. The navigator soon observed that the commander was behaving erratically. The commander engaged in strange exchanges with flight control and flew the aircraft in a highly dangerous manner. He hallucinated a landing strip and prepared to land. The navigator succeeded in convincing the commander not to land, but to return to the home base. The plane eventually was landed safely, but only after several abortive attempts during which the commander and the navigator fought for control of the stick.

These two instances of psychiatric episodes in a confined space underscore the simple wisdom expressed by Hartman and Flinn in 1964, "there are many advantages to having two men in a space vehicle, except when one of them is psychotic" (p. 63).

It is not clear that confinement increases the frequency of psychological disturbance, although it is reasonable to expect that it does. It is known that psychological disturbances in confinement are not unusual: Six mentally disturbed persons were removed from the American Antarctic bases during the International Geophysical Year (Law, 1963); and on two cruises on the submarine Triton, approximately 5% of the crew of 137 were seen for some form of psychiatric disorder (Serxner, 1968). These crewmembers suffered either from anxiety or from depression and were treated primarily with drugs. The submarine Triton also had one full-blown psychotic episode. The victim, a chief petty officer, experienced migraine headaches followed by hallucinations. The man was treated with phenothiazine and barbiturates by the medical officer, but was kept as much as possible in his normal activities, accompanied at all times by other chief petty officers. Significantly, this man's disturbance centered on his job, the inadequacy he felt in his performance, and the slights he felt, both from his superior officers and from his subordinates, in being excluded from decisionmaking sessions.

Nature and cause— It has been observed (S. Perry, 1976) that a person experiencing a severe psychological disturbance will show impairment in at least one of four functions: thinking (disordered, as in schizophrenia); perception (distorted, as in paranoia); mood (elevated, as in mania or despondent, as in depression); and impulse (violent, as in hebephrenic excitement or frozen, as in catatonic stupor).

Impairment in mood expressed as depression is the most commonly observed form of psychological disturbance. Depression is frequently reported in confined groups (S. Smith, 1969), with depressive symptoms particularly marked among those isolated for long periods of time, such as Antarctic wintering-over parties or Polaris submarine crews (Earls, 1969). Depression is marked by lack of ability to concentrate, feelings of worthlessness, irritability, weight loss, hopelessness, guilt, etc. With depression, the individual suffers a diminution in relations with other people and with the outside world in general.

Akiskal and McKinney (1975) show that the various models of depression reflect one or more of five dominant schools of thought: psychoanalytic, behavioral, sociological, existential, and biological. Among the models based on the psychoanalytic school are those that describe depression as anger turned inward (see Freud's Collected Papers, 1950) or those that equate depression with a negative cognitive set (Beck, 1974). Behavioral models would include the theories of low positive reinforcement (Lazarus, 1968; Lewinsohn, 1974a, 1974b) and theories of learned helplessness (Seligman, 1974; Seligman, Klein, and Miller, 1976), where individuals perceive themselves to have no control over the events affecting their lives. The sociological school views depression as resulting from a loss of status (Bart, 1974), whereas the existential school focuses on the loss of the meaning of existence (Becker, 1964). The biological models emphasize the roles of genetic or chemical factors in the onset of depression (Coppen, 1967). In considering several suggested etiologies of depression, Blaney (1977) has noted that "it may well be that each element is sufficient to lead to depression but that none are necessary" (p. 218).

Akiskal and McKinney (1975) posit an integrative psychophysiological model of depression. In this view, various chemical, experiential, and behavioral processes translate into the functional impairment of advanced depression or melancholia at the reinforcement center at the diencephalon. According to Akiskal and McKinney, once the individual reaches the melancholic phase, the disruption in functioning becomes "biologically autonomous" of the conditions which preceded it.

Contributing factors— Whatever the primary cause of a disturbance, it has been found that events, or series of events, can aggrevate or trigger a psychological episode.

Events— Since the early work on life change scaling at the University of Washington (see, e.g., Holmes and Rahe, 1967), a positive correlation has been demonstrated between major life events and subsequent physical and psychiatric disturbances.[6] (For a review of this work, see Rahe and Arthur, 1978.) These findings are in overall agreement with descriptions of the generalized stress response as outlined by Selye (1976). Although both welcome and unwelcome events can extract a toll, it is the negative event, and usually the "exit" event (e.g., the death of a loved one, marital separation, the loss of a job, etc.) that gives rise to the greater problem and that has a particular relationship to the occurrence of severe depression (Paykel, 1974).

In extended spaceflight, one could readily see how conditions on-board could contribute to psychological disturbances, and especially to depression. For instance, as noted above, depression is sometimes interpreted as anger turned inward. Under normal Earth conditions, one might expect some individuals to turn their anger inward and other individuals to direct their anger toward others. In confinement, confrontation is studiously avoided. It is reasonable to assume, then, that even normally outward-directed individuals could be subject to depression in confinement. Other conditions, such as lack of control over the environment, loss of support of family members, or the loss of status also could contribute to a depressive response.

For future space travelers, an especially worrisome form of depression is one following the relational breakdown between the individual and the group. As described in chapter V (see Compliance, Conformity, and Independence), this breakdown could result in ostracism (Schachter, 1959) or in the quasi-psychotic symptoms associated with "long-eye" (Rohrer, 1961). It is significant that even the adoption of group norms may not fully protect a crewmember from social ostracism. Strange and Klein (1973) have found that separated groups tend to have at least one isolated member. It seems clear that all effort must be made to ensure that ostracism, "long eye," and related phenomena do not occur in space.

---

[6] Recent work emphasizes the role of minor as well as major events in physical and psychological illness (Lazarus, Richard S.; Little Hassles Can Be Hazardous to Health. *Psychology Today*, July 1981).

Environmental adaptation— Presently at least, some psychiatric conditions are thought to represent a failure of the individual to adapt to the environment. Investigation of the stress response (Selye, 1976) reveals a range of pathologies, both physical and psychological, that can follow an inadequate adjustment to one's environment. The stress literature provides some guidelines as to which factors might influence the adaptation process.

Andrews, Tennent, Hawson, and Vaillant (1978) suggest that events which have significance to the individual can lead to stress and that the ability to deal with stress is related to patterns of coping, to personality characteristics, and to social support. Patterns of coping refer to those particular ego defenses used to deal with stress. Vaillant (1976) has found that suppression, sublimation, humor, and altruism lead to a healthy adjustment; defenses such as fantasy, projection, or passive aggression can lead to disruption in the individual's relationship to reality, to others, or to his or her own conscience. In terms of personality characteristics, Andrews et al. suggest that anxiety proneness, self-esteem, and perceived ability to control could be important predictors of adjustment. In support of the personality hypothesis, Kobasa, Hilker, and Maddi (1979) found that, among management personnel, those who have a sense of commitment, those who feel they have control over their own lives, and those who seek novelty and challenge remain relatively free of illness, even under considerable stress. The third determinant of the Andrews et al. model relates to the environment in which the individual operates. Although it is frequently suggested that the quality of social support is more important to successful adjustment than the quantity of support, Andrews et al. (1978) note that we have very little understanding of what constitutes quality support and, by implication, very little understanding of what constitutes an adequate social support system. However, we can say with certainty that crew interaction will be a powerful determinant of general adaptation since, as Selye (1974) has pointed out, the stress of living with one another is among the more significant causes of distress.

Based on the Andrews et al. model, one would advise that any attempt to analyze adjustment to space should include: the significance of the event(s) to the individual, the possible multiplier effect of the individual's personality, the effectiveness of the individual's coping patterns, and the support system available.

Selection— The first defense against the occurrence of a psychological disturbance in space is personnel selection. Psychological assessment of astronaut candidates has historically been part of the medical evaluation. In the early space program, a fairly extensive battery of tests was used, along with interviews and observations, to arrive at a judgment concerning the candidates' abilities to withstand the rigors of space. Included in this battery were standard personality tests such as the Rorschach, the Thematic Apperception Test, and the Minnesota Multiphasic Personality Inventory, along with various tests of general intelligence and special aptitude. (For a review of the psychometric procedures used in the early space program, see Ruff and Levy, 1959; Hartman and McNee, 1977.)

At the time of preparation for the Shuttle program in 1977, ten years had passed since astronaut candidates had been selected. Psychological testing procedures were reassessed and the decision was made to drop the clinical battery, relying instead on psychiatric interviews, situation testing, and general observation. For the Shuttle, two psychiatrists with differing interview styles formed the primary team for the psychological screening of candidates. The recommendations of these psychiatrists became a part of the medical evaluation submitted to the Space Medicine Board for final selection. The same psychological evaluation techniques were used for the crew and for mission specialists.[7]

The NASA decision to drop the clinical battery portion of the psychological selection procedure was based on findings that the earlier psychometric data had contributed little to the selection process. In this connection, it should be remembered that standardized psychological tests are intended to select out individuals (i.e., to identify those with potential pathology). Yet with a large pool of prescreened candidates, NASA's primary task was to select in individuals (i.e., to identify those with high levels of physical and emotional health). Selecting for positive attributes is clearly the more difficult task. C. Perry (1965) found that, for a group of well-qualified applicants, there was greater agreement among evaluators on which candidates were (comparatively) less qualified than on which candidates were more qualified. In the future, as manned

---

[7]It should be noted that NASA selects only NASA astronauts. The Department of Defense will select its own payload specialists for Shuttle flights. European payload specialists who are part of the cooperative NASA-European Spacelab Program are selected through the European Space Agency, which employs its own selection procedures (Shapland, D. J., J. DeWaard, and G. Nichols; Choosing ESA's First Astronaut. ESA Bulletin No. 13, 21-28, May 1978).

spaceflight attempts to accommodate a wide variety of space travelers, techniques may be needed to select out individuals whose psychological makeup could lead to problems in space. However, for the foreseeable future, the need will remain that of identifying unusually sound individuals.

Just what is a mentally sound person? One suggestion comes from the work of Vaillant (1976) and of Lindemann (1979), both of whom link emotional stability with the concept of maturity. Lindemann describes a mature person as one who is able to perform the required tasks, meet ordinary stresses of life without disintegration, operate without making others sick, and adapt his or her own perceptions to reality.

The concept of maturity provides at least a starting point in the search for positive qualities associated with space adaptation. An interesting aspect of the maturity/adaptation relationship is the observation that maturity is frequently linked to a balanced approach toward people and events. As Lindemann (1979) notes, specialization, by consuming so much of an individual's energy, tends to be at odds with such balance. The need for special skills and for high levels of emotional stability will present a challenge in selection for extended spaceflight.

Treatment— If a person should experience psychological problems in space, the requirements of other crewmembers as well as the needs of the individual should be taken into account in selecting a treatment strategy. Because of the special interdependence of crewmembers, procedures must be developed and an understanding reached before flight as to just when and what kinds of measures are to be employed to control aberrant behavior. On Earth drugs are commonly used in treating both acute psychotic states and many chronic conditions. It can be expected that drugs will play a similar role in space. The following are some other possibilities for avoiding or dealing with potential psychological disturbances in space.

Psychotherapy— The value of psychotherapy has long been debated in medical circles. However, there is a growing literature indicating the efficacy of psychotherapy when used alone or in conjunction with drug therapy (Marshall, 1980). For psychotherapy to be carried out in the standard fashion, a trained therapist would have to be aboard the spacecraft. Pope and Rogers (1968) found that having a psychiatrist along on an Arctic mission provided a vehicle for crewmembers to vent their hostilities, averting psychological and emotional problems. However, having a therapist aboard a space flight may not always be an option.

An interesting alternative is the use of two-way communication for remote counseling. In recent years, various experiments involving mediated therapy have been tried. These experiments range from totally impersonal computer counseling (where the individual interacts with a preprogrammed "therapist"), to telephone therapy (where the client speaks to a trained person while retaining his or her anonymity), to two-way video phone or closed-circuit TV sessions (where a client and a therapist meet in a one-to-one relationship, separated by the medium of contact). Since these situations are less intimate than the face-to-face encounter, and give the patient considerable control over the interaction, they frequently are preferred by clients (Lester, 1974). Although all of these techniques show promise, it is not well understood which therapeutic qualities can be mediated and which cannot. Depending on the flight, another barrier to remote counseling could be the delays involved in long-distance communication.

An option to the use of professionals, either aboard or remote, would be the selection of crewmembers with therapeutic-like qualities, trained in helping skills. Carl Rogers has identified characteristics which he believes are essential for such an individual. He or she should possess congruence (stability, personal integration), empathetic understanding, and the ability to convey unconditional positive regard (C. Patterson, 1980). The concept of training for supportive behavior has been gaining acceptance in the last decade. Using models by Carkhuff (1969) and others, Egan (1975) outlines an approach for training paraprofessionals. Egan's techniques borrow from learning theory, social influence theory, behavioral modification, skills training, and problem solving techniques. Significantly, Egan sees the small group as fundamental to the development of skilled helping behavior. This and related work appear extremely promising for addressing the needs of spacecrews on extended missions.

Awareness training— Awareness, sensitivity, and related training may be useful in avoiding relational problems or in dealing with such problems when they arise. Although extensive experience has been gained with these approaches in everyday situations, we need to understand the potential benefits and limitations for confined groups. In one case in which such training was given in preparation for confinement, the crew reported that it was not only helpful, but the most valuable part of their training program (Dunlap, 1968).

Hypnosis— Serxner (1968) has reported some success in using hypnosis to treat psychological disorders during underwater cruises of the submarine Triton. Sharpe (1969) has suggested various ways in which hypnosis might prove useful in the space environment. However, to date, hypnosis has received little serious attention as a therapeutic device for space. Yet, as Orne (1959) points out, one of the characteristics of a hypnotic trance is the "potentiality for experiencing as subjectively real, distortions of perception, memory, or feeling based on 'suggestions' by the hypnotist rather than on objective reality" (p. 297). Although such techniques would have to be used judiciously, hypnosis might help to reduce anxiety or to direct attention to specific tasks to be performed under stress. Empirical evidence supports the notion that hypnosis could be useful in training individuals to sleep when desired, and to awaken in an alert condition (Dorcus, 1961). This area of research deserves attention, particularly as it offers a viable option to drug therapy.

Meditation and exercise— As discussed in chapter III, exercise has been found to have positive therapeutic effects in helping an individual maintain a sense of vitality. Of late there has also been a surge of interest in meditation, with practitioners attesting to an increased sense of calm, greater control over their lives, and higher levels of awareness.

In discussing the relationship of both meditation and exercise to well-being, Schwartz, Davidson, and Goleman (1978) draw a distinction between cognitive distress ("I have difficulty concentrating") and somatic distress ("My heart beats faster"). Their research with meditators and exercisers shows that although both groups report similar levels of upset, exercisers report less somatic, and more cognitive, disturbance than meditators. Although far from clear, a possible conclusion of this research is that techniques such as meditation and exercise could be employed to prevent or to combat particular forms of distress. The roles of meditation and exercise in addressing specific distress symptoms need to be explored.

Other— Techniques now being used to control negative reaction to stress include progressive relaxation, autogenic training, and biofeedback. Progressive relaxation involves tensing and releasing muscle groups to teach the body how to relax. Autogenic training uses self-suggestion to help individuals gain voluntary control over bodily functions. Biofeedback employs instrumentation to tell the individual how well he or she is progressing in attempts at self-regulation. It would be desirable to know how these techniques could best be used, either individually or as part of an integrated therapeutic program.

A common method of relieving psychological upset is to engage in distracting and rewarding activities. Although there is considerable overlap in the kinds of activities people engage in, there is also considerable individual difference (Rippere, 1977). As discussed in chapter III, effort should be made to match activity options to the preferences of particular spacecrews. However, in space, activity options will necessarily be limited and space travelers will probably have to learn new distraction/relaxation techniques. We do not yet know how successful such substitutions will be.

In terms of future spaceflight, the value of interacting with and caring for pets should be considered. Of late, pet therapy is receiving a great deal of attention, particularly as it applies to socially and emotionally isolated individuals.

Transcendant Experiences

The general question of consciousness alteration or transcendant experiences in space has been raised frequently, and several episodes have occurred that heighten interest in this question. On Soyuz 7 the flight engineer reported glancing at the Earth below and hearing the sounds of a dog barking and of a child crying (Leonov and Lebedev, 1975, p. 135). On Soyuz 26 the commander apparently decided to make an unauthorized extravehicular excursion. This in itself is surprising; more surprising is the fact that he failed to attach his space suit to the restraining device and was saved from being cast into space only by the quick action of his companion, who managed to grab him just before he floated away and return him to the vehicle (Oberg, 1981). One might argue whether this latter episode represents an altered state of consciousness or simply overexuberance. Whatever the explanation, such episodes should not be summarily dismissed. Mostert (1974) reports on a phenomenon found among sailors in which young men, otherwise apparently healthy both physically and mentally, answer the "call of the water" and quietly slip into the sea. The call of the water could have a direct corollary in the "call of space."

Clark and Graybiel (1957) describe a phenomenon in which jet pilots experience a pronounced feeling of physical unreality and of separation from the Earth. For some, this phenomenon was accompanied by feelings of exhilaration and exalted powers; for others the sensations were unpleasant and fear-inducing. Clark and Graybiel call this experience "breakoff," and conclude that the effect occurs most frequently when a pilot is alone, is flying at high altitudes, and is relatively uninvolved with the details of flight. A phenomenon which can result in flight personnel feeling dissociated from Earth (not subject to its laws) should be of concern to astronauts and mission planners.

Sours (1965) tested the hypothesis that the experience of breakoff is related to emotional and personality disorders. He found a higher incidence of breakoff experiences among aviators with positive psychiatric findings (presence of symptoms) than among other naval and marine jet aviators. Among those with psychiatric symptoms, the breakoff phenomenon generally was anxiety-producing and unpleasant. Sours concluded that the breakoff phenomenon probably precipitates an anxiety reaction in susceptible individuals. Benson (1973) examined 78 aircrewmen referred for clinical assessment because of disorientation in flight. About 40% of these men had experienced the breakoff phenomenon, and in all cases the experience was one of unease and apprehension. Since in this study breakoff was experienced by helicopter and by fixed-wing pilots, it appears that high altitude, although a contributing factor, is not necessary to the effect.

While the etiology is thought to be different, a similar phenomenon has been observed among divers and others operating under the sea. This phenomenon is usually termed "nitrogen narcosis," although it has been observed with various gas mixtures. Like the breakoff phenomenon, it is sometimes accompanied by a feeling of general euphoria or intoxicated-like state, whereas some individuals become "aggressive, irritable, insolent, and fussy" (Adolfson, 1967, p. 43). Among the changes that have been noted are a diminishing of cognitive function, a decrease in audio perception, a deterioration of motor performance, and amnesia (Behnke, Thomson, and Motley, 1935; Adolfson, 1967; Bennett, 1969; Miles, 1969).

It appears that the breakoff phenomenon and nitrogen narcosis have elements in common, and both bear an intriguing resemblance to general sensory isolation effects. If breakoff and similar phenomena are found to be associated with, or aggravated by, isolation and sensory deprivation or monotony, we might expect at least some related effects to occur in extended spaceflight.

Much more needs to be known concerning the cause of breakoff and related phenomena. The physiological correlates of altered perceptions or unusual behaviors, possibly related to atmospheric pressure as well as environmental triggers, need to be understood. The suggested relationships between personality characteristics and both the incidence and the effects of breakoff and related phenomena also need to be explored.

Substance Abuse

One does not usually think of substance abuse in relation to spaceflight; the problem of access would seem to preclude this concern. Yet Shurley, Natani, and Sengel (1977), among others, believe that substance abuse is a potential problem in space. It is unlikely, although not impossible, that there will be problems with illegal drugs. More likely would be the abuse of drugs brought aboard for medical purposes, or of accepted recreational drugs.

All spaceflights carry a medical kit containing pharmacological agents for pain, for motion sickness, and for other uses. Two questions related to the use of medicines in space need to be addressed. The first involves the interaction of drugs and weightlessness. As discussed in chapter II, we have little information on how commonly used drugs might affect individuals in a weightless environment. More specifically, and in terms of the subject matter of this chapter, we have no first-hand information on the potential effects of drugs that might be used to control psychotic episodes in space.

The second and equally important question concerns how drugs themselves might precipitate a crisis in space. Among drugs to which spacecrews might find ready access are stimulants and tranquilizers. On occasion, both categories have been associated with psychiatric symptoms. If a drug gives rise to abnormal behavior (e.g., by releasing psychological controls) the delicate balance of the isolated and confined spacecrew could be disturbed.

Alcohol is presently the only drug fully accepted for recreational purposes in American society. In fact, alcohol is so much a part of our culture that its use is considered the norm. Only when alcohol intoxication results in familial or occupational disruption do we consider alcohol use a problem. With this level of cultural acceptance, alcohol could find its way into space when resupply arrangements are in place and when storage capacity is plentiful. In its extreme form, the use of alcohol has been linked to dependency. The user fears rejection or failure and uses alcohol, quite literally, to "feel no pain." His use of alcohol makes him disdained by others, which reinforces the cycle of dependence, fear of failure, and rejection. Alcohol abuse can result in physical distresses such as tremors or "shakes," in vomiting, and sometimes in seizures. It can also accompany or mimic the symptoms of psychiatric disorders.

Specific data on the use of alcohol in confinement are limited. In many isolation studies, alcohol has not been available; in other situations where alcohol or other drug use could have been observed, authors are surprisingly mute. One report of alcohol use (aboard a supertanker) appears similar to the better uses alcohol finds under Earth conditions (Mostert, 1974). Here, "pourout" or the evening happy hour was apparently a very important, though very controlled, occasion, used to promote relaxation and good fellowship. However, we know that alcohol has been widely, and apparently less efficaciously, used in the Antarctic.

In one laboratory study of confinement, tobacco and alcohol were employed as experimental variables (Rogers, 1973). In this study subjects were confined for 5 days under either high- or low-decor conditions. In the relatively plush surrounding, only about half as much was spent on tobacco and alcohol as was spent under the more austere conditions. With high decor, the largest percentage of the leisure activity budget was spent for music. With the more spartan surroundings, the importance of music fell, and tobacco and alcohol became the top priority items in the leisure budget. These findings are suggestive of the importance alcohol could play in a confined and limited environment.

We need to understand more fully the impact of drugs used for medicinal purposes, especially as they influence personality change. In terms of the use of drugs for recreational or nonmedicinal purposes, we need to understand how such drugs might be employed in space. We know that isolation and confinement lead to increased emphasis on food, and probably heighten oral needs generally. In space, stimulation is low and reports of monotony and boredom are to be expected in all but the briefest flights. It is reasonable to assume that substitute methods of gratification will be sought, perhaps including the use of alcohol or other diverting drugs. Alternately, the self-censuring which restricts competitive interaction in confinement may also preclude the use of drugs. It would be helpful to understand how people, given the time and opportunity, would choose to use drugs in a confined and isolated environment, and how such use could exacerbate, or perhaps relieve, the problems of living in space.

Grief

Grief can be experienced in response to any significant loss, such as the loss of status, the loss of income, or the loss of a valued possession. However, grief is most profoundly experienced in the loss of a relationship. This loss may be temporary, as occurs in sporadic separations, or it may be permanent, as in divorce or death. Grief in its extreme form (e.g., in response to the death of a loved one) is among the most profound of all physiological and psychological stressors. One might wonder what the response to such a loss would be in the confines of space.

Bereavement and recovery— Averill (1968) distinguishes between mourning behavior, which is determined by the customs and mores of the society, and grieving, a ubiquitous reaction to loss which is found in all human and in many animal societies. Grief is marked by such psychophysiological reactions as fatigue, sleep disturbance, loss of appetite, choking or shortness of breath, empty feeling in the abdomen, lack of muscular strength, apathy, and change in the level of activity. Numerous grief researchers have noted that, although grief is relieved by other relationships, the behavior during grief is antithetical to the formation or continuance of such relationships. Typically, the grieving individual withdraws from other relationships or is antagonistic to sympathetic others. The individual experiences a loss of warmth in relationship to others and responds to their offers of concern with irritability and anger.

In the normal course of events, grief reactions follow a well-defined sequence. In the most general terms, this sequence can be described as shock, followed by despair, and eventually by recovery. Some researchers divide the stages of grief more finely. For instance, Ramsey (1977) distinguishes phases of shock, denial, depression, guilt, anxiety, aggression, and reintegration, with considerable overlap among phases.

Behaviors associated with loss and the resultant grief are well documented (Vachon, 1976). Although mania has been reported as a response to bereavement (Rickarby, 1977), depression is the more common reaction. However, behavior during the initial stages of bereavement may not mimic the inactivity usually associated with depression. Lindemann (1979) reports that depression following a crisis or loss is often accompanied by heightened (though non-directed) activity, increased speech, or increased hostility. An episode involving such hostility which occurred in an isolated group is reported by Shurley (see San Francisco Chronicle, Sunday Punch, Jan. 14, 1982, pp. 2 and 5). A member of a South Pole expedition who had recently received word of the death of his father got drunk and went on a rampage, smashing dishes and attacking other members of the expedition. Injuries were limited to gashes and bruises; however, it was several hours before order was restored. If alcohol or drugs were to be available in space, problems associated with grief responses could be aggravated.

In some cases bereavement can threaten the life of the grieving individual. Durkheim (1951) has called attention to the incidence of suicide among widows. However, the relationship between grieving and mortality goes beyond the incidence of suicide. In a review of the data on conjugal loss, Jacobs and Ostfeld (1977) found a significantly elevated mortality rate among survivors due to a variety of causes. This elevated rate was found to be more extreme for younger than for older persons, and for men than for women.[8]

The activity involved in recovering from a significant loss has been termed "grief work." Grief work is a staged process in which the survivor accepts the painful emotions associated with the loss, reviews the variety of experiences shared with the lost person, and gradually rehearses and tests new patterns of interactions and roles to replace those that are lost (Lindemann, 1979). Time for recovery varies with the severity of the loss. Many therapists consider that grief lasting a year or more is within normal limits.

Most students of bereavement agree that experiencing grief is essential to recovery (see, e.g., Hodge, 1972). Because of the extreme pain of grief, many individuals employ any and all defenses to avoid it, resulting in delayed or distorted reactions. A treatment that is employed in therapy sessions with those experiencing pathological grief reactions is to break down their defenses and to force them to live through the pain of their loss. Such therapy is extremely distressing to patient and therapist alike.

---

[8] Lindemann (1977) notes that one of the distorted reactions to grief is the adoption of symptoms associated with the last illness of the deceased. Perhaps related is the tendency noted by Parkes et al. (Parkes, C. M., B. Benjamin, and R. G. Fitzgerald; Broken Heart: A Statistical Study of Increase of Mortality Among Widows. *British Medical Journal, 1,* 740–743, 1969) for bereaved widows to die of the same illness that claimed their spouses.

From the standpoint of spaceflight selection, Ramsey's (1977) analysis contains a clue as to who might pose special bereavement problems. This author draws a correlation between phobias and extreme depression following bereavement, both being suggestive of individuals whose inclination is to avoid confrontation and to escape from difficult situations. The relationship between phobic behavior and distorted grieving requires testing.

Since there is consensus that little can be done to help a person who is experiencing normal grief, space planners must concentrate on ensuring that, in the event of a loss, the grief of a crewmember follows a normal course. Just what is a normal course of grief for a space traveler can only be speculated upon. If a loved one at home should die, it can be assumed that the space traveler has already passed through some anticipatory grief in the process of prolonged separation (Moss and Moss, 1973); but a loss under these conditions can be expected to be accompanied by extreme feelings of helplessness and guilt. And, if bereavement occurs at a time when the person is confronted by important tasks, or when there is a necessity to maintain the morale of others (conditions likely in space), that person may postpone normal grief reaction for a prolonged period of time (Lindemann, 1977).

Death aboard the spacecraft— A death aboard the spacecraft can be expected to have a profound effect on the crew. Yet, as spaceflight continues to expand, it seems likely that a death will eventually occur away from the home planet. When this happens, crewmembers will be required to deal with both the physical and the psychological demands of the situation. If the person who dies is central to the operation, his or her passing can be expected to have dramatic repercussions throughout the system. In addition, there will be the factor of grief. Although one might expect that the grief associated with the loss of a crewmate would be less distressing than grief in response to the loss of a family member, this may not be the case. Lindemann (1977) points out that the strength of a grief reaction is directly related to the intensity of the interaction with the deceased before death and that this interaction need not have been positive. The interaction of crewmates may or may not be positive, but it is likely to be intense. These factors suggest that the death of a crewmember would be an extremely traumatic event, affecting the surviving crewmembers on many levels, both practical and emotional.

Cohen (1976) provides insight into the aftereffects of such a loss. At the sudden death of one of their members, a therapy group first reacted with shock and sadness. This period was brief and rapidly was replaced by normal group activity. However, the group remained resistant to any activity which related to their deceased member. Three attempts to introduce a new member to the group met with failure. Cohen concludes that the unresolved feelings of responsibility for, and guilt concerning, the death of the lost member were sufficiently strong that the group members were unwilling to risk accepting a new member. A similar rejection could occur in space if a new crewmember were introduced to replace a deceased member.

Homicide and suicide— Chapter I has described how tendencies to act out are dealt with in confinement. Oversimplified, such tendencies are consciously suppressed. When they do occur, they appear in altered forms, such as the "pinging" or taunting behavior observed among members of the submarine Seawolf (Ebersole, 1960), or the loud cries and shrieks displayed during runs on the SAM two-man space cabin (Cramer and Flinn, 1963). Confined groups appear to be painfully aware of the costs of allowing violent tendencies to surface, and seek to avoid confrontation, even if this means curtailing activity. The reverse side of this relationship has also been observed. Mostert (1974), reporting on life on a supertanker, notes that whenever the level of energy was allowed to rise, as during pool games, the activity turned violent.

Much work needs to be done to determine how acting-out tendencies can be dissipated safely in extended spaceflight. However, it seems unlikely that a serious physical attack by one crewmember upon another would occur in any but the most prolonged space missions. (The issue of the special juridical requirements of space is considered in chapter VIII.)

The likelihood that a crewmember would injure himself or herself does seem significant, and the possibility of a suicide occurring on an extended flight should be considered. Breed (1972) provides insight into the characteristics that might help identify a potential suicide. Noting the convergence of his research results and those of Miller (1967) and Miller and Goleman (1970), Breed suggests that the typical suicide victim is a rigid person of high commitment, who, on meeting serious failure, experiences shame and eventually becomes socially isolated. Such individuals, like all committed persons, set high standards for themselves and identify strongly with their goals. However, because they are rigid, they are unable to shift their goals. Since they see society as incapable of supplying needed regulation, they respond with excessive self-regulation. Breed reports that many of the suicide cases he studied were described as overly neat, meticulous perfectionists who gave scrupulous attention to personal tradition. After Neuringer (1964), Breed uses the term "brittleness" to describe the affective and cognitive orientations of these individuals. Failure is identified as a central factor in suicide, with shame the perfectionist's response to failure. Failure-shame act as the precipitating force which mobilizes the person toward the suicide response.

The final component in the suicide syndrome is social isolation. The individuals perceive negative reactions or labeling from those around them. This process is so painful to them that they withdraw from their present associations. These individuals do not seek new contacts, since they believe such contacts will also condemn them. Such persons become isolated and are unable to receive the legitimization and validation necessary for any individual to function. Significantly, this description of the would-be suicide does not rest on the concept of mental illness. Although the individual is depressed, the depression is less important than his or her inability to consider other solutions.

As applied to space, Breed's analysis suggests some selection guidelines. The inflexible, single-goal-oriented individual described is extreme, but is recognizable as a hard-working, self-directed, high achiever. These are qualities that would lead an individual into, and make him or her a valued member of, a space mission. But it is fair to say that, given the pressures of space, these qualities are not enough. For extended spaceflight, individuals must be competent and committed; they must also be capable of dealing with the inevitable failures in a nondestructive manner.

Again, the possible compounding effects of alcohol and drugs must be considered. It is perhaps significant that the one homicide that has occurred in recent history among isolated teams developed from a dispute over the distribution of wine (Time, Sept. 28, 1970). Excessive alcohol use has been related to depression, with high rates of suicide being recorded by alcohol abusers (Zimberg, 1976). Among drugs, barbiturates are the most commonly employed agents in suicide attempts (Robbins, Angus, and Stern, 1976).

It is axiomatic that a threat of suicide probably means serious intent. There are other signs that a person is contemplating or has decided to make a suicide attempt. Among these signs are making plans to give away or actually giving away valued possessions, and the sudden and unexplained onset of calm in an individual who has been distraught. Space travelers should be taught to recognize the significance of these behaviors.

Crisis Intervention

Crisis intervention techniques seem particularly well suited for dealing with the immediate responses to crises that could arise in space. (For a description of the theory and methodology of crisis intervention, see Aguilera and Messick, 1974.) Crisis intervention has evolved from the community health medicine field and therefore is structured to operate in an environment with limited resources. Crisis intervention recognizes that there are patterns of behavior in many crisis situations, and focuses attention, not on the psychodynamics of the person, but rather on helping the individual to deal with the immediate problem.

Steps in crisis intervention are

1. Assessment of the problem and the person's response to it. Although the cause of the individual's distress may be obvious, the therapist must take into consideration the likely responses of the individual.

2. Planning of therapeutic intervention. At this stage the therapist explores with the individual the disruption to the person's life and the lives of others, as well as what strengths, coping skills, and supports the individual may be able to bring to the situation.

3. Intervention. Here, the therapist aids the individual in understanding the event, in expressing his or her feelings about it, in exploring and testing coping strategies, and in replacing some of the roles disrupted by the loss.

4. Reaffirmation of the progress made. The final stage reviews the strategies that have been successful and helps the individual plan for the future (Morley, Messick, and Aguilera, 1967).

The goal of crisis intervention is a rapid return of the individual to a state of equilibrium. The length of time usually required for intervention is about 4-6 wk (Jacobson, 1965). Most important to the requirements of space, this approach to crisis intervention is designed to be carried out by non-health-professionals. Some or all crewmembers could receive intervention training as part of their overall orientation.

Implications for Space

A crisis in space could arise from any of a number of psychological and/or relational disruptions occurring within the spacecrew. Of the various psychological factors that could menace the integrity of the crew, depression poses a particular threat. Of those factors that could lead to a depressive response, the relationship of the individual to the group appears the most volatile. Being ostracized from a group that is itself isolated can be expected to result in severe, and potentially unbearable, pressures on a space traveler. Ostracism must be actively planned against. One suggestion is the use of a "buddy" system, where each individual is assigned a partner whose responsibility it is to understand that person's perspective and to defend him or her, if needed, to the larger group. However, the more important need is to understand more fully the factors that contribute to the occurrence of ostracism.

Closely related to the issue of psychological well-being is the question of how well an individual performs a work assignment. Even in the enriched environment of Earth, the relationship of an individual to his or her work takes on special significance (Breed, 1972). Rushing (1968), Maris (1969), and Richman and Rosenbaum (1970) all have noted the primacy of job failure in the syndrome of suicide. In isolation, work takes on an added significance. We can reasonably expect that if job-related failure should occur in space, it would place a severe strain both on the individual and on his or her relationship to the rest of the crew.

SUMMARY AND CONCLUSIONS

The occurrence of an accident in space can be expected to result in fear reactions. Research on responses to fear suggests several questions of relevance to extended spaceflight. Although the various measures of fear response (physiological, self-reports, and performance) are positively correlated, the precise relationship among these measures is not well understood. The dynamics of the fear response itself also deserve attention. Several authors (Noyes and Kletti, 1976; W. Smith, 1966) have found a lack of understanding or a denial to be the first response to threat. We need to understand how prevalent such denial is, the time dimension associated with denial, and the relevance of training for reducing this period of inactivity.

Fear arousal follows prescribed patterns. In general, good performance has been associated with early arousal; that is, arousal in advance of the threatening event (Fenz and Epstein, 1967, 1969; Fenz and Jones, 1972). From these data Epstein (1967) has postulated a generalized pattern of emotional response in which inhibition rises steeper than excitation, allowing the person to be in control when he or she must deal with the threat.

An hypothesis has been advanced that fear arousal is at a maximum when a decision must be made. Another hypothesis might also be considered. Some data suggest that fear responses are most pronounced at those points when something is likely to go wrong, whether or not there is anything one can do about it. For instance, we know from the work of Fenz and his collaborators that the arousal of experienced jumpers is very low just before they emerge from the aircraft. Yet Ouchida (in Levanthal and Lindsley, 1972) has found that experienced jumpers report themselves to be just as fearful as novice jumpers in the few seconds before their parachutes open. Just how peak fear relates to personal decision or transition points needs further evaluation.

Experiments on sports parachuting show that experience is an important predictor of both fear response and performance. However, the relationship between experience and training has not been determined. Other training questions concern the value of generalized stress training, the feasibility of training individuals to direct their attention outward or inward (e.g., to external or internal events), and training individuals in communication and coordination skills to function in a crisis situation.

The role of personal characteristics such as birth-order and gender eventually may prove useful in understanding reactions to severely threatening situations. However, available birth-order data are not sufficiently consistent to have predictive value, and gender studies have provided only preliminary conclusions.

Not all methods of dealing with stress are equally beneficial. It has been shown that blaming or scapegoating is a common method of dissipating stress after a crisis has occurred. Yet, blaming in the space environment could be costly. The first step to control this behavior is to determine the conditions under which such behavior occurs in isolated and confined groups, and the forms this blaming behavior takes. Conversion reactions such as shared illnesses are other undesirable methods of relieving tension following a crisis. To aid the selection process, one might explore the relationship between conversion reactions and personal styles such as general rigidity or an inflexible approach to work obligations.

Research involving realistic threat situations often has ethical implications. Over the years, attitudes about this type of research have shifted markedly toward emphasizing the rights of test subjects. Although few would argue with this development, it is doubtful that future research involving fear responses will reach the levels of realistic simulation that has occurred in the past. Future studies of fear reactions will be restricted either to manipulations of low levels of fear, or to piggyback observations of events in which fear is a naturally occurring byproduct. In order to contribute to the research needs in this area, we should make ourselves aware of those situations in which information about fear responses can be gained without subjecting the individuals involved to additional stressors.

One of the broad challenges of the future is to begin to understand which factors will foster psychological health in the space environment. From the selection perspective, we need to identify those characteristics that are particularly well suited for dealing with crises, as well as with other conditions of space. Tests which could help identify the highly stable individual would be useful in reaching this goal. We also need to identify, and find methods of dealing with, those situations which could precipitate an on-board crisis. In this regard the work role commands particular attention. Since there are few ways of diffusing the potentially devastating effects of failure in space, emphasis must be placed on assuring that every member of the crew can succeed. In this regard we need to examine both the advantages and the disadvantages of work role specialization, and, more broadly, to understand the range of skills, both personal and professional, an individual must have to be integrated successfully into a crew.

We can assume that psychological health relates to how the individual deals with stress. Andrews et al. (1978) have shown stress reactions to be related to three factors: personality characteristics, patterns of coping, and the availability of social support. The personality characteristics they suggest as pertinent (anxiety proneness, self-esteem, and expectations of control), among others, should be examined. However, both patterns of coping and social support are concepts that, though imperfectly understood, may be more fruitful avenues of investigation. What constitutes effective patterns of coping in the space environment? What are the social supports available in space? And, what are the factors associated with quality social support?

It is possible that, in spite of efforts to promote all aspects of health, psychological incidents could occur in space. In treating such incidents, we need to understand the effects of pharmacological agents in the weightless environment. More importantly, we need to explore nonpharmacological options. Particularly intriguing are the possibilities of group therapies aboard the spacecraft and remote counseling between the spacecraft and the ground. The latter poses questions concerning the adequacy of various mediated systems (audio, audiovisual, computer) in terms of feasibility, efficacy, privacy needs, etc. (see chapter VI).

In terms of personal techniques for maintaining emotional balance, the work of Schwartz et al. (1978) suggests some interesting possibilities. Basically, these authors describe two types of anxiety response — somatic and cognitive. These investigators found that people who exercise have lower somatic anxiety than people who meditate (although it is unclear whether somatic anxiety is reduced or cognitive anxiety elevated for exercisers). Follow-on studies should include all groups (those who exercise and meditate and those who do neither) and should explore how factors which relate to the mode of anxiety influence the choice of activity. Nor is it clear that the two modes identified constitute the most significant split. Schalling, Cronholm, and Asberg (1975) have found psychic (cognitive) anxiety to be more prominent in introverts than in extroverts, whereas somatic anxiety did not distinguish along this personality dimension. This finding suggests that there may be other equally valid ways to view anxiety. The exercise-meditation data are intriguing, since they suggest a relatively direct problem/therapy relationship. However, we are far from the point where we can confidently prescribe a particular kind of activity to relieve a particular form of anxiety.

Other events that could trigger a crisis within the crew are transcendent experiences such as the breakoff phenomenon, or drug-induced behavioral change. Significant questions include the special role of space physiology in perceptual and behavioral changes. The death of a family member or crewmember could precipitate a crisis. In the area of grief management, we need to be sure that selection procedures exclude potentially pathological grievers. A suggested area of research toward this end would include an investigation of the relationship between phobic behavior and grief avoidance.

A question of particular relevance to the mental well-being of space travelers concerns how much they should be told of negative events involving relatives and friends at home. Although this question has surfaced regularly, it has never been addressed in a systematic fashion. When incidents arose during Polaris submarine cruises, the decision to inform or withhold information from a crewmember was left to the ship's captain. This solution was judged not altogether acceptable by the crewmembers affected (Weybrew, 1980, personal communication). A decision of this kind already has been required in space. On Soyuz 6, Commander Romanenko was informed by the visiting crew that the father of his companion, Cosmonaut Grechko, had died. Romanenko decided to keep the news from Grechko until they were safely back to Earth (Oberg, 1981). No mention is made of Grechko's later response on learning of this decision. Intuition suggests that the duration of a mission should be an important variable in the decision to inform or to withhold information. If a mission is relatively brief, crewmembers might be willing to suspend their right-to-know, assuming the probability of a serious negative event to be acceptably small. However, if crewmembers were to be away for long periods of time, even negative information might be preferable to ignorance. Just what is acceptable to crewmembers, and where along the time continuum the shift occurs (assuming it does) needs to be determined.

The discussion presented here is necessarily ground-based. It uses our experiences on Earth, modified by our guesses about how space would influence our expectations, to anticipate particular problems which might arise, treatments which could be useful, and questions which suggest themselves as important. This approach should be supplemented, whenever and to whatever extent possible, by careful observation of the space experience.

Artist: Mitchell Jamieson

VIII

ORGANIZATION AND MANAGEMENT

INTRODUCTION

An understanding of small group dynamics is essential for planning a nonsolitary mission of any size. However, as crew size increases beyond that point in which each crewmember can interact with each and every other crew member on a face-to-face basis, additional, higher-order social variables become salient. Interpersonal relations are supplemented by intergroup relations, and organizational-level phenomena must be taken into account. For present purposes, organizations may be defined as large, complex social entities which are composed of interdependent subgroups. In the present chapter, we consider means for motivating and coordinating the space organization's members and subgroups to achieve overall mission goals.

Relatively small crews can achieve the necessary degree of coordination through professionalization. This involves identifying a select group of individuals whose particular blending of abilities, motives, and interests provide the right combination for the exhaustive training associated with the profession (Perrow, 1979). As a result of years of formal and informal preparation, the recruit acquires a backlog of knowledge and the skills deemed appropriate for the profession. After overcoming a series of hurdles culminating in a major *rite de passage* (receiving pilot's wings; being introduced to the press as a trained astronaut, etc.), the professional is deemed capable of exercising independent judgment while remaining coordinated with coworkers.

Different members of a profession share a body of knowledge, expectations, and common skills (Perrow, 1979). When confronted with a particular problem or situation, the professional's training permits rapid comprehension of the problem, retrieval of relevant knowledge, speedy identification of admissable courses of action, and prompt response. If teamwork is required, years of training have taught each person his or her place. Thus, for example, small teams of astronauts can work effectively with only minimal communication. Under the professional mode of organization, learned expectations coordinate different people's activities.

As crew size expands it will become increasingly difficult to achieve the necessary degree of coordination through professionalization, for this mode of organization will become too expensive, too time-consuming, and too unwieldy. Truly large-scale missions will necessarily depend upon modes of organization which require less selectivity in choosing astronauts and less extensive training than is presently the case. Some of the burden for social organization must shift from internalized rules common to members of a profession to external social constraints.

Traditionally, a distinction can be drawn between the planners and managers who design a space-capsule social system, and the crew that puts the system into operation. Although this distinction is convenient for purposes of discussion, design and implementation are not necessarily separate. Weick (1977) notes that since the processes of design and implementation tend to clarify and elaborate one another, the implementers (in the present case, the spacecrew) should be heavily involved in both the formulation and modification of plans. From Weick's perspective, organizations may be treated as self-designing systems, or systems in which users contribute to the design or redesign.

There are two major advantages to self-design. Although of value in any setting, these advantages become pronounced when time or distance separates the planners and implementers. First, self-design ensures that the implementers' abilities, needs, and perspectives are taken into account. Through self-design, idiosyncratic characteristics become an integral part of the overall social system. Second, self-design facilitates prompt and appropriate response to unanticipated or changing conditions.

Discussing the value of self-design, Weick cites an incident involving the overprogramming of astronauts' time by means of a relentless barrage of computerized instructions. (This incident will be described fully later in this chapter.) The pace set by the computer proved difficult or impossible for the crew to maintain, and distracted them from activities which did not appear on the computer's list, but which should have been given high priority. Weick ventures that general instructions coupled with a list of desirable projects would have better accommodated the crew's personal requirements and the efficiency with which they attained various mission goals. Weick also notes that even if the crew had displayed enthusiasm for carrying out computerized instructions, latitude for discretionary action would have remained important. Otherwise, the crew's progress could have been halted by a computer malfunction.

The basic goals of self-design are a system which is *pliable* in the sense that it form-fits its members; *responsive,* in the sense that it can react appropriately to changing conditions; and *proactive,* in the sense that it contains the seeds for self-development or improvement. Although there are no hard and fast rules for establishing self-designing systems, there are certain general guidelines. These include (1) blurring the distinction between planners and implementers; (2) incorporating latitude for improvisation; (3) devising mechanisms which the crew itself can use to solve problems and improve its own procedures; (4) encouraging crewmembers to broaden their sights to encompass an expanding array of alternatives and options; and (5) promoting questioning attitudes.

SPACECREW STRUCTURE

Organizational structure refers to social influence patterns and social rules which help determine crewmembers' typical reactions to one another. Organizational structure is external to the individual, and at a high level of abstraction. Structural factors constrain individual behavior. They encourage dependability in performance, and, according to classical theory, make it possible for one person to substitute for another without a major disruption in the functioning of the whole (Weber, 1947; Graen, 1976; Perrow, 1979).

Structure is reflected in the distribution of authority and tasks. Structure defines organizational positions and specifies their relationships to one another. Associated with each position is a social role; that is, a set of expectations that define certain behaviors as essential, certain behaviors as admissible, and certain behaviors as unacceptable.

Power Structures

Power structures refer to the social influence patterns that regulate individual conduct and have the potential of coordinating activities within and between groups. The multilevel hierarchical or pyramidal form of power structure is the most common form in large governmental and industrial organizations today. This is evident in the military or paramilitary command structures which have typified past space missions and which may carry forth to future missions as a matter of convenience, tradition, or choice. A number of common assumptions support the perpetuation of this form: (1) there must be a strong advocate of the sponsoring agency's interests aboard; (2) only a single individual's decision could be made fast enough to cope with certain situations; (3) crewmembers would prefer this type of arrangement because they are used to functioning in hierarchies; and (4) the average crewmember will easily understand a form of organization that approximates one that he or she has frequently encountered on Earth. These assumptions, however, are open to question. Not all potential crewmembers will be used to functioning around the clock in formal hierarchical structures; autocratic decisions, although speedy, may be resisted by the group; and a form of organization that is readily understood does not necessarily mean that it will be readily accepted. Thus, after reviewing some of the implications of the multilevel hierarchical control model for future space missions, we will consider two alternatives: the community democracy model, and the labor relations model (Whyte, 1967).

Multilevel hierarchical control model— Hierarchical control structures require that individuals holding positions at one hierarchical level be able to direct or influence the activities of individuals at lower hierarchical levels. Social power refers to the maximum influence that person A is capable of exerting on person B. Multilevel hierarchical control structures may thus be conceptualized in terms of the differential distribution of social power such that persons at each level have greater power than do persons at successively lower levels.

Bases of social power— According to French and Raven (1960), there are several bases of social power. Each of these bases may be eroded under conditions of extended-duration spaceflight.

Legitimate power arises from social norms and internalized values which dictate that person A is entitled to influence person B and that person B is obligated to accept influence attempts by person A if these attempts fall within a prescribed range of conditions. As usually conceived, this form of power is based on person A's and person B's mutual acceptance of the organizational norms and values. On a space mission, legitimate power may be maximized through (1) a careful specification of authority relations; (2) the creation of a strong value system within the sponsoring agency; and (3) the use of selection and training procedures which ensure that the individual astronauts subscribe to the organization's value system. Prolonged separation from Earth, however, may undermine the expectations and prescriptions that provide a basis for legitimate power. As Haythorn (1970, p. 164) notes:

> Leadership and other behavioral patterns are clearly determined to some degree by role expectations and behavioral prescriptions of larger segments of organizations and societies than are fully represented in the small group. When groups are isolated from contact with (such larger segments) these prescriptions and expectations cannot be as frequently and strongly reinforced as they normally are . . . . Leadership under such circumstances is unable to rely as strongly on formal role relationships and must depend more on the individual capabilities of the men to whom leadership is assigned.

Uncovering the means for mitigating loss of legitimate power under conditions of extended-duration spaceflight is of primary importance. One type of research which could contribute to this area would focus on techniques for maintaining crew allegiance to the Earth-based authority and the formally designated leaders. For example, social rituals, tangible reminders of allegiance, and symbols of rank might accomplish this. Other research might examine the feasibility and consequences of establishing legitimacy through means other than appointment by the Earth-based authority. For example, if legitimacy were controlled by the crew, legitimate social power would flow from proximal rather than distal sources, a change which could circumvent some of the problems identified by Haythorn (1970). Indeed, some findings suggest that leaders with internally conferred legitimacy have an advantage over leaders with externally conferred legitimacy. Comparisons of elected and appointed leaders suggest that the former are perceived as more influential and more qualified and are less likely to promptly lose influence following failure (Hollander and Julian, 1970, 1978). Thus, it will no doubt prove useful to select future spaceflight managers from among the ranks of former crews, perhaps even through the nomination and/or vote of the crew itself. A third possibility for future research involves systems under which legitimacy flows from both internal and external sources.

Reward power and coercive power refer to person A's abilities to affect person B's outcome or level of satisfaction. Such power is based upon the control of rewards (such as pay increases, recognition, and advancement through the hierarchy) and punishments (such as fines, censure, and termination). Although on-board managers can be invested with reward and coercive power, these forms of social power may also be undermined as Earth becomes remote. For example, an augmented or docked paycheck may lose significance in a microsociety where regular currency is not used, and it is difficult to "fire" someone who cannot be replaced and who has no place to retreat. The problem is thus one of finding incentives and disincentives that are effective under conditions of isolation, confinement, and risk, and ensuring that these remain under the control of legitimate occupants of leadership roles.

Expert power flows from competence. Person A is said to have expert power over person B to the extent that person B perceives person A as an exclusive possessor of superior knowledge or skill. Leonov and Lebedev (1975) believe that astronauts must have total confidence in their leaders' expertise. To help achieve this, astronauts can be chosen for important positions on the basis of their expertise, or invested with expert power by means of special training programs. In addition, steps can be taken to help experts maintain and enhance their skills under conditions of extended-duration spaceflight. For example, computer simulations of systems dysfunctions can allow experts to exercise their problem-solving skills, and telecommunicated courses and teleconferenced or on-board seminars can contribute to professional development.

Although expert power is considered fairly robust (Katz and Kahn, 1978), it, too, may be eroded under conditions of extended-duration spaceflight. First, a certain amount of cross-training may be necessary to prepare back-up personnel. Since expert power implies an exclusive access to knowledge, cross-training will serve to dilute it. Second, McCarthy (1979) hypothesizes that leaders risk loss of expert power whenever they make foolish statements, even when such statements regard matters that fall outside of their acknowledged area of expertise. Presumably, the risk of a display of ignorance is increased in isolation and confinement, conditions which may promote protracted superior-subordinate discussions on far-ranging topics. Self-imposed isolation on the leader's part may be too terrible a price for the leader to pay. Thus, a major problem is one of finding ways for leaders to maintain credibility without remaining highly isolated from the rest of the crew.

Additionally, there is referent power, or power that is based on liking or identification with another person. A prerequisite for referent power is that the leader have "membership character" in the group; that is, he or she must be perceived by the other astronauts as "one of us." Crew leaders can be chosen on the basis of their appeal, and training procedures may increase their attractiveness. However, as missions progress, leaders may have to take unpopular actions that would reduce their overall level of attractiveness. Furthermore, in accordance with the reinforcement model of interpersonal attraction (Byrne, 1971), it is hypothesized that the negative affect generated by isolation, confinement, and risk may become associated with the leader, again to the detriment of the leader's referent power. Once again, then, there is a potential loss of social power under conditions of extended-duration spaceflight. In this case we must look for ways in which leaders can inspire continued affection from the crew under conditions of isolation and confinement.

In summary, legitimate power, reward and coercive power, expert power, and referent power make it possible for individuals at one level of the organization to coordinate and direct the activities of individuals at a lower level of the organization. Our research task is to understand the causes and the rates of erosion of each form of power under conditions of extended-duration spaceflight, and to identify those techniques which could reverse or retard such erosion.

Centralization of authority— Authority is centralized to the extent that decisionmaking power is concentrated within the hands of one or a limited number of individuals. In a highly centralized authority structure, managers retain decisionmaking power and expect subordinates to operate within relatively narrow latitudes. In a highly decentralized authority structure, managers delegate power downward to lower-level managers and expect subordinates to demonstrate initiative and independence.

The optimal degree of centralization of authority depends upon a complex array of interrelated factors (Wexley and Yukl, 1977). For the most part, centralization is desirable to the extent that (1) there are good communications with subunits; (2) higher-level managers have access to staff specialists or other information sources which are not readily available to lower-level managers; (3) higher-level managers have the same knowledge of local conditions as do lower-level managers; and (4) decision speed is only a minor consideration.

As missions become more remote from Earth, involve more people, and last longer, decreased centralization may gain the advantage. Under tomorrow's spaceflight conditions, communication with Earth may become erratic or delayed; the crew is likely to contain an increasing number of staff specialists; and the crew commander's knowledge of local conditions will surpass that of Earth-based personnel located hundreds of thousands or millions of miles away. In addition to promising efficiency under tomorrow's spaceflight conditions, decentralization promises two long-range advantages (Perrow, 1979). First, it will free the time of higher-level leaders on Earth to deal with long-range issues. Second, it will help train or prepare lower-level leaders aboard the spacecraft for higher-level responsibilities. The alternative to decentralization, maintaining a high degree of central control, requires a solution to communication problems and ensurance that common or superordinate goals will be maintained by both flight and ground personnel.

Hierarchical positions and leadership tasks— To some extent, organizational charts and rule books make it possible for selected individuals to lead. But structural factors themselves cannot ensure an optimal level of organizational functioning (Katz and Kahn, 1978). Such factors must be supplemented by the exercise of judgment and the application of human-relations skills. By themselves, organizational charts and rule books do not make adequate allowance for (1) the juncture between the organization and the (external) environment; (2) changing environmental conditions; (3) internal dynamics and development; and (4) the needs of participants to be treated as unique individuals.

On missions involving few people, there may be only one formal leader. On missions involving scores or hundreds of people, there may be many leaders ranging from low-level supervisors to the person in overall command. The extended discussion of small-group leadership presented in our group dynamics chapter is certainly pertinent to leadership within the organizational setting. But, suggest Katz and Kahn (1978), the demands placed upon leaders change as a function of hierarchical level. Low-level supervisors, middle management, and top-level leaders require different intellectual and human-relations skills.

Lower-level leaders work within the framework of preexisting policies and procedures. The intellectual requirements are technical skills (for example, those regarding equipment use or organizational procedures) and the human-relations requirements (concern with equity and fairness when rewarding and sanctioning subordinates).

Intermediate-level leaders supplement and piece out preexisting policies and procedures. The intellectual requirements are an understanding of the subsystem and a two-way perspective which encompasses superiors and subordinates. The human-relations requirement is an ability to integrate primary and secondary relationships; that is, to deal with people as unique personalities and as role occupants.

Higher-level leaders create policies and procedures. The intellectual requirement is an understanding of the system as a whole and its relationship to interfacing systems. The human-relations requirement is charisma; that is, a magical aura or halo which is derived from followers' emotional needs, and from bold and imaginative acts of leadership.

Competent leadership at the top level will become increasingly important as missions encompass ever-increasing spans of time. The authority structures, roles, norms, and other organizational variables which are set prior to mission departure may or may not fully withstand the passage of time. Although dramatic events could force reorganization shortly after a mission's departure, and space settlements could preserve traditions for generations, we anticipate that the status quo will be preserved on relatively brief missions, but that organizational change will be likely on missions which involve extended periods of time. Future top-level leaders would thus do well to master the techniques of organizational development; that is, planned organizational change (French and Bell, 1978).

Isolation, confinement, and risk are expected to increase the demands on leaders at all levels. As mission planners and managers undertake leadership selection and training, consideration might be given to the intellectual and human-relations requirements of each leadership level. For example, with respect to selection, one assessment device might examine candidates' tendencies to adopt individual, subsystem, or systems perspectives. Such a test might consist of case histories that can be interpreted at different levels.

Another research issue involves the effects of physical remoteness on the performance of top-, intermediate-, and low-level leaders.

Leadership level and communications media may have interactive effects on leadership effectiveness. For example, a physically remote low-level leader, who communicates with subordinates via teletype, may find it difficult to convey a high level of social support to subordinates, and may hence find it difficult to gain their cooperation. Alternatively, a physically remote high-level leader who is rarely or never actually seen by subordinates may find it relatively easy to preserve charisma.

Community democracy and labor relations models— According to Whyte (1967), certain organizational models, such as the multilevel hierarchical or bureaucratic model, are so common that it becomes difficult or impossible to envision alternatives. Yet no single model will prove optimal for all purposes. The bureaucracy has many strengths (Perrow, 1979), but it also has many weaknesses. For example, in many multilevel hierarchies, downward communication is impaired because leaders at each level feel compelled to pass messages through subordinates rather than directly to those for whom the message is intended. Upward communication is impaired, not only because such messages also pass through many levels, but because they may be distorted to achieve compatibility with subordinates' perceptions of their superiors' desires. Within bureaucracies, decisions may be seen as imposed from above, and hence fail to be endorsed or implemented by the organization's rank and file. In space there will be another problem: each crewmember will be so expensive to maintain that extremely careful consideration will have to precede the addition of yet another bureaucratic level.

Perhaps the best-known alternative to the bureaucratic model is the community democracy model (Whyte, 1967). According to this model, each member of an organization participates in the decision-making process. Participation may be organized along one-person-one-vote or discussion-to-consensus lines. Under this model, authority is deemphasized and negative sanctions may be applied to people who fail to make appropriate inputs. This model rarely prevails in truly large social units, but is closely approximated in the New England town meeting (Whyte, 1967).

The community democracy model is consistent with American democratic ideology, and the decisions that result, at least when reached through consensus, are likely to be accepted and implemented by the membership. However, the community democracy model is difficult to apply when the group is large and when it is necessary to reach prompt decisions. Not every member of a large group can make informed and useful inputs; not every member of a large group cares to be consulted on each and every issue.

The community democracy model has particular disadvantages when it is bogus or only partially applied. Sometimes, for example, the group is allowed to reach minor decisions but not allowed to vote on "hot" or consequential issues. Sometimes, the rank and file is openly channeled toward a certain decision, or discovers later that its efforts have been negated or reversed by a higher authority. Such misapplication of the community democracy model wastes time and energy, invites disillusionment and cynicism, and undermines morale.

An alternative to the bureaucratic and community democracy models is the labor relations model (Whyte, 1967) which openly combines both authority and participation. One group, the managers, has the power to make and enforce decisions. The other group, the rank and file, has the right to suggest changes, influence decisions, and protest decisions that have already been made. The managers and the rank and file take each other seriously, and bargaining and negotiation surround the decisionmaking process. On the one hand, this model acknowledges conflicts of interest and, according to Whyte, makes it possible for such conflicts to be dealt with in realistic ways. For example, in the labor relations model the rejection of the rank and file's demands invites protest, but not disillusionment or cynicism. On the other hand, there is always the risk that the conflicts inherent in this form of organization will exceed acceptable limits.

Each model has some advantages and disadvantages. The point is that no one model is necessarily the best for all future missions, and that rather than relying on the best-known models, planners and managers would do well to carefully evaluate both familiar and novel alternatives. The crew size and the nature of the crew (e.g., military or civilian) are among the many variables that need to be taken into account.

Work Roles

Kanas and Fedderson (1971) observe that four types of task or work roles would seem to be important on missions of any appreciable size. These roles are here labeled flight-operations roles, scientific-investigative roles, environmental-support roles, and personnel-support roles. To this list should be added a fifth type of role, the production role, which will become important on certain types of missions. On small missions, we have seen "doubling up," when a given individual performs more than one work role. On future missions, expanded crew size may give rise to the option of having many crew members able to perform each type of role.

Flight-operations roles involve command, navigation, flight engineering, systems monitoring, and telecommunications. Historically the first to develop, such roles remain essential to any space mission. Thus far, commanders have been drawn from among flight operators, but this might not always be the case (Helmreich et al., 1980). That is, during these early years of spaceflight, all other considerations are insignificant when compared with the problem of getting there. In the future, "getting there" may become routine and top leaders may require skills which are useful to attain other mission objectives. Thus, the top leader of an orbiting work station may be a professional manager rather than a pilot.

Scientific-investigative roles involve research tasks. People who perform such roles are expected to generate new data that have relevance beyond the immediate flight. Although perhaps not essential for any one mission, in the aggregate scientific-investigative roles help justify space exploration.

Environmental-support roles involve the management of supplies and the maintenance of facilities. Environmental-support roles are essential for a mission, and on small missions they are easily combined with flight-operations functions. On truly large missions, people who perform such functions might range from the equivalent of a quartermaster to crewmembers who perform very routine facilities-maintenance tasks.

Personnel-support roles include promoting physical and mental health, attending to the psychological needs of crewmembers and attempting to maintain the morale of the crew as a whole. Astronauts who fill such roles will serve what Katz and Kahn (1978) refer to as the maintenance function. On small missions, personnel-support requirements may be met by a physician who is well trained in interpersonal relations (Kubis, 1972). Large missions may involve an elaborate personnel-support system that includes physicians, psychologists, and personnel administrators.

Production roles will be prominent on missions that pursue explicit industrial or economic goals. Included in this category would be roles related to the manufacture of products in space and the exploitation of energy or other resources. On certain future missions, the proportion of crewmembers who perform production functions may be larger than the proportion of people who perform all other functions combined.

Most writers assume that crewmembers should perform well-planned and well-learned work roles (Berry, 1973; Helmreich, et al., 1979a; Helmreich et al., 1980; Kubis, 1972; Leonov and Lebedev, 1975; Natani and Shurley, 1974; W. Smith, 1966). Expressing the views of many authors, Berry (1973a, p. 1142) notes:

> ... in order to ensure that crew interaction is orderly and does not become a source of friction, roles must be strictly defined. The assignation of specific roles has been a feature of past space flight missions and will be a feature of future ones. As space crews grow larger and mission length increases, organizational structure will become even more important....

Among the continuing concerns of mission planners and managers are (1) defining articulated task roles which collectively satisfy the mission's behavioral requirements, and (2) ensuring that each crewmember learns and performs his or her role. Although the careful specification of roles is often a necessary and desirable goal, there are reasons to hypothesize that this goal should be pursued with some moderation. Succumbing to the temptation to prescribe roles to the smallest detail could have at least three adverse consequences. First, the sharp definition of roles can discourage functional as well as dysfunctional out-of-role behaviors. (For example, Leonov and Lebedev (1975) report that two Antarctic explorers refused to help put out a fire because it "wasn't part of their job.") Second, stringent behavioral prescriptions for individuals can generate psychological reactance; that is, threaten the person's sense of freedom and autonomy and provoke grumbling and defiance (Brehm, 1966). Third, the minimum level of performance that is prescribed tends to define the maximum level of performance that is obtained (Katz and Kahn, 1978). Further definition is required to identify the optimal degree of role specification over a range of space missions.

Prescribed and emergent roles— Roles may be imposed by higher authorities (prescribed roles) or they may arise less formally in the course of social interaction (emergent roles). Emergent roles typically complement and supplement prescribed roles, but occasionally the two will conflict. Four conditions encourage emergent roles (Katz and Kahn, 1978); the first two of these conditions will be particularly pronounced in space.

First, prescribed roles may not make adequate allowance for people to satisfy their personal needs. For example, well-being requires some behavioral variety. Rigid role prescriptions limit behavioral options. Conditions of isolation and confinement can intensify this problem, since, under any circumstances, crewmembers will have a limited opportunity to enjoy social variety.

Second, prescribed roles are devised on the basis of the anticipated and known. They are likely to prove inadequate in light of the unanticipated or unknown. For example, a polar expedition described by W. Smith (1966) was intended to have highly defined work roles, with each person assigned specific duties such as driving or taking geological measurements. However, an icy blast through the SnoCat's floorboard made sustained driving unbearable, and the geological measurements were extremely difficult for one person to obtain. This led to a breakdown of preassigned duties. Men took turns driving, and offered the geologist assistance. Because spacecrews, like other adventurers, are likely to encounter the unanticipated or unknown, emergent roles become likely.

Third, roles are prescribed on the basis of conditions that are in effect at a given point in time. A change in conditions may make certain roles obsolete, force the rapid development of new roles, or require one person to perform another's functions. For example, the death or disablement of a crewmember could force another crewmember to assume the deceased crewmember's duties. To cover such contingencies, some degree of redundancy must be associated with key roles.

Finally, prescribed roles cannot fully cover even routine situations (Perrow, 1979) For example, crewmembers may discover that they have to respond to gradations and distinctions which the official plan ignores. To illustrate, a flight-operations officer and a personnel officer may enjoy the same official status or rank. Yet the latter may enjoy greater informal influence if he or she is responsible for performance evaluation or for the assignment of duties.

Defining space crew roles, then, requires a sensitivity to socioemotional as well as to task requirements; the inclusion of options and alternatives to cover unanticipated, unknown, and changing conditions; and an acceptance of the fact that not every contingency can be fully covered. Useful in this regard would be procedures for identifying those situations for which predefined roles are necessary, those situations for which predefined roles are desirable, and those situations which need not be covered by predefined roles. Another important research topic is identifying the conditions under which emergent roles are likely to complement and supplement (rather than undermine) prescribed roles. We hypothesize that such incompatibilities are likely to be minimized when (1) selection and training yield a crew with a high degree of commitment to official structures, prescribed roles, and overall organizational goals; (2) prescribed roles are flexible in the sense that they can be adjusted in response to changing conditions; and (3) prescribed roles provide latitude for discretionary activities and the development of voluntary socioemotional relationships.

Role overload— Role overload exists when a position carries a combination of tasks that are sufficiently burdensome that they cannot be competently performed without undue stress. Attempts to adjust to overload, such as omitting certain tasks or back-logging tasks, result in inefficiency and a lowering of performance standards; in addition, overload increases wear and tear on the individual performing that role (J. G. Miller, 1960, 1978). Thus, care must be taken to ensure that individual crewmembers are not overloaded.

Combinations of roles, in particular, may prove especially burdensome for individual crewmembers because the extent of the duties associated with one role may not be fully taken into account when the other role is formulated. Managers may underplay the burdens of each role in an attempt to induce crewmembers to accept the combination. Thus, crewmembers who accept multiple roles may not be fully aware of the demands that each role will place on their energy and time. If aware of potential overburdening, crewmembers may nonetheless accept multiple roles because of a high need to achieve, a "can do" self-image, real or imagined long-term career considerations, or a lack of assertiveness to decline. Managers may be sorely tempted to assign heavy combinations of duties to each crewmember, because weighted against the addition of new crewmembers are engineering considerations and immense costs.

Overload conditions appeared in the course of Spacelab Mission Development Test III (SMD III). This project was a ground-based simulation of a series of life sciences experiments conducted by Ames Research Center and Johnson Space Center. Personnel from Ames Research Center planned and prepared experiments which were then delivered to Johnson Space Center, where they were carried out in the context of a 7-day spaceflight simulation (Helmreich et al., 1979a).

For many of the people involved, the addition of heavy SMD III requirements to normal responsibilities resulted in an onerous workload. First, commitment to SMD III made it difficult to fulfill administrative and other routine obligations at the two Centers. Second, principal investigators often felt frustrated, for their commitment to SMD III made it difficult to carry on with their usual activities. Third, administrative duties made it particularly difficult to complete high-quality scientific research; that is, the combination of managerial and scientific-investigative roles proved particularly burdensome (Helmreich et al., 1979a).

Certain procedures may help minimize the problems associated with role overload. For example, Helmreich et al. (1979a) suggest that some of the difficulties that managers and workers encountered in the course of SMD III could have been reduced by (1) ensuring that both had a clear understanding of the extent of their multiple obligations; (2) ensuring that both set appropriate priorities among multiple obligations; (3) improving communication among occupants of adjacent roles; and (4) minimizing people's exposure to irrelevant communications. Despite such progress, important topics for future research include more definitive specification of the conditions that lead to, and ameliorate, role overload during the course of a space mission.

Role-related conflicts— A certain amount of conflict may be expected among occupants of different spacecrew roles. One potential form of conflict, that of role conflict, is associated with expectations and demands centering around the roles themselves. Such conflicts occur when the person's expectations are inconsistent with the expectations of people in related roles, the person is subjected to inconsistent demands from people in different interlocking roles, and the person is assigned two or more roles which prescribe incompatible behaviors.

Role conflict causes frustration, vacillation, and deteriorating performance (Wexley and Yukl, 1977). Although it is doubtful that role conflict can ever be eliminated, careful analysis and planning followed by a high degree of training should help reduce the potential for role conflict. Direct tuition, observational learning, and role-playing are among the techniques that are available for ensuring that crewmembers understand the behaviors expected of them and maintain appropriate expectations concerning each other's behavior.

Another potential source of conflict is the different perspectives of the people who are likely to occupy the different roles. Systematic differences in the personal backgrounds and hence the values and interests of different role occupants may generate on-board frictions (Doll and Gunderson, 1971; Natani and Shurley, 1974). Although the conflicts usually involve minor issues, such as musical selections (Doll and Gunderson, 1971), they may be blown out of proportion as the term of isolation and confinement continues. Perhaps the most serious incident recorded involved sailors on an Antarctic research vessel who threw overboard the scientists' 2-yr collection of biological specimens. The sailors did this to make more room for beverages in the ship's freezer (Bluth, 1981). It is important to learn more about how people with noncomplementary personalities and interests gravitate toward interfacing roles. Remedies may be sought in personnel selection or in role redefinition.

The fact that some roles are more important than others also has the potential for generating conflicts. For example, some of the military test pilots who initiated the conquest of space have expressed nonsupportive attitudes toward the "hyphenated astronauts" who joined to play scientific and other nonpiloting roles (Cunningham and Herskowitz, 1977; Wolfe, 1979). In the future, environmental support roles may be assigned low status by other crewmembers, and line officers who are expected to risk their lives in extravehicular activities might view staff officers as second-class citizens. We might expect crewmembers to see colleagues who have the less important jobs as not fully carrying their own weight, and, over time, the colleagues may lose their self-esteem. Certainly, one task is to identify and establish those conditions under which each crewmember recognizes the other crewmembers' importance. Perhaps the critical variable here is perceived fairness; that is, a conviction that everyone is carrying his or her weight and receives rewards that are appropriate, given the level of his or her contributions (Adams, 1965; Walster, Berscheid, and Walster, 1973; Leventhal, 1976). Planners and managers need to know more about perceived fairness and equity under conditions of isolation, confinement, and risk.

Yet another threat is that people within a role category will form factions or cliques and mark boundaries that discourage constructive interaction with members of other cliques or with nonaffiliated individuals. Factions may show a certain amount of prejudice toward one another, or attempt to reach special-interest goals that are incompatible with mission goals. For example, scientific-investigative personnel might argue in favor of a dangerous, but curiosity-satisfying, change in course. Thus, another research task is to find means to cope with the dangers of factionalism. Techniques are needed to detect and combat prejudice, promote social interaction and communication, and encourage far-ranging friendship networks.

Allowing crewmembers to (sequentially) perform many different roles may discourage the formation of cliques along work lines and offer other benefits as well. Shurley et al. (1977) suggest that role rotation and personnel exchange can (1) allow crewmembers to gain other crewmembers' perspectives; (2) foster tolerance for behaviors that are perceived as idiosyncratic, but which are actually due to structural or role variables; (3) encourage mutual problem-solving; and (4) help meet some of the needs frustrated by the abandonment of everyday roles. To this roster of benefits we add that role rotation and personnel exchange should (5) promote a breakdown of invidious status distinctions; (6) underscore the importance of each role for the integrity of the overall mission; and (7) provide welcome variability in a relatively unchanging environment.

Role rotation, of course, is not a panacea. Many roles require a high degree of knowledge and skill. It is unrealistic to expect individuals to be competent in several highly technical areas. Moreover, unless carefully controlled, role rotation could contribute to role overload or to role conflict. For example, the expectations which persist from yesterday's role may make it difficult to perform today's role.

In summary, many of the most important research and planning issues surrounding tomorrow's space missions have to do with defining and articulating crewmembers' roles. Although it is tempting to specify roles in great detail, an extremely high degree of specification can yield dysfunctional consequences. Emergent roles will accompany prescribed roles; an important research topic is identifying the conditions under which the former will complement and supplement, rather than conflict with, the latter. Further research and planning issues include preventing role overload, minimizing role conflict, minimizing conflict among the interests and needs of people who are drawn to certain types of roles, fostering the perception that different roles have appropriate balances of obligations and rewards, and preventing the rise of role-related factions or cliques.

Normative Structures

Social norms or mutually shared expectations about attitudes and behaviors are another structural determinant of individual conduct and interpersonal coordination (Hackman, 1976). In addition to prescribing appropriate ways of behaving, norms prescribe means for dealing with individuals whose methods or levels of performance deviate to some degree from those considered acceptable by the group. Normative structure refers to the entire set of norms that is in use by a group or an organization. Normative structures do not provide a framework for all potential behaviors; only those that the group considers relevant to its functions and purposes (Hackman, 1976). In most settings, people who work together do not also live and play together, with the result that the norms tend to be limited to job-related activities. In spaceflight and other isolated and confined settings, the normative structure is likely to encompass not only those activities that are directly related to the job, but to living and recreational activities as well.

On American spaceflights, the crew's normative structure will in part be derived from our society's norms, in part derived from NASA's norms, and in part derived in the course of interaction within the crew. In some cases the attitudes and behaviors encouraged by society or by NASA may be preempted by norms that emerge in the course of interaction within the group. Typically, such emergent norms support lower levels of accomplishment than those which are officially prescribed, and support ways of proceeding which ignore or perhaps violate the official rules. From the outsider's (or mission control's) perspective, emergent norms are based less on standards of excellence than on personal convenience, and less on organizational objectives than on peer-group compatibility. From the "insider's" (or crewmember's) perspective, emergent norms reflect first-hand experience regarding what actually can be accomplished in a given situation, and a realistic concern for getting along with one's peers. As in the case of prescribed and emergent roles, official and unofficial norms can complement, supplement, or conflict with one another.

A group that maintains an elaborate and clearly defined normative structure is expected to function smoothly. But norms that are too rigid or too strictly enforced can have dysfunctional consequences. As noted at several points in our discussions of group and organizational variables, social structures require some flexibility to encourage innovative behaviors, allow social variety, and minimize the dangers of ostracism.

It is important to gain a better picture of spacecrew norms and their likely rationales. One concern is identifying dysfunctional norms that impose excessive restrictions on those behaviors that do not threaten social stability or detract from attaining mission objectives. Another concern is identifying conduct areas that appear to require social constraints but which the existing normative structure leaves uncovered. A third concern is identifying and dealing with conflicting norms that may be maintained by the different groups that participate in a mission.

In summary, spacecrews, like other groups, will develop shared expectations regarding appropriate attitudes and behaviors, and means to ensure that these expectations are confirmed. Because astronauts rest and play as well as work together, crew norms are likely to extend into many different areas of human activity. Some of the norms that emerge from within a spacecrew may conflict with the norms that are maintained by the mission sponsor. We see some value in mapping and monitoring spacecrew norms and devising means for changing normative structures when necessary.

MOTIVATION

Since space missions are run by human effort, planners and managers must take human motivation into account. Conditions must be established to ensure that participants join and remain with a mission, perform their roles dependably, and respond spontaneously and appropriately to unanticipated situations (Katz and Kahn, 1978).

Much of our present knowledge regarding human motivation has to be reexamined before it can be applied effectively in isolated and confined settings. The dominant theme in this section is that some of the conditions that motivate and satisfy people on Earth are either unavailable or are substantially altered in environments such as those likely to be found in space. However, some writers (Law, 1960; Natani and Shurley, 1974) have noted that exotic environments offer certain inducements which are not found in most other locations. Specifically, polar outposts, subaquatic dwellings, and space-capsule societies may offer an environment (1) that tolerates total immersion in work, or "workaholism"; (2) in which people are likely to be judged more on the basis of their performance than on the basis of their appearance, age, race, and similar qualities; (3) that is socially uncomplicated; (4) that is financially uncomplicated; and (5) that offers adventure.

To effectively shape behavior, rewards and punishments must be connected to performance (L. Miller, 1979). This linkage presupposes criteria for evaluating performance. According to Lawler (1976), evaluative criteria must be firmly linked to organizational purposes and goals, clearly and unambiguously expressed, set at levels which at once recognize organizational requirements and human capabilities, and accepted as fair and equitable by the people to whom they apply.

Certain variables associated with extended-duration spaceflight may make it difficult to establish evaluative standards which satisfy such criteria. Many of the tasks involved will be peculiar to space. Furthermore, performance levels are likely to be affected by weightlessness, somatic dysfunction, and other correlates of extended-duration flight. It may thus be very difficult to set appropriate performance standards. For example, Cooper (1976) believes that some of the antagonism and irritability expressed by crewmembers of the third manned Skylab mission, Skylab 4, reflected performance standards which were too high because they did not take weightlessness and related variables into account. The general remedy in such situations is to allow the subjects of the evaluation to help set the evaluative criteria (Lawler, 1976; Weick, 1977). Unfortunately, at the onset, few workers will have had sufficient experience in space to identify appropriate performance standards.

We might expect highly motivated, self-confident people of the type likely to volunteer for space missions to set unrealistically high performance standards. Later failure to meet these standards (perhaps brought about by unanticipated conditions) could prove to be a major blow to self-esteem and morale. The relationship of personality factors associated with volunteering, levels of aspiration, and the impact of success and failure under conditions of isolation and confinement are themselves important topics for further study.

Rewards

Work provides people with two kinds of satisfactions or rewards. First, there are extrinsic rewards (such as monetary compensation) that have value outside of the immediate work situation. Second, there are intrinsic rewards (such as a growing sense of competence) that come from the performance of the work itself. Both are important for attracting workers and ensuring high levels of performance. However, there are individual differences in responsiveness to extrinsic and intrinsic rewards. In Antarctica, for example, scientific-investigative personnel tended to seek intrinsic satisfaction in the venture; Navy support personnel tended to seek extrinsic satisfactions in the form of extra pay and accelerated promotions (Kanas and Fedderson, 1971; Natani and Shurley, 1974).

Extrinsic rewards— The two most prominent extrinsic rewards are pay and social recognition. However, the reward value of each may be diminished under the conditions of extended-duration spaceflight. For example, pay might not be collected for months or years. In some cases, crewmembers may never return to Earth, with the result that they will not enjoy (firsthand) social recognition from the folks back home.

Pay is satisfying because it provides basic necessities, because it provides luxuries, and because of other reasons. Space capsules, polar camps, military bases, ships at sea, prisons, and other environments which hold people captive on an around-the-clock basis must provide them with basic necessities independent of their levels of contribution to the group. Furthermore, space environments will not be able to support large stores of luxury items. Some of pay's immediate utility may be lost under the conditions of extended-duration spaceflight because pay will not be necessary for survival and will not purchase many luxury goods. Pay will not be rendered valueless, however, because people can and often do work toward very distant financial goals. Laboring in an environment in which necessities are provided and there is little opportunity to squander money offers an excellent opportunity to amass a small fortune.

Thus, alternative economic systems should be explored. Small items, such as candy bars and cigarettes, are known to influence the behavior of adult workers in normal organizational settings; when distributed by supervisors small reports have symbolic as well as material value (Lawler, 1973; L. Miller, 1979). Under conditions of isolation and confinement, relatively minor luxuries and conveniences may assume high value, and a prolonged shower, the use of a private room, or increased access to telecommunications or recreation equipment may prove to be powerful incentives. One possibility for consideration is the establishment of a token economy within the space-capsule microsociety. Certain behaviors might be rewarded with tokens which can later be traded for the few available luxury goods or services. Thus, an immediate reward system would supplement the long-term reward of accumulated pay. We hypothesize that such a system would have to meet stringent requirements of fairness in the crew's eyes; otherwise, it could undermine morale. That is, the rewards gained by one crewmember must not be offset by the frustration and embarrassment of other crewmembers. It is possible that such an incentive system, even if scrupulously fair, could prove disruptive in a closed society.

At present, space travelers receive substantial recognition from their coadventurers, from family and friends, and from society. But the conditions of extended-duration spaceflight may militate against a steady stream of social recognition. Communication with people on Earth is likely to be limited, society may become bored with missions which only rarely have news to report, and heroic welcoming ceremonies upon return are likely to become infrequent as space travel becomes routine.

Helmreich and his associates (Radloff and Helmreich, 1968; Helmreich et al., 1980) predict that both the costs of space travel (defined in terms of risk and discomfort) and the rewards of space travel (defined in terms of social recognition) will decrease as technological and other factors conspire to make space travel safer and more routine. However, they also predict that the *rate* at which the costs may be expected to decline is not likely to be as fast as the rate at which the rewards will decline. At some time in the future, the risks and discomforts of life in space are likely to become disproportionate to the rewards.

However, there may be certain steps which could help retard or offset declining social recognition. One step would be active publicity campaigns to advise the public of the importance of individual missions. Another step would be to increase recognition from people within sponsoring agencies such as NASA to offset decreasing recognition from society at large. A third possibility is to alter the basis for recognition. For example, less emphasis might be placed on reaching and returning from the destination and more emphasis placed on scientific, technological, production, and leadership achievements during the missions' course.

Intrinsic rewards— Intrinsic rewards or satisfactions are those which derive from activities themselves. Crewmembers are likely to value work. However, there may be a discrepancy between the high level of importance crewmembers accord work and the lesser extent to which work activities will occupy their time and absorb their energy (Kanas and Fedderson, 1971; Kubis, 1972; Law, 1960; Natani and Shurley, 1974; Rohrer, 1961).

Work is likely to be important to crewmembers for several reasons. First, crewmembers are likely to be selected on such bases as intelligence, drive, and an interest in achievement, and they are also likely to have strong interests in the technical or business end of the mission. Such people can be expected to place an especially high priority on work. Second, work and work roles are likely to be of great importance to crewmembers because the work role is the one everyday role that is likely to be maintained in an isolated and confined group (Rohrer, 1961). Astronauts may be expected to remain active in their occupational roles as pilots, physicians, and systems engineers, but to relinquish other roles such as spouse, parent, and scout leader. Under conditions of extended-duration spaceflight, the work role may be the only everyday role that remains intact.

But although crewmembers are likely to place a high priority on work, future flights (unlike past flights) are likely to be marked by periods of relatively light task demands. Although the craft will have to be staffed to handle times of peak activity, there will be long stretches during which there is very little meaningful work to do. In the Antarctic (Natani and Shurley, 1974) and on sailing ships (Day, 1969), low work load is associated with low morale.

Certainly, a major research issue is finding suitable meaningful work to maintain interest on long, and potentially tedious, missions. Leonov and Lebedev (1975) suggest including among the crew people who have expertise in areas that other crewmembers would like to learn about; hence, a space mission could become, in part, an educational activity. "Think tanks," consulting activities, and other knowledge industries, and possibly even effort-intensive light industries (electronics and the like) could also reduce the proportion of leisure time and increase the proportion of productive time, thereby helping to satisfy work-oriented individuals.

One set of conditions under which intrinsic satisfactions are likely to be high is when work assignments are congruent with personal interests. Underlying theories of vocational selection is the principle that people's interests (preferences for classes of activities) will in part be served or frustrated by the person's work setting and the nature of his or her job. When there is a good match between the person's interests and the job requirements and rewards, that person should find the job intrinsically satisfying. A good point of departure for future research in this area is Holland's (1959, 1973, 1976) theory of vocational selection. According to Holland, people tend toward one or more of six personality types: realistic, investigative, artistic, social, enterprising, and conventional. Different occupations, suggest Holland, tend to cater to one of the six personality types. Holland's basic argument is that attempts should be made to discourage square-peg people from ending up in round-hole jobs, but as Campbell (1974) notes, Holland argues this in a very sophisticated way. People may have mixed interests and occupational environments may have mixed characteristics. Thus, a person won't simply succeed in one type of occupation and fail in all the rest; some occupations should be satisfying, some dissatisfying, and some in between. Similarly, a given occupation doesn't require a specific kind of person. Many people may be able to do a tolerable job, but some people can be expected to do a better job than others.

Holland's theory of vocational selection has become the basis for the formerly atheoretical Strong-Campbell Interest Inventory (SCII). This self-report test consists of 325 items arranged into seven parts; essentially, these involve indicating preference for various school subjects, occupational alternatives, and types of people. Respondents' answers are compared with normative data collected from workers in different fields. The SCII has good psychometric properties (in particular, test-retest stability) and might eventually be extended to help select people for different roles in missions of different sizes and durations. Certainly, a point of departure might be gathering normative data from people who have adapted, to varying degrees, to conditions of isolation, confinement, and risk.

According to Hackman and Oldham (1974), many jobs can be designed in such a way as to maximize intrinsic motivation. These authors identify five job dimensions which have been shown, again and again, to affect worker commitment and job performance. First, jobs are intrinsically motivating to the extent that they provide workers with the opportunity to engage in varied activities and to exercise a range of different skills. Second, jobs are intrinsically motivating to the extent that they have task identity; that is, they result in a whole and visible (as compared with partial or obscure) piece of work. Third, intrinsic motivation is likely to be high when a task is significant in the sense that it affects the physical, psychological, or social well-being of other people, or clearly contributes to the attainment of group goals. Fourth, jobs are likely to be intrinsically motivating to the extent that they provide the worker with autonomy, that is, the opportunity to assume responsibility and exercise individual discretion. Finally, jobs which make some provision for workers to gauge their progress or performance tend to be more satisfying than jobs that do not provide performance feedback. Thus it would be useful to know more about designing spacecrew tasks to incorporate the features that are known to promote intrinsic motivation.

Also of interest here is Kahn's (1973) work-module form of organization. This is implemented by first determining the shortest length of time that is economically feasible and psychologically meaningful to work at a given task, such as navigating, analyzing data, or cooking. For purposes of illustration, let us assume this unit of time to be 2 hr. Time task units define work modules. From the overall perspective, a large-scale mission might consist of thousands of modules involving scores or hundreds of crewmembers performing hundreds of tasks. Under conventional forms of organization, missions might consist of a certain number of shifts or watches, each of which requires repetitive activities on the individual worker's part.

Under the work-module system, a crewmember would be allowed to qualify for several different kinds of tasks (such as navigating, analyzing data, and working in the galley) and then construct his or her own schedule using the requisite number of modules. For example, one crewmember might choose two modules of navigating, one of analyzing data, and one of working in the galley to satisfy the requirements of an 8-hr shift. Still another might change job content by day of the week. Moreover, Kahn's system could provide a crewmember with the opportunity to vary the way he or she distributes work in the course of the overall mission. Thus, within the limits established by the individual's qualifications and the organization's needs, crewmembers could, in effect, construct their own jobs.

Sanctions

Although it will be possible to "select out" individuals with histories of criminal or delinquent behavior, increasing crew size and heterogeneity will be accompanied by an increased likelihood of deviant behavior (Shurley et al., 1977). As noted in our earlier discussion of group dynamics, peer-group pressures will help keep individual crewmembers in line, as, under conditions of isolation and confinement, rejection and ostracism are severe penalties indeed. However, it is likely that informal sanction systems will need to be supplemented by formal sanction systems which include standards or rules, policing mechanisms, sets of graduated punishments, and procedures which govern the application of these punishments.

At present, the applicability of criminal law to space is spotty and inconsistent. The activities of military personnel in space are covered under the Universal Code of Military Justice. However, for civilian personnel, the law is still evolving. A recent addition to the Special Maritime and Territorial Jurisdiction Act (Amendment 6 to Title 18, Section 7 of the U.S. Code) extends the federal government's jurisdiction in criminal matters to cover events occurring within the space vehicle. This amendment dramatically increases the number of activities for which space travellers can be held accountable. However, even this amendment leaves many questions unanswered. For instance, it does not extend jurisdiction to those activities which occur outside of the spacecraft, nor does it resolve international jurisdictional issues.[1] More germane to the present discussion, it is difficult to project how this or any set of statutes can satisfactorily address the broad problems associated with enforcement in long duration spaceflight.

In space, one can expect legal and moral standards to evolve which are supportive of both the safety and success of the mission. On the one hand, the mission's legal and moral standards are likely to reflect the values of the sponsoring society. On the other hand, these standards must make accommodation for the likely behavioral impacts of isolation, confinement, and risk.

One possibility, with respect to enforcement, is a security force with rotating membership. This arrangement would have the advantage of encouraging the acceptance of attitudes that would help minimize "police-citizen" schisms. The identification of appropriate policing procedures is also an important topic for future research.

---

[1] For a full discussion of the applicability of 18 U.S.C.A., 7 (West Supp., 1981) to spaceflight, see Robbins, Karen, The Extension of the United States Criminal Jurisdiction to Outer Space, Santa Clara Law Review, *23(2)*, Spring, 1983, 627-661.

For instance, certain techniques which are accepted on Earth (such as the use of electronic "bugs" and human informants) may prove intolerable in the confines of the outer-space habitat.

Sanctions include social censure, material deprivation, and confinement. Social censure includes an oral dressing down or a written letter of censure. As mentioned above, such stated disapproval may be a powerful force in the closed environment of space. Material deprivations include fines and the withdrawal of luxuries or privileges. Under conditions of isolation and confinement, fines may have little significance if pay has low immediate utility, but the withdrawal of luxuries or privileges may be a major deterrant because few such satisfactions exist. Confinement includes restriction to quarters or imprisonment. Planners may be very reluctant to devote valuable living space to a brig, and find it tempting to think of imprisonment as a rather informal arrangement. However, unlike censure and material deprivation, imprisonment serves an immediate security function as well as deterrant and retribution functions. Other forms of protection from the unrepentant or the seriously deranged include tranquilizing drugs and physical restraints.

Establishing a set of procedures for applying such sanctions in space is fraught with many knotty problems. It will be essential to protect individuals from vigilante approaches to justice. However, time-consuming and expensive legal systems such as those found in western societies may be virtually impossible to duplicate except when there are extremely large crews, and a multitiered set of appeals procedures will not be possible (except, perhaps, through telecommunications with Earth). On some missions, it may prove possible and desirable to postpone final judgment until after return to Earth or home base, but on some missions security or justice may demand immediate confinement or punishment. (It should be noted that the prospects of a long trial upon return to Earth may pose more personal and social turmoil than somewhat crude but immediate legal proceedings.) It is not clear how crewmembers will react to sanctions in space, and there is little basis for predicting subsequent reintegration into the crew.

In summary, reward systems which are reliable on Earth may prove difficult to implement in space organizations, or, if implemented, may prove ineffective. Mission planners must identify rewards that will attract and retain mission participants and ensure dependable performance. Pay may have relatively little value in space; it is important to continue evaluating token economies and other economic systems which could prove effective. As missions become routine, social recognition from society at large is likely to decline; it is hypothesized that it may be possible to offset this decline by increasing social recognition from other sources. A major research problem is identifying suitable activities to maintain interest on long and tedious missions; educational and industrial activities are salient possibilities. Further research aimed at increasing intrinsic satisfactions should include extending such techniques as the Strong Campbell Interest Inventory into the realm of space jobs, seeking improved job designs, and evaluating work-module forms of organization. Another major task is establishing a legal system which is simple, gives rise to few jurisdictional disputes, and at once preserves the interests of mission security, justice, and customary civil rights.

EXTERNAL RELATIONS

Organizations exist within larger social frameworks. Although space missions will be separated from Earth by ever-increasing amounts of distance and time, they will nonetheless be influenced by Earth's social systems, including mission-control systems. First, spacecrews are likely to have internalized many of the values, goals, and norms of these systems. Second, spacecrews are likely to be dependent upon these other systems for legitimation and for economic support. Third, spacecrews are likely to maintain (sometimes erratic) communication with many of these systems. This section reviews some of the ways that space missions are likely to interact with their social environments; that is, social systems or subsystems which are external to the boundaries of the flights themselves.

The amount of influence that the sponsoring agency, home community, or other Earth-based organization is likely to have on a mission is expected to vary as a function of four general factors. These are the degree of overlapping membership in the Earth-based organization and the crew, the stage of the mission, the autonomy of the mission, and communication links. For example, we hypothesize that NASA or another sponsoring agency is likely to maintain a high degree of influence on a mission (1) when crewmembers are drawn from NASA's ranks and endorse NASA's goals; (2) relatively early in a mission, before crewmembers have psychologically disengaged from home, and relatively late in a mission, as crewmembers begin to anticipate their return home; (3) when the authority structure is centralized, with the Earth remaining the site of central control; and

(4) when there are good lines of communication between the spacecraft and Earth.

Boundary Roles

Certain positions serve as interfaces between one organization or subsystem and another. Individuals who occupy such positions are referred to as "boundary role persons" (Adams, 1976). In space, boundary roles will be filled by certain crewmembers and by selected members of mission control.

Boundary role persons (BRPs) serve as agents for purposes of bargaining and negotiation; that is, they attempt to influence the behavior of other persons and organizations whose priorities may differ from the priorities of their own organization. As the target of influence attempts, BRPs are forced to depict the interfacing organization's interests to their own constituencies.

Because BRPs have close dealings with representatives of other organizations, they are likely to be closely monitored by their constituencies and to be expected to display openly their loyalty and adherence to group norms (Adams, 1976). As negotiating agents, BRPs affect the outcomes that each organization experiences.

BRPs contribute heavily to the view that each organization has of the other. That is, as representatives of their respective organizations, BRPs have considerable control over the image that their organization represents to interfacing organizations. Similarly, much of what people learn about other organizations they learn from their associates who serve as BRPs. In effect, each BRP is a gatekeeper who regulates the flow of incoming information about other organizations and the flow of outgoing information about his or her own organization. Thus, BRPs affect the kinds of attitudes and views that each organization develops of the other.

Each BRP is subject to multiple influences. These include (1) the norms of his or her own organization or constituency; (2) the interfacing organization's norms, which limit the interfacing BRP's opportunities to negotiate; (3) the BRP's own attitudes, needs, and predispositions; and (4) the interfacing BRP's attitudes, needs, and predispositions (Adams, 1976).

To a large extent, the quality of the relationship between a spacecrew and mission control will depend on the behavior of the people who occupy the interfacing boundary roles. It will be incumbent upon spacecraft and ground-control communicators to avoid impasses resulting from conflicting norms, to arrange cooperative ventures with equitable outcomes, and to display appropriate emotions.

There are many important questions surrounding the creation and functioning of the spacecraft communicator, the mission-control communicator, and other boundary roles. How should boundary roles be defined, given the requirements and goals of each mission? What kinds of selection and training procedures will ensure that boundary roles are satisfactorily performed? How do electronic communications media, such as two-way radio or television, affect the interaction of BRPs? What are the stresses and long-term psychological effects of occupying a boundary position?

Interorganizational Conflict

A salient interest of mission planners and managers is maintaining good relationships with people aboard a spacecraft and others with whom the latter must interact, both within and outside of NASA. Although, as we shall soon see, conflict can serve some useful purposes, relationships that are considered good are not marked by hostility and divisiveness. Certainly, one important relationship to preserve is that between a spacecrew and mission control. As the sponsor of the mission, mission control seeks to maintain its authority and achieve its goals. Mission control also seeks to maintain cooperative relationships with the spacecrew. Yet there is increasing evidence that externally based authority may be difficult to sustain and that spacecrew/mission-control interactions can become marked by misunderstanding and friction. First, many studies of isolated and confined groups have found hostility directed toward external authorities (Kanas and Fedderson, 1971; Kubis, 1972). Second, during SMD III, scientific personnel expressed negative attitudes toward planners and managers (Helmreich et al., 1979a). Third, actual spacecrew/mission-control interactions have occasionally assumed a negative form. These involve an incident in which the Apollo 7 crew removed biosensors (Leonov and Lebedev, 1975), and a potentially more serious incident involving a Skylab crew (Bluth, 1979; Cooper, 1976, 1979).

According to Cooper (1976, 1979), the cordiality which characterized the relations between the first two manned Skylab missions (Skylab 2 and 3) and mission control did not characterize the relations between the third manned Skylab mission (Skylab 4) and mission control. Instead, the relationship between Skylab 4 and mission control was strained for approximately the first half of the mission. According to Cooper (1976), the crew was given to "blistering language," displayed a high degree of "grumpiness," and reacted to external authority with "ridicule, hostility, and exasperation." The climax came when, according to Cooper, the Skylab 4 crew "rebelled" by taking a day off. After the "rebellion," the relations between the crew and mission control improved, and so did the crewmembers' performance.

Not everyone shares Cooper's view of these incidents. According to Bluth (1979), many of the "blistering comments" about the living conditions were in direct response to a habitability questionnaire. Rather than symptomatizing general irritability and peevishness, the critical comments and complaints reflected candid assessments of faulty equipment. Bluth agrees that the astronauts did take a day off. However, she characterized the work slowdown as a legitimate attempt on the commander's part to put a "hold" on activities pending a clarification of instructions and establishment of a suitable work pace. However, there is some agreement that the Skylab 4 crew appeared to be under an unusual degree of pressure, and that there was an unanticipated degree of friction between this particular crew and mission control.

One important irritant to the Skylab 4 crew was an overprogramming of the astronauts' time (Bluth, 1979; Cooper, 1976, 1979; Weick, 1977). To keep the astronauts occupied and to make good use of the crew's very expensive time in space, planners had assigned the astronauts a seemingly endless string of tasks. Although the pace might have been maintained if the astronauts had been on a short-duration flight, they were aloft for a 3-month mission and hence required a different balance between work and nonwork activities.

A second possible source of friction between Skylab 4 and mission control may have been the latter's response to the astronauts' unsuccessful attempts to hide illness (Cooper, 1976). This episode can be viewed as the result of a conflicting interpretation of who is in control — the crew commander or mission control. The commander decided against advising mission control that the crew had experienced motion sickness (perhaps to avoid bad publicity), but the commander's decision was criticized and effectively reversed by authorities on Earth. The incident was, quite possibly, a source of chagrin and embarrassment which set an unpleasant note on which to begin a long stay in space.

Third, unfulfilled expectations might have contributed to the Skylab 4/mission-control conflict. Clearly, there were problems with some of the Skylab facilities. Frustrations and disappointments which are relatively minor under normal conditions tend to be exaggerated under conditions of isolation and confinement. The mission planners, as the agents perceived responsible for the unpleasant state of affairs, were the likely targets for subsequent aggression.

Most analyses of spacecrew/mission-control conflicts state or imply that the conflicts involved are motivational conflicts or conflicts of interest. The emphasis on such motivational determinants of conflict reflects conflict theory and research as a whole. However, cognitive factors, or differences in the ways in which people make judgments, can also breed conflict. Research by Hammond, Brehmer, and their associates suggests that cognitive conflicts can be particularly pernicious and difficult to resolve. Furthermore, what begins as an intellectual disagreement can develop full-blown motivational and emotional components (Hammond and Brehmer, 1973; Hammond, Stewart, Brehmer, and Steinmann, 1975; Brehmer, 1976).

Cognitive conflicts reflect differences in the data and rules that form the bases for individual judgments on social or other issues which do not have one objectively correct solution. When different people draw upon different sets of experience (data), and apply different means of analysis and methods of decisionmaking (rules), they may render conflicting judgments. Each person sees his or her judgment as accurate and conflicting judgments as faulty. The other person's judgment may be viewed as reflecting duplicity or sinister motives, rather than reflecting a different perspective on the problem. Rational discussion is unlikely to eliminate the conflict, because social judgments are only quasi-rational. That is, people cannot fully identify all the data that influenced their decisions, describe the weights and combined rules that were used to treat the data, or explain the procedures which transformed the results of their private analyses into public behaviors.

Available research (Brehmer, 1976) suggests that people involved in cognitive conflicts may eventually recognize that other people are using different mixtures of experience and analyses as bases for their judgments. Following such recognition, they are torn between serving the interests of accuracy and factuality as they see it by maintaining their present views, or altering their approaches to reduce the conflict. Generally, subjects in cognitive conflict experiments prefer accuracy to conflict reduction. If they do try to reduce the conflict, they are likely to abandon their own positions before they understand the conflicting positions, and to make a series of trial-and-error judgments which makes them look very inconsistent, if not outright capricious. This impression of inconsistency itself serves to perpetuate the conflict. Continued interaction may slowly reduce cognitive conflict, but it is unlikely to result in its eradication (Brehmer, 1976).

The implications of the cognitive conflict research for spacecrew/mission-control interactions are clear. Personnel at each location may be drawn from different backgrounds and traditions and are certainly privy to different data. For example, compared with mission control, Skylab 4 crewmembers had first-hand knowledge of the effects of weightlessness, the inconvenience posed by the misplacement of equipment, and the possibilities of unexpected scientific opportunities. Given such considerations, the generally modest level of conflict associated with spacecrew/mission-control interactions is far more remarkable than the one or two flare-ups which provide the exceptions to the general rule.

The Skylab 4 experience raises a number of issues. The first is identifying an optimal balance between work and nonwork activities for missions of varying lengths. To reduce the threat of overprogramming, mission planners and managers, first, might maintain a high degree of sensitivity to the socioemotional requirements of prolonged space flights. Second, planners and managers might incorporate principles or organizational self-design by further involving experienced astronauts in the planning of tasks and activities and by incorporating scheduling flexibility so that crews can readily modify their timetables on the basis of conditions encountered in space. Third, it might prove useful to conduct a path analysis of attitudes toward supplies, equipment, and living conditions on the one hand, and attitudes toward mission planners and managers on the other. It is not clear whether opinions regarding supplies and equipment are causes of, or symptoms of, spacecrew/mission-control conflicts. Another problem is identifying procedures for ensuring that crews and mission control have shared interpretations of each other's motives and behaviors. It is particularly important to find ways to make criticisms constructive and to minimize such consequences as embarrassment. Finally, there is the wide-open area of cognitive conflict. Future research might explore ways that differing parties can become privy to each other's data or better grasp each other's analytical processes and decision rules.

Models of Conflict Management

Thomas (1976) points out that whereas theories in such areas as international relations and race relations tend to focus almost exclusively on the destructive aspects of intergroup conflict, organizational theorists tend to view limited amounts of conflict as beneficial. First, limited conflict may contribute to an optimal level of arousal which in turn has beneficial effects on performance and satisfaction. In some cases, at least, conflict has been credited with alleviating depression (Leonov and Lebedev, 1975). Second, conflicting views may have to be presented before it is possible to identify or synthesize the best solution to a problem. Third, conflict can draw attention to problems which would otherwise go unremedied. For example, in the absence of conflict, the Skylab 4 crew might have remained saddled with a work schedule which undermined their morale and impaired the overall quality of their performance throughout the entire flight. Fourth, conflict is often instrumental, in the sense that it helps the contending parties reach important goals. Such accomplishments are not necessarily at the other party's expense, since conflict may simultaneously force the two parties to identify mutual goals or to enter into mutually beneficial arrangements.

Because conflict has functional as well as dysfunctional consequences, contemporary organizational theorists are less devoted to eliminating conflict than to managing or limiting its expression. Highly relevant in this regard are two complementary models developed by Thomas (1976). The first of these, the structural model, is concerned with the conditions which shape or channel conflict. It suggests certain constraints which decrease the likelihood that destructive forms of conflict will be initiated. The second model, the process model, addresses specific conflict episodes. It offers suggestions concerning how conflict can be minimized or managed once it has been sparked. Stated another way, the structural model is oriented toward prevention; the process model is oriented toward remedy or cure.

The structural model of conflict— The structural model of conflict is built upon four variables. These are behavioral predispositions, social pressures, incentive structures, and rules (Thomas, 1976).

Behavioral predispositions refer to attitudes, needs, personality traits, and the like. As noted in our chapter on group dynamics, attitudinal dissimilarities, competitive needs, own-gain or relative-gain motivation, incompetence, and socially devalued personal qualities breed conflict. Planners and managers must not only be sensitive to the compatibility of individuals who comprise the crew, they must also be sensitive to the compatibility of the BPRs who, through telecommunications, link spacecrews with mission control.

Social pressures are of two types. The first type, constituent social pressure, flows from the groups which the parties in the conflict represent. Typically, notes Thomas, constituent social pressures are directed toward competitive stances, although the reasons for this are not always clear. The other type, ambient social pressure, flows from outsiders, and includes larger social systems than those which encompass the conflicting parties. Ambient pressures tend to channel the conflict generated by constituent pressure into socially acceptable forms.

Certainly, a major problem confronting space mission sponsors is to ensure that potentially conflicting subsystems, such as space crews and ground crews, are exposed to constituent or ambient pressures which limit destructive conflicts. There are basically three techniques for fostering conflict-limiting norms. The first is selection. Specifically, crew candidates can be screened to select those who subscribe to conflict-limiting norms. Second, training and indoctrination can promote conflict-limiting norms. Third, reward systems can be geared to foster peaceful solutions to problems.

The influence of NASA over the norms of crewmembers, however, may be expected to diminish as missions become larger and prolonged. First, increased crew size forces decreased selectivity. Second, poor or intermittent communication may make it difficult to maintain a high degree of ambient social pressure. Finally, as noted in the earlier discussion of control systems, certain rewards may lose force under conditions of extended-duration spaceflight.

An example for maintaining ambient social pressure during long-duration missions in the future has been given by the role of U.S. Consulates in maritime affairs during the era of sailing (Day, 1969). Such consulates, available in many if not most major foreign ports, provided important contacts for ships and crews that had been away from the U.S. mainland for a period of years. The consulates buttressed the captain's authority and provided crews with redress against the captain's excesses. In addition, ship, captain, and crew were offered protections, services, and supplies that were not available at sea. In the distant future, space consulates located on large satellites or even planets might serve an analogous function, both to mediate disputes within the crew and between the crew and the ground.

The third element in the model, incentive structure, refers to the distribution of rewards following cooperative and noncooperative transactions. Of crucial importance here are the conflicts of interest that occur when two or more subsystems (for example, the crew and mission control) pursue mutually exclusive goals. As noted in our discussion of group dynamics, conflicts of interest may be minimized or eliminated by superordinate goals; that is, goals which are of overriding importance to both groups or factions (Sherif et al., 1961). Thus, an essential part of the design process is identifying and incorporating goals which can be shared by all subsystems (and by each individual within a given subsystem) and which override separatist or special-interest goals.

Finally, rules and procedures refer to laws, customs, conventions, and the like which govern ongoing negotiations. Presumably, such rules and procedures will be established through the efforts of international courts of law, the home community, the sponsoring agency, and, on a less formal basis, the contending parties themselves. Decision rules provide advance codes regarding specific conflicts of interest; procedural rules provide guidelines for bargaining, negotiation, and reaching resolutions in instances not covered by the decision rules. To the extent that such rules are effective, each party accepts the outcomes or constraints imposed by the rules and bears the expense in terms of decreased discretionary power. Reliance on rules generates less hostility than the exercise of coercive power, a common response to conflict situations (Thomas, 1976).

Space organizations, like other organizations, thus require sensible sets of rules for conflict containment and management. Some of these rules must offer advance resolutions to likely conflicts of interest, and others must govern the process of resolving conflicts of interest when prejudgment has not been made. Particularly important procedural rules will be those governing third-party intervention, including mediation (the use of a third party to bring the two contenders together) and arbitration (the use of a third party to render a decision). Rule enforcement, mediation, and arbitration from a distance are also important topics for future research.

The process model of conflict— Thomas' (1976) process model focuses on specific conflict episodes. It involves five variables: frustration, conceptualization, behavior, others' reactions, and outcome. These variables represent sequential events. The pattern of these events will determine the likelihood and intensity of subsequent aggressive episodes.

Perceived frustrations by one party initiates the conflict episode. Frustration, in this model, refers to any interference that makes it difficult for the party to satisfy a need or reach a goal. Crewmembers, for example, may perceive ground-based personnel as frustrators if the latter authorizes low-grade or faulty supplies or equipment, overburdens the crew with endless tasks, or interferes with the crew's normal pattern of social relations. Similarly, ground personnel may perceive the crew as frustrating if the crew fails to follow their instructions.

Conceptualization refers to the manner in which the frustrated party views the perceived frustrator. Conceptualization subsumes many variables such as magnitude of frustration, importance of frustration, implications of frustration (inadvertent or deliberate), and the framing and evaluation of possible responses.

The parties' conceptualizations of the conflict are, according to Thomas (1976), an immensely important and unexplored determinant of the conflict episode. Presumably, contending parties conceptualize frustrations in different ways, and this can add to the conflict. For example, the frustrating party may view the frustration as accidental and of small magnitude. The frustrated party may view the frustration as deliberate, of great magnitude, and fraught with far-reaching consequences. The frustrated party may then react in a way that it considers appropriate, but which the frustrating party would view as an overreaction. Presumably, understanding each contender's view of the situation — and helping each contender understand the other side's view of the situation — will help minimize and resolve conflicts. There are a number of unexplored research issues regarding conceptualization of conflict in the restricted environment of space. We would hypothesize, for example, that breadth of perspective, empathy, and other personal qualities would be particularly important to nondestructive conceptualizations under conditions of limited communications.

One party's behavior and the other party's reactions constitute interaction sequences which can escalate, contain, or deescalate the conflict. Specific hypotheses can be derived from Gouldner's (1960) norm of reciprocity, which suggests that a conciliatory move on one party's part is likely to be matched by similar actions on the other party's part. Although intended for application at the intersocietal level, Osgood's proposal for defusing intersystem conflict may be appropriate at the interorganizational level. Osgood's plan is called Graduated Reciprocation in Tension Reduction or GRIT (Osgood, 1962). He suggests that tensions typically build up slowly over time as a result of repeated incidents involving implacability and threat. Tensions can also be reduced slowly over time through a series of conciliatory actions. These actions, Osgood states, should be graduated and reciprocated. That is, the initial gestures should be relatively minor, and further overtures should await a conciliatory response on the other party's part. Small steps make it possible to maintain face and a sense of security while attempting to transform the opponent into a partner.

Finally, there are the outcomes of the conflict. One outcome is an integrative agreement or true resolution of the issue to each party's satisfaction. If both parties can be satisfied, the stage is set for subsequent cooperation. Domination, compromise, accommodation, and other outcomes are likely to leave residual feelings of frustration, hostility, and mistrust. Such feelings breed stereotypes, augment perceptions of incompatibility, and in other ways set the stage for future conflict (Thomas, 1976). Thus, although a conflict appears to be over, there may be an aftermath which sets the stage for additional trouble. Effective action after a conflict may thus minimize the chances of recurrence (Caplow, 1976).

In summary, the causes and cures for intersystem conflicts are an important area for future research. Because within certain limits conflict can serve useful organizational functions, conflict containment and management are more appropriate (as well as realistic) goals than are complete prevention or elimination. Possible points of departure for future research include Thomas' structural model of conflict, which is oriented toward prevention, and Thomas' functional model of conflict, which is oriented toward containment or cure.

Reassimilation

Upon return from a mission, crewmembers will need to be reassimilated into their home communities. In the future, when spaceflight is routine, tickertape parades, public appearances, and other acknowledgments will be reserved only for astronauts returning from the most heroic of missions. The welcome awaiting the bulk of tomorrow's space voyagers will be comparable to that accorded an average sailor returning from a peacetime voyage, rather than to that accorded a victorious admiral returning from war.

Clinical studies and surveys suggest that separation brought about through military service or other work-related duties can be highly disruptive to the family that remains behind (Greiff and Munter, 1980; Isay, 1968; Pearlman, 1970). Clearly, the remaining family member who bears the brunt of the negative consequences of separation is the spouse; almost invariably this has been the wife. Separation confronts her with three sets of challenges. First, and most obviously, there are feelings of loneliness and sexual frustration. The marital obligation prevents, or at least imposes a barrier to, the formation of compensatory relationships. Second, she must assume responsibility for areas that were formerly in her husband's domain. For example, she may have to take over family finances, keep the home and car in good repair, and so forth. Third, she must assume full responsibility for parenting; serve as both a male and female role model, make all decisions regarding the children's education and welfare, act as disciplinarian, and so forth. The same general challenges would have to be faced by a husband who remained at home.

The impact of separation on family life involves three phases. Prior to departure there is likely to be a period of bickering and general marital dissatisfaction. During the separation, the person who is away is likely to experience guilt about missing important family occasions (birthdays, anniversaries, graduations, and so forth) and about overburdening his or her spouse with abandoned duties. The person at home is likely to experience anger, resentment, and depression; depression, in particular, has been repeatedly noted among submariners' wives (Isay, 1968; Pearlman, 1970). Following the separation is a painful period of readaptation. Now is the time when hostilities and resentments are freely expressed. At this point, the wife is expected to abandon her hard-won abilities to make and enforce important family decisions. The better her performance as family manager during the period of separation, the harder it may be for her to reassume a shared or secondary leadership role within the family. Both partners are confronted with the task of redefining spousal and familial roles. Each spouse may have expectations which make awkwardness and conflict difficult to bear. For example, the person who has been away may feel that a hero's welcome is in order, and both may feel that the reunification should be a time for joy and wedded bliss. The discrepancies between such high expectations and bleak realities may exacerbate interpersonal conflicts and feelings of guilt.

The impact of a mission on an astronaut's family is important not only in its own right, but because it could affect the astronaut's motivation and morale. Specifically, the stresses and conflicts associated with departure, guilt over abandoning one's familial obligations, recognition that one is generating resentment in one's spouse, and the painful necessity of having to revise major life roles upon return to Earth could discourage qualified and desirable people from applying for service as an astronaut, and make it difficult to retain astronauts whose training and experience represent very sizable investments. According to Greiff and Munter (1980), the costs of separation are playing an increasing role in people's decisions *not* to accept important new corporate assignments.

Consequently, we consider it important to examine the effects of impending separation, separation, and reentry on the astronauts' relationship to his or her family and upon the family itself. It would be useful to identify the factors that are associated with a successful adjustment at each stage, and the kinds of resources that might be of use. For example, during the early years of spaceflight, most astronauts' families lived in the same community and could provide support for one another while the husbands were away. This type of close-knit social support network might not be practical in the future, but some sort of useful programs and resources might be devised. Another possibility, of course, is to send entire families into space.

SUMMARY AND CONCLUSIONS

As crew size expands, organizational-level phenomena will gain in importance. In this chapter, we addressed some of the issues related to the formation and operation of organizations in space. Organizational structure is defined by the differential distribution of authority and tasks.

Hierarchical control requires that individuals at one hierarchical level have the power to influence individuals at subsequently lower levels. French and Raven (1960) have identified several bases for such power, each may be undermined in the course of an extended-duration space mission. The norms and prescriptions which give rise to legitimate power may not be sufficiently reinforced in space. The satisfactions and sanctions that underlie reward and coercive power on Earth may prove unavailable or unreliable in space. On extended-duration missions, the decline of unused technical skills, cross-training, and the mounting likelihood of appearing foolish may undermine expert power. Finally, interaction under conditions of isolation, confinement, and risk may decrease interpersonal attraction and erode referent power. Further research is required to assess and bolster the robustness of each form of power.

It is expected that crews will become increasingly autonomous as they involve more people, last longer, and travel farther from Earth. If an acceptable level of Earth-based or centralized control is to remain possible, communications problems must be solved, and new means must be identified for setting bounds on the behavior of crewmembers.

Large-scale missions will involve several levels of leaders. Low-level leaders, intermediate-level leaders, and high-level leaders perform different tasks and exercise different intellectual and human-relations skills. One important research issue is identifying procedures for selecting leaders at each level. Another is examining the effects of mediated communication on the performance of low-, intermediate-, and top-level leadership roles.

Predominant among the work roles anticipated in space are flight operations, scientific-investigative, environmental support, personnel support, and production roles. The organizational design process must include ways to define these roles and ensure that each is satisfactorily performed. It is hypothesized that a moderate degree of specification is preferable to a high or low degree of work-role specification.

Prescribed roles do not always satisfy socioemotional requirements or cover unanticipated contingencies. For such reasons, emergent roles are likely to arise in the course of a mission. Ways must be found to ensure that emergent roles do not conflict with essential aspects of prescribed roles. We hypothesize that incompatibilities will be minimized when crewmembers show a high degree of commitment to the overall organizational goals, when prescribed roles can be adjusted in response to changing circumstances, and when prescribed roles provide latitude for discretionary socioemotional activities.

Role overload occurs when people are assigned positions with excessive demands. Role overload causes personal wear and tear and performance deterioration. A clear understanding of obligations, a sense of priorities, open communication channels, and the minimization of exposure to irrelevant communications are expected to retard or prevent role overload.

There are many forms of social conflicts that are associated with work roles. Some conflicts are generated by the expectations and demands that are associated with the roles themselves. Others are generated by the personalities and interests of people who are likely to be drawn to different roles. The perception that some roles are more important than others and the formation of factions or cliques among people within different role categories are additional sources of conflict.

Social norms or shared standards of conduct impart regularity to behavior. In isolated and confined settings, normative structures may extend beyond job-related activities and encompass living and recreational activities as well. Research in this area might include identifying norms that impose unnecessary limits on individual discretion, identifying important conduct areas which lack appropriate normative constraints, and dealing with potential conflicts among the norms maintained by the different groups or subsystems that are involved directly or indirectly in a mission.

Inducements and deterrents which prove reliable on Earth may be unavailable or unreliable in space. Under conditions of extended-duration spaceflight, pay may lose value as an inducement. Additional research is required to determine the subjective value of pay to spacecrews, and to develop supplementary economic systems. Token economies that meet stringent fairness criteria may prove appropriate. Helmreich and his associates (Radloff and Helmreich, 1968; Helmreich et al., 1980) have suggested that the rewards accorded to astronauts will decline faster than the costs of participation; this decline might be offset by new sources and forms of rewards.

Many knotty issues surround the creation of formal sanction systems. Considerable effort will be required to establish a legal system that is simple, has clear jurisdictional boundaries, and is consistent with the requirements of security and the traditions of social justice and individual rights. Further research is required to identify suitable policing mechanisms and to find appropriate forms of restraint and punishment.

Spacecrews interact with mission control and other external or environmental organizations. Boundary-role persons, who serve as interfaces between the spacecrew and mission control will play a major role in setting the tone of such interactions. Boundary-role persons are subject to multiple influences and serve, in effect, as gatekeepers who regulate the flow of information. The evolution of tomorrow's space organization would be facilitated by a greater understanding of boundary roles. Of particular importance are the tasks of identifying selection and training procedures for boundary-role occupants, examining the effects of mediated communication on boundary-role performance, and understanding the psychological and social consequences of prolonged boundary-role tenure.

Observations of many isolated and confined groups, including spacecrews, suggest friction between crews aloft and mission control. Possible contributing factors include overprogramming of the astronauts' time, a jurisdictional dispute, unfulfilled expectations concerning the quality of life in space, and cognitive conflicts. Mission planners are thus confronted with the issue of identifying an optimal balance between work and nonwork activities on missions of varying lengths, finding means to ensure that crews and mission control have shared interpretations of each other's behaviors, uncovering the link between negative attitudes toward physical facilities and hostility toward mission planners and managers, and helping mission control personnel and flight personnel to better understand each other's decision-making processes.

Organizational theory suggests that conflict can have functional, as well as dysfunctional, consequences. The goal, then, is not to prevent all conflicts, but to prevent destructive conflicts. Two models devised by Thomas (1976) provide useful conceptual frameworks for conflict containment and management. The first, or structural model, involves behavioral predispositions, social pressures, incentive structures, and rules, and is proposed as a preventive model. The second, or process model, involves perceived frustrations, conflict conceptualizations, behavioral interaction sequences, and outcomes, and is proposed as a containment or curative model. Particularly salient research problems are those which center around conflict conceptualization, containment, and de-escalation strategies, and the inculcation of superordinate goals.

An astronaut's departure may have an adverse impact on his or her family. During the period of separation, the astronaut may experience some guilt, and his or her spouse may experience anger, resentment, and depression as a result of loneliness, frustration, and the assumption of additional family responsibilities. Upon the astronaut's return there may be a prolonged and awkward period of reassimilation. We need to know more about the effects of separation on the astronaut and on his or her spouse and children. Resources must be developed to help all parties cope with the separation, and to facilitate the process of family reunification.

Artist: Susan Kaprov

IX
SUMMARY AND RECOMMENDATIONS

PURPOSE

Our overall goal has been to understand and plan for human psychological and social adjustment to space. To do this, we have attempted to assemble and integrate information that bears on human performance capability, psychological health, and social and organizational adaptation as they relate to space, and to indicate those areas in which additional research could further ease the Earth/space transition. Although there have been observations, analyses, and experimental studies that bear upon the human aspects of extended spaceflight, the connections and interrelationships among these strands of evidence have not been satisfactorily explored. Our purpose, then, was to seek these connections and, by imposing some order on them, examine holistically the issues of humans in space.

Past attempts to address the overall needs of space travelers have generally involved assembling articles covering a range of topics from a number of well-known specialists. The primary advantage of this approach is the high level of expertise brought to each individual topic; the limitation is the lack of an overall perspective. Since our purpose is to provide an overall perspective, our approach is one of integration rather than compilation. Although the present work builds on, and we hope extends, the earlier work, breadth is necessarily purchased at the expense of depth, and our treatment is that of the informed generalist rather than the topic specialist.

The goal of promoting a better understanding of human adjustment to space subsumes three major subgoals. The first subgoal is to provide a conceptual outline for organizing the personal and interpersonal dimensions of spaceflight. Although in a few cases our ordering is somewhat arbitrary, our chapter headings and subheadings represent our solution to this problem.

Our second subgoal is to achieve some semblance of balance between two extremes which presently dominate discussions of man in space. At one extreme are the experimental efforts which examine a single or very small number of variables. At the other extreme are the sweeping, often Utopian, views which pay only limited attention to the results of behavioral research. We sought to strike a balance by taking a more comprehensive view than the former approach, and a more research-oriented view than the latter. To this end, we reviewed an abundance of literature, including many studies which were not previously identified as pertinent to the human side of life in space. In addition to the volumes compiled by topical experts mentioned above, we drew from scholarly books and journals covering a broad range of disciplines, and to a significant extent, from the report literature. Where information was unavailable in published form, personal communication was sought.

Our third major subgoal was to examine the needs of humans in space, and to suggest research areas that could contribute to our understanding of how to satisfy these needs. Our assessments of needed research are presented in the various chapters, and in the section entitled Directions for Future Research, we outline what we feel are, overall, the most salient research issues.

CHAPTER SUMMARIES

This section presents an overview of the findings presented in the preceding eight chapters. As noted in chapter I, this overview emphasizes areas in which significant problems are likely to arise; that is, those situations which could, under certain conditions, pose a threat to the well-being of the spacecrew or to the success of the mission.

Chapter I, LIVING IN SPACE, presents an introduction to issues of human adaptation to space and outlines the basic assumptions underlying our approach to this topic. This chapter also provides an overview of the basic psychological and interpersonal consequences of isolation and confinement. These conditions often have been found to result in impaired intellectual functioning, motivational decline, somatic disturbance, psychological changes, and social tensions, the latter being reflected in a tendency to withdraw from fellow confinees and to react in a hostile manner toward outsiders. Withdrawal among isolated and confined individuals frequently is accompanied by a pattern of depression which peaks approximately two-thirds of the way through the confinement experience.

Chapter II, BEHAVIORAL AND SELECTION IMPLICATIONS OF BIOMEDICAL CHANGES, examines behaviors which may change as a result of physiological alterations coincident with space travel. Although physiological responses to space and to environments simulating space are relatively consistent, behavioral measures taken under the same simulation conditions generally have failed to demonstrate consistent effects. However, the possibility of genuine behavioral effects cannot be excluded. The present lack of consistent response differences may reflect an insufficient sensitivity of measuring instruments or the presence of characteristics in the experimental environment which permit the individual to focus on the task in a way that would not be possible in the more demanding environment of space.

There appears to be a number of important psychophysiological variables that relate to an individual's ability to adapt to spaceflight. For example, space sickness or space adaptation syndrome affects about one-half of all space travelers, primarily during the early days of a mission. Earth-based studies, although far from conclusive, suggest that psychophysiological correlates, including age, gender, and personality traits, could be predictive of an individual's susceptibility to space sickness. Sensory conflict theory continues to be helpful in directing research efforts in the general understanding of why space sickness occurs and how to simulate it on Earth. Psychophysiological correlates have proven helpful in determining adaptability to other spaceflight conditions. For example, one intriguing line of evidence suggests that the highly athletic individual may present no advantage in withstanding the effects of weightlessness and may even be at a disadvantage compared with an average, healthy individual. Similarly, there is evidence to suggest that older individuals may handle certain types of physiological stress better than younger ones. These and related findings could have important selection implications.

In chapter III, HABITABILITY, we note the need for a shift in emphasis from merely sustaining human life in space to maintaining a high quality of life in space. To accomplish this shift, it is necessary to broaden our perspective to include not only such obvious issues as environmental integrity and safety, sanitation, and nutrition, but also such subtle issues as environmental richness, temperature and humidity, and compatibility among crewmembers. The use of leisure time is a significant habitability issue. Confined individuals tend to place heavy emphasis on assigned work and little emphasis on recreational opportunities. When recreation is sought, it tends to be passive in nature. Although exercise is needed to meet many of the health requirements of space, exercise programs have not been pursued enthusiastically by space travelers.

Privacy, for the group as well as for the individual, has been identified as a significant issue for extended spaceflight. Many of the privacy mechanisms used on Earth are lacking in space. Of the privacy devices that remain to the space traveler, manipulation of the interior space and decor and training in the regulation of the sharing of information about oneself appear to offer the greatest hope for privacy protection.

In chapter IV, PERFORMANCE, we consider demands placed on spacecrews to accomplish mission objectives. Work schedules are influenced by workload demands, sleep requirements, stress, and many other variables. In space, work capacity can be altered by changes such as increased metabolic requirements, and by other conditions of weightlessness such as difficulty in moving from one location to another. Both difficulty in sleeping and the necessity for sudden awakening can contribute to performance decrement. Individuals show performance decrement throughout the day; therefore, the circadian rhythms themselves have implications for establishing work schedules. Desynchronosis, the result of changes in circadian rhythms produced by artificial environments, can contribute to performance decrements in space. Changes in work and sleep schedules can produce symptoms similar to those of jet lag. In space, when performance decrements occur, fine motor skills appear to be more negatively affected than gross motor or cognitive skills.

Aside from the earliest days of Mercury and Vostok, spaceflight has been a shared venture. In chapter V, SMALL GROUPS, we note that a clear understanding of interpersonal dynamics is important for mission success. Groups whose members lack compatibility have been found to have lowered morale, increased stress, decreased motivation, and, when the opportunity is present, a high rate of member loss. In assembling a crew, it is necessary to consider how the qualities of different people interplay. Complementary needs, similar values and attitudes, and a strong but noncompetitive work orientation are among the qualities that are thought to contribute to social compatibility. Groups can be aided by the inclusion of individuals who are personally attractive, competent, and who can respond both to the demands of the task and to the needs of others. People who are similar to each other appear to have a compatibility advantage, at least in the short term, but bringing together individuals differing in age and social and cultural backgrounds, and of both sexes is likely to enrich the confinement experience over time.

Leadership is presumed to be a major factor in the effectiveness of spacecrews. Although leadership is often thought of as a quality of the individual, leadership depends on three sets of characteristics: characteristics of leaders, characteristics of followers, and characteristics of situations and tasks. Leadership is associated with two types of activities: task and socioemotional. Autocratic decision making appears to have an advantage when expertise is concentrated in the hands of the leader and when speed in decision making is important. Democratic procedures can lead to better decisions when expertise is distributed across group members and when it is desirable for all to have a high degree of commitment to the decision.

There are many ways in which spaceflight conditions are likely to affect dynamics within small groups. Historically, members of isolated and confined groups have had difficulty containing interpersonal conflict without resorting to social withdrawal. Direct training in interpersonal relations may be of some help in this regard. In isolated and confined groups, conformity and compliance pressures are likely to be high; this can make it difficult for individuals to offer creative solutions to problems. Conformity pressures can result in the rejection of the nonconformist, a result which may be unacceptable under spaceflight conditions. Established groups would seem to offer some advantages for long-duration missions. Ideally, a crew entering space would be sufficiently "old" that its members will have achieved a high degree of interpersonal coordination, but sufficiently "young" that they will not have become bored with one another. Space stations and other space environments that are likely to have crews of rotating membership will have to grapple with the problem of assimilating new members into the crew.

In chapter VI, COMMUNICATION, we note some of the ways that conditions of spaceflight influence the transmission of information from person to person. Within the spacecraft, propulsion and life-support equipment are noisy and could interfere with verbal communication. Nonverbal communication may be hampered by distortion of facial expressions and distancing cues associated with weightlessness. Among the most significant communication issues for space are those involving contact between people aboard the spacecraft and people who are on the ground or aboard another space vehicle. Mediated communication systems must be developed to meet the needs of the crew throughout an extended mission. Under Earth conditions, mediated systems are generally used to handle formal, businesslike, and routine exchanges, tasks which they do well. In space, mediated systems will have to handle all communications needs — personal and subtle, as well as impersonal and straightforward. Systems designs for space must take into consideration people's emotional requirements as well as their requirements for information exchange. To this end, the use of video is promising.

Among the problems of formal communication networks are barriers to information flow. One such barrier is the tendency for individuals at all levels to impede the upward flow of negative or disconcerting information. In addition, some strategically located individuals known as gatekeepers have particular power to influence the flow of information throughout the group or organization. In space, heavy reliance must be placed on formal communication arrangements, for instance between the spacecrew and mission control, since informal communication networks will be difficult or impossible to establish.

Precipitous events affecting either the physical safety or the psychological well-being of the crew are discussed in chapter VII, CRISES IN SPACE. The spacecrew must be prepared to deal with both physical threats and the emotional responses to such threats. Some evidence suggests that fear can be managed by learning to control the timing of the fear experience. Effective handling of a threatening situation will require that organizational arrangements and procedures be in place prior to the occurrence of the event. Following recovery from a threat, it is possible that, in space as elsewhere, blaming or scapegoating will occur.

Other occurrences that could cause a serious disturbance on-board include a severe psychiatric event or the death of a crewmember or a loved one at home. Isolation and confinement may contribute to depression and other forms of psychological distress and, on rare occasions, psychotic episodes have been observed in environments having some of the characteristics of space. In maintaining the mental health of space travelers, success in the work role and maintaining good relationships with other crewmembers will be of primary importance. Also of concern are the immediate reactions to the death of another and the predictable, but disruptive, reactions associated with grief.

In chapter VIII, ORGANIZATION AND MANAGEMENT, we note that large spacecrews will take on the characteristics of complex organizations. Informally evolved means of achieving interpersonal coordination will be supplemented, if not replaced, by externally imposed rules and regulations, and structural considerations will come to the fore. Social structure is reflected in the distribution of authority and tasks. The multileveled hierarchy is the most common form of authority structure. However, alternate models exist and their application to space deserves careful consideration. The Russians report success in using hierarchical methods to make determinations regarding the technical and performance aspects of a mission; however, they found democratic methods to be more appropriate for making determinations regarding life within the spacecraft. Centralization of authority is desirable to the extent that central authorities have good communication with remote units and a good understanding of their situation. However, as missions last longer and become more remote from Earth, the effectiveness of Earth-based, centralized authority may decline. Truly large crews will require leadership at many different levels. Middle-level and top-level leaders may require managerial capabilities above and beyond those required to successfully lead the relatively small crews associated with the space missions of today.

Future spaceflights are likely to be marked by pressures toward increased specialization. Among the work roles that are likely are flight operations, scientific-investigative, environmental support, personnel support, and production roles. Formal roles that are specified in advance by mission planners and managers will tend to be supplemented or even replaced by emergent or informal roles. Role-related conflicts could emerge. Some roles may be seen as more important than others, and occupants of certain roles may form blocs or factions which are antagonistic to other groups. A suggestion for extended spaceflight is that individuals be rotated across roles; among other benefits, this practice could foster tolerance by breaking down status distinctions and helping crewmembers to gain each other's perspective. Role rotation also could encourage creative problem solving and introduce welcome changes of pace into an otherwise monotonous routine.

In chapter VIII, we also consider the motivational forces that attract people to organizations and encourage dependable role performance. Certain extrinsic rewards are expected to lose force under tomorrow's spaceflight conditions. Pay may lose incentive value, since money cannot be spent freely in space. Furthermore, as space travel becomes common, another important extrinsic satisfaction — social recognition — will decline. Little is known about how sanction systems might work in space, but it seems likely that the usual punishments (fines, incarcerations, etc.) will not be practical or effective. To some extent, the expected decline in the availability of extrinsic satisfactions might be offset by increases in the availability of intrinsic satisfactions. Such satisfactions tend to be high when work is available that taps personal skills, provides autonomy, and is consistent with both personal and superordinate goals.

Many isolation studies have revealed overt hostility between crewmembers and external monitors or authorities. Although there is general agreement that conflict can have a functional aspect, damaging conflict between organizational units such as a crew and mission control needs to be prevented or at least contained. Persons who serve as interfaces between the crew and other groups or organizations play a critical role in determining the presence or course of intergroup hostility.

DIRECTIONS FOR FUTURE RESEARCH

A major impetus for the present book has been to identify some of the salient unanswered questions regarding psychological functioning and social dynamics during extended spaceflight. All of the research questions identified are directed toward ensuring that crewmembers are able to maintain high levels of personal and social adjustment while performing well under the unremittingly challenging circumstances of outer space. Stated in the most general terms, the task for extended spaceflight planners and managers is to match people, environments, and tasks; that is, to devise a system consisting of compatible and mutually supportive environmental, technical, and social subsystems.

There are three basic methods for aligning people, environments, and tasks. One method is to select people who are presently fit to meet the mission's requirements. The choice of short, lightweight astronauts during the era when spacecraft size had to be kept to an absolute minimum is an example of this personnel selection approach. Of course, physical characteristics provide only one basis for selection; technical skills and social compatibility are other bases. In the past it has been possible to devote a great deal of time to selecting each individual. To process or at least screen large numbers of applicants and to compose large crews, new testing techniques need to be developed.

A second method is to modify people's biological, psychological, or interpersonal reactions to gain greater consistency with environmental, technical, and social requirements. The use of pharmacological agents to combat space sickness, training programs to develop skills in the use of equipment, and direct training in human relations are examples of this conditioning or training approach. To prepare large numbers of astronauts, we will have to identify conditioning and training procedures that are efficiently and effectively applied to entire groups.

The third method for improving the compatibility between people and life in outer space involves structuring environments in such a way that they gain greater consistency with their human users. Increasing or reorganizing interior space and simplifying the operation of scientific apparatus illustrate this method. Often overlooked, but in the same vein as environmental engineering and ergonomics, is the formulation of rules and procedures and the structuring and pacing of tasks.

In practice, we cannot afford to narrow our sights to any one technique for adjusting people and spaceflight environments. We must rely, instead, on a mixture of selection, conditioning and training, and engineering procedures to produce a smoothly functioning overall system.

In the present summary, we will first review research issues which are of interest to future spaceflight in a general sense. We will then review some of the research questions which arise, or which gain salience or complexity, as mission duration increases and as crews become larger and more heterogeneous. Next we will consider some conflicting perspectives and recommendations and the kinds of research which might reduce or eliminate some of these conflicts.

Finally, we will outline research issues which we believe to be particularly deserving of consideration because little attention has been paid to them in the past.

General Research Issues

As our review of the literature demonstrates, over two decades of laboratory studies, field studies conducted in environments bearing some resemblance to space, and, of course, experience in space itself have added greatly to our abilities to forecast people's reactions to various spaceflight conditions and to take steps to upgrade the quality of human life in space. Despite these advances, our experience of space remains somewhat limited, and many important questions remain unanswered. Some of these questions are basic in that they pertain to missions of differing specifications. One group of basic questions centers around psychophysiological reactions to space. For example, it may be necessary to develop a new pharmacopoeia that takes weightlessness and time frames induced by clocks, electric lighting, and other artificial *Zeitgebers* into account. In this era of the Space Shuttle, space sickness remains a problem and an enigma. The underlying mechanisms are not well understood, and age, gender, physiological variables, personality variables, and many other possible predictors have yet to be fully explored. Despite the lack of clear theoretical supports, simulation studies suggest that biofeedback may be useful for controlling motion sickness on Earth, but the procedure remains untested in space. If subsequent findings confirm the value of biofeedback for controlling space sickness, further research may be in order to explore the possible value of biofeedback for controlling blood pressure, heart rate, and other psychophysiological responses in space. Optimal biofeedback training procedures would then have to be devised. The hypothesis that people who are smaller, older, and not unusually physically fit have certain advantages in space is also of interest; more information is needed on the dynamics of these relationships. We need to identify which specific exercises are best suited for space as well as to identify methods of increasing commitment to exercise regimens. Tailoring activity programs to the personal preference of the individual astronaut and including reliable methods for tracking one's progress or skill development are among the possibilities that might be explored.

Among the other psychophysiological questions that one might raise are those pertaining to (1) how individuals differ in susceptibility to desynchronosis, (2) how work schedules might be adjusted to coincide with biological clocks, and (3) how aids might be employed to combat desynchronosis. The effects of artificial gravity also need to be explored. For example, would the performance benefits that might result from a 0.25-g artificial gravity justify the expense? Although a fair amount of research has been devoted to the effects of spaceflight conditions on psychomotor performance, very little is known about the effects of these conditions on cognitive performance. The combined effects of weightlessness, desynchronosis, isolation, confinement, risk, and so forth on intellectual functioning is a salient and important research area.

Additional questions need to be asked about perception, motivation, and other psychological processes. For example, a certain amount of routine work is required to better understand color and brightness sensitivity, dark adaptation, and taste and odor thresholds and tolerances under spaceflight conditions. Noise annoyance thresholds, particularly those related to unrelieved noise sources, also need to be examined. The questions regarding human motivation are numerous and complex. One central research issue of particular importance in interplanetary flight involves work-oriented people who must spend prolonged periods with little to do. Another central research area involves identifying ways of increasing both extrinsic and intrinsic rewards. Traditional rewards such as pay will probably lose value in space, so supplementary or alternative means of compensation need to be studied. Extrinsic rewards in the form of social recognition are likely to decline as spaceflights become more routine, and techniques for offsetting this decline should be identified. Among the techniques that could increase social recognition are publicity campaigns, increased within-group recognition, and altering the bases for recognition from the journey itself to achievements in scientific or technical areas or in organizational leadership. The attainment of mission goals will no doubt be a powerful reward; since these goals may be remote in time, we need to explore the potential benefits of interim goals. Techniques that would produce increased congruence between tasks and personal interests, increased variety in tasks performed by an individual, and increased individual responsibility have been proposed as methods of increasing intrinsic satisfaction. Task structure and design and its relationship to workers' personal needs is a large and potentially fruitful area for future research.

In addition to unanswered questions about work in space, there are unanswered questions about leisure in space. Food seems to be an important gratification in exotic environments. We need to understand how food might be used, or perhaps abused, to fulfill psychological and social needs in space. Individuals in remote environments show a preference for passive forms of recreation. What is the basis for this preference, and under what conditions will active, and perhaps even competitive, pastimes be enjoyed?

Continuing isolation, confinement, deprivation, and danger are likely to place extreme stress on crewmembers, and there are many unanswered questions regarding their mental health. Little is known, for example, about the personal resources that may be required to cope with emergency situations in outer space. Additional research is required to identify characteristics of people who have high stress tolerances and to devise training procedures to increase people's abilities to control stress. Dissociation, or out-of-body experiences, sometimes occur when a disaster is imminent. A hypothesis that has been offered here is that this phenomenon relates to the perception that the individual has no control over the event which is threatening him. Testing this or competing hypotheses could aid the process of identifying possible countermeasures.

The general role of objective, perceptual, and cognitive factors governing the arousal, control, and dissipation of fear is an area with potentially important research implications. Means of dealing with specific crises pose another set of questions. How do isolation, confinement, and risk affect the grieving process? How will surviving crewmembers cope with a death aboard the spacecraft? It is of paramount importance to uncover the means for maintaining the level of personal functioning and social coordination that is required for a crew to successfully cope with such disrupting events. Russian research suggests that crews that are "homeostatic" (i.e., crews whose different members show synchronous psychophysiological responses and behaviors) may be particularly well equipped to cope with highly stressful and demanding situations. This suggestion, if confirmed, could have a direct impact on selection strategies.

There is evidence that "blaming" behavior follows a crisis, and there is some suggestion that the second or third in command is a likely target of this blame. We need to examine blaming and particularly scapegoating and their implications for life in space. Psychotherapy under spaceflight conditions is a wide-open research area; peer support strategies, the use of on-board mental health professionals, and therapy via telecommunications systems are of particular importance.

For the most part, space travel is a group activity. A major concern, therefore, is to compose crews which are compatible in the sense that the different individuals can work with and enjoy one another. We suspect that the research conducted to date only begins to tap the factors that are related to compatibility. First, we need to identify means for assessing interpersonal and technical competence. Second, we need to expand our search for personality characteristics that prove desirable for spaceflight conditions. In some cases these traits might be dual traits which are variable in their expression. An example of such a trait is androgyny; androgynous individuals can adopt either the instrumental or the expressive role. Another desirable dual trait, we propose, is ambiversion; that is, the ability to be social and outgoing or self-sufficient and contained, depending on the demands of the situation. Still other dual traits might be identified which could lead to selecting individuals particularly suited to life in space. The intermeshing of different people's needs also deserves further study. Research should include groups larger than two or three individuals, and the number of needs considered should be expanded beyond that of present research. From the opposite perspective, it would be useful to explore further the consequences of incompatibility.

In search for complex bases for social compatibility, we should not ignore the simple fact that some people possess mannerisms which other people find annoying. Development of an annoyance questionnaire would help eliminate people who are sensitive to too many human frailties as well as people who possess too many irritating qualities. In conducting research on interpersonal compatibility, it should be kept in mind that crewmembers' perceptions of compatibility may be more important than the actual intermeshing of underlying personal qualities.

Communication is the basic process which underlies the coordination of different people's activities. Reduced atmospheric pressure and the environmental conditions associated with weightlessness can interfere with both verbal and nonverbal communication. The effect of spaceflight conditions on direct personal communication requires further exploration, and the means for improving the signal-to-noise ratio need to be identified.

Tremendous demands are placed upon spacecrew leaders. We suspect that many of the forms of social power that make it possible for an individual to direct and coordinate others will erode under future spaceflight conditions. Additional research is needed to assess the decline of various forms of social power and to find counteractive measures. Most of the existing accounts of leaders in space-like environments focus on the leader's characteristics, but contemporary approaches to leadership encompass follower characteristics and situational factors as well. Research on leadership under spaceflight conditions should be expanded to include these additional variables. Some leadership requirements may be satisfied by people not designated as leaders; other leadership requirements may be satisfied by formalized rules, constructing equipment in such a way that certain forms of misuse become unlikely, and other techniques. It would be useful to know more about the optimal distribution of leadership functions for various mission categories. We also need to research the advantages and disadvantages of various autocratic and participative decision-making procedures, given different crews, tasks, and environments. Future research on leadership in space-like environments should not rest exclusively on the traditional hierarchical model, but reflect instead a wide range of perspectives and possibilities.

A certain amount of compliance and conformity are required for people to work together effectively. However, compliance and conformity can militate against innovation and reduce personal satisfaction. There are many unanswered questions regarding the optimal balance between social control and individual expression. For example, under spaceflight conditions, the constraining effects of conformity on innovative problem solving may be particularly severe, and we should seek ways to ensure that good ideas will be aired. Aggression, hostility, and interpersonal conflict must be kept within acceptable limits without resulting in other disruptive behaviors such as psychological withdrawal. We see some real value in exploring the effects of direct training in human relations, the use of on-board specialists in human relations, and external arbitration and mediation conducted through some form of teleconferencing arrangements.

As yet no legal system has been developed that is simple enough to work in space, but which preserves some of Western society's traditional standards and safeguards. However, acceptable sanctions must be found. Under spaceflight conditions, ostracism resulting from either violating the group's informal rules or from breaking the law is likely to have a devastating effect on the deviant and on the rest of the crew. Other methods must be sought to control unacceptable behavior and to allow for reintegrating transgressors into the group. Similarly, we must find methods of integrating newcomers into an established crew.

For each space mission, planners are confronted with the task of defining and articulating a set of roles which is comprehensive but which leaves crewmembers with some opportunities and options. Role-related research issues include identifying means for preventing role overload, preventing the rise of role-related factions or minimizing the conflicts if such factions do arise, and fostering perceptions that different roles have appropriate balances of obligations and rewards.

Most spacecrews are far from being completely isolated; there is some communication with mission control and, oftentimes, with family and friends. A critical problem is improving our understanding of spacecrew/mission-control dynamics. There are many subsidiary issues: the degree to which mission control should be privy to various crew communications and activities, symmetry in spacecrew/mission-control access, the distribution of power and authority, controlling information in such a way that information overload is prevented and proper priorities are maintained, spacecrew/mission-control rapport, and the role of people who serve at the boundaries of the interfacing systems. Additional research is also required to better understand how interacting through telecommunications systems affects relationships with mission control and with family and friends, and how these relationships with external parties in turn affect interaction within the crew. Among the possible telecommunications modes are one-way and two-way audio, video, and alphanumeric or computer systems. These modes will vary in terms of difficulty of use and in terms of impact on interpersonal relations. We need to know more about the benefits and limitations of these different communications systems in the space application. Finally, there are questions regarding the reintegration of crewmembers into their home communities following an extended stay in space.

Extended Spaceflight Variables

In chapter I, we described extended spaceflight as marked by significant increases in crew size, crew heterogeneity, and mission duration. Increased size increases the pool of skills aboard, as well as the friendship options, and also introduces variety into social life. However, increasing crew size can also introduce new problems. Large crews may be more difficult to organize (requiring, perhaps, bureaucratic rules and regulations), and may be more likely to give birth to disruptive cliques. Increased crew size can result in lowering the social recognition which can be accorded to each member, and may have an adverse effect on individual motivation. Larger crews also increase demands placed on leaders and may require multiple layers of leadership.

As with large crews, crews composed of a heterogeneous mix of individuals offer the opportunity for varied social relationships and varied work roles, but can also introduce the opportunity for interpersonal conflict, for factionalism, and for prejudice and stereotyping. Furthermore, heterogeneous crews can result in miscommunications and can increase the difficulty in achieving interpersonal coordination.

Increasing the duration of spaceflight is likely to increase the general level of social tension aboard and to decrease the effectiveness of traditional rewards and sanctions. In addition, very long flights may exacerbate the drift away from the values and controls of the sponsoring society, and increase the need for local authority.

In summary, increasing crew size, heterogeneity, and mission duration are likely to yield a complex array of effects, some of which may undercut the quality of life in space, and some of which may actually enhance it. Let us turn now to some of the research issues specifically associated with the variables of size, heterogeneity, and duration.

Size— Increasing crew size raises the salience of virtually all of the research issues identified under the General Research Issues section of this chapter. First, the decreased selectivity required to staff large or multiple crews is likely to yield astronauts who are less resistant to some of the pressures and problems of extended spaceflight than are astronauts of yesterday and today. Hence, it becomes more important to find ways to reduce the pressures and to prevent or solve the problems. Second, if attached to each astronaut is some probability of an adverse occurrence (i.e., a psychotic episode, a criminal act, serious illness, or death), then the greater the number of astronauts who venture into space, the greater the likelihood of occurrence of an adverse event and the greater our need to know how to deal with such an event.

Increasing crew size also generates new research issues. The interior space needs of crews of different sizes are not well understood; in fact, it is not clear whether the amount of additional space required per person increases or decreases as crews become large. Increasing crew size can also affect motivation if, for example, associated with increased crew size is increased task specialization.

Additional research is required to better understand the interrelationships among such variables as crew size, activity range, and motivation. In addition, we need to know more about the effects of increasing crew size on the formation and development of social relations. For instance, attitudinal differences among crew members appear to have a negative influence on compatibility of small crews, but may be an enriching factor among larger crews. At what point does this shift occur? Or, more generally, under what conditions are attitudinal differences tolerated and even welcomed?

Other significant size-related questions affect organization and management. At what crew size do potentially disruptive factions begin to emerge and how can differences among groups be minimized? Also, what crew sizes require multiple levels of management, and how should supporting communications networks be organized to accommodate this end?

Heterogeneity— As crews are selected from various backgrounds, plans and procedures that are predicated on studies of technically trained, white males will lose precedence, and research will be required to identify the effects of age, various backgrounds, and various mixes under spaceflight conditions. For example, we need to know more about the effects of such variables as age, gender, educational background, and other differences on psychophysiological processes: the arousal and control of fear, deconditioning tolerances, and space sickness susceptibility. In addition, we need to understand how various individuals and combinations of individuals differ in terms of habitability requirements such as internal space needs and recreational preferences. There is suggestive evidence that among isolated individuals, diverse groups engage in more active leisure than do homogeneous groups; this evidence requires substantiation.

A possible negative aspect of heterogeneity of the group occurs in perceptions of and responses to crowding. Privacy invasions by friends appear to be relatively rare, even in high-density situations. This may be because friends usually share values as well as interests. If shared values form a basis for privacy protection, heterogeneous groups may have particular difficulty in preserving privacy in space. The relationship of heterogeneity of the group to perceptions of crowding and privacy should be investigated.

We also need to understand the effects of increased heterogeneity on social interactions within the crew and between the crew and mission control. For example, what attitudes and interaction patterns will prevail when crews are composed of both genders and span a range of ages and ethnic backgrounds? Cultural heterogeneity raises many special questions about conflict, performance, and morale. The possible effects of cultural heterogeneity on communications are particularly deserving of study. Since both direct and indirect communications may be more difficult in heterogeneous crews, we need to learn more about what communications mechanisms work reliably across cultural and language barriers, and what level of redundancy is needed to ensure that communication does not break down.

Duration— Increasing mission duration increases the significance of some of the basic research issues identified earlier. First, increasing the amount of time in space increases the wear and tear on the human organism. Conditions which can be tolerated on a short-term basis may not be well tolerated on a long-term basis. Second, even as increased crew size increases the probability of an untoward event, so does increased mission duration. For example, a natural death in a small group becomes relatively likely if we consider a sufficiently long span of time. We must point out, however, that there are complex interrelationships among the many variables associated with spaceflight. Thus, we might expect desynchronosis to be a greater problem on short flights than on long flights, because people are likely to resynchronize to an artificial environment over time.

New questions emerge when we consider missions of longer duration than those which have already been mounted. The onset of space sickness is time-dependent and there is evidence that space sickness is a passing phenomenon, but the time line of the disturbance and the conditions of recurrence are not well understood. Related research issues involve long-term habituation to weightlessness, and the possibly different effects of drugs in long-term compared with short-term missions.

As in the case of crew size and heterogeneity, mission duration is expected to affect habitability requirements. The amount of interior space that suffices on brief missions may prove inadequate on extended missions. Also, we need to understand human adaptation to an unusual three-dimensional world. If space travelers can learn to adapt to an existence lacking a vertical orientation, we will need to plot the course and process of this adaptation. There are questions about the use of food in space; we need to understand how the use of food will change over time, and to determine the potential evolving roles of alcohol and other "recreational" drugs. Eventually, spaceflights will be long enough to encourage a relaxed living style; to plan for such flights we need to determine what kinds of environments (decor, lighting, sound, etc.) are supportive of life in space, and how demands (e.g., for increased complexity) change over time.

A reasonable hypothesis is that boredom will become a greater problem as missions gain in length, and that simple, repetitive tasks may be particularly unpopular on extended-duration missions. As work becomes routine, we might expect space travelers to look for challenges elsewhere. A significant question is whether use of leisure time becomes (or could be directed into) more active pursuits as spaceflight lengthens.

Other concerns are the effects of increased mission duration on interpersonal relations. We need to determine how anxiety can be diffused in space without invoking behaviors that would be harmful to the individual or to group relations. There is evidence that in confinement people tend at first to overshare personal confidences and then, over time, to psychologically withdraw from others in the group. We need to determine the time line of this process, as well as the techniques that can be introduced to offset it. We also need to determine how individuals who have themselves withdrawn, or who have been ostracized by others, can be reintegrated into the group. There are suggestions that group processes in space will lead to a high level of conformity and a lessening of creativity over time. This area needs to be explored. Other questions involving intragroup functioning on long-duration spaceflights concern how leaders emerge in space and how leadership roles change over time.

Finally, and perhaps most crucially, we need to examine the effects of the passage of time on the relationships between the spacecrew and mission control. We anticipate a general drift away from the control of the sponsoring society. In addition, there is a point in many confinement experiences at which hostility toward, and rejection of, outside authority is displayed. We need to determine how this process is affected by the duration of flight and whether factors, such as known length of stay, influence intergroup attitudes. For instance, does the length of the flight affect the crew's willingness to have bad news from home withheld? Other crew/ground-control questions concern how to maintain communication when the flight is sufficiently long to render the mission effectively independent of the sponsoring society.

Competing Perspectives

In this section we will point out some questions which arise because different analysts, reviewers, or researchers have approached a problem with different underlying assumptions; because they have formulated conflicting interpretations of a consistent data set; or because the data themselves have been inconsistent.

One frequently expressed assumption is that selection and training techniques used in preparation for spaceflight are so highly sophisticated that issues of personal and social adjustment become almost irrelevant. We believe that this assumption begs the question and is potentially harmful to the cause of extended spaceflight. Selecting a small number of individuals from a large pool of candidates is considerably easier than selecting a large number of individuals from a shrinking pool of candidates. But, more to the point, we would argue that the success of past flights reflects the ability of space planners to effectively implement information supplied by researchers and practitioners. It does not mean that the space program has any proprietary formulae in these areas or that information on which selection and training decisions are made is complete, or even adequate. In the relationship between the space mission and the human, there exists a gap between what we know and what we need to know; this gap can only widen with extended spaceflight.

Looking closely at the selection area itself reveals differences in the way analysts view the task. Do we, for instance, select on the basis of task requirements (resulting in a crew of specialists) or on the basis of social requirements (resulting in a crew with a more generalist orientation)? And what, specifically, are the preferred social characteristics of space travelers? Here again we see differences of opinion. Some observers have argued that those with little need for, or skill in, social exchange would do well in space; others feel that a more responsive personality would be the appropriate choice. Characteristics of the flight are likely to assume importance in determining preferred characteristics of crewmembers. For instance, one could imagine that a narrowly focused individual would do well on a short flight with intense work requirements, whereas an individual with considerable flexibility would be required for a long-duration mission with periods of less concentrated work. Whatever the resolution of these and similar questions, the selection process will continue to suggest varied and potentially conflicting requirements.

Psychophysiological research presents many interesting puzzles. Some researchers have shown very positive results using biofeedback to control symptoms of motion sickness, yet others indicate that such training, in theory, should not work at all. If both areas of investigation are correct, one is left with the question of what is being trained in the former instance and how it relates to motion sickness. This question is not tied exclusively to motion sickness. For instance, biofeedback has been used to treat such common symptoms as headaches, yet there is little or no understanding of the etiology of these headaches or how biofeedback relieves them. Other psychophysiological discrepancies concern the interrelationship of physiological characteristics, personality traits, and susceptibility to motion and space sickness.

In the perceptual area, too, there are inconsistencies to be resolved. In the early days of spaceflight, sightings by astronauts gave rise to the belief that vision was improved by weightlessness. Subsequent analysis of the details of the viewing situation showed space acuity to be within normal Earth limits. Later Russian experiments indicated that certain aspects of vision, rather than being improved, are degraded in space. As with the reports of improved vision, these reports are likely to be in error. However, our knowledge of how various aspects of vision are affected by weightlessness and other spaceflight conditions remains sketchy.

Evidence on sleep in space has also been inconsistent. Sleep has been poor during most flights, yet there is some evidence from Skylab that sleep may actually be better and more efficient in space than on Earth. One can easily construct plausible explanations for either of these conclusions; still, the actual effects of the space environment on sleep need to be determined.

Once agreement has been reached that a problem exists, observers with different perspectives may prescribe conflicting strategies for solving it. For instance, hostility among isolated and confined individuals tends to be quite predictable and to follow a consistent pattern. Some believe that the nature of such conflict should be explained to space travelers, along with suggestions on how to respond. Others argue that such warnings would constitute self-fulfilling prophesies that would further add to the burden of adapting to life in space. Similarly, some advisors propose that on-board "gripe" sessions should be held as a means of managing hostility in space. Other advisors, relying on the observation that isolated and confined individuals circumvent even mild encounters, suggest avoiding confrontations, since they may be damaging to the stability of the group.

Conflicting recommendations can also be found in the area of leadership. Some observers believe that leaders best maintain their positions of authority by psychologically distancing themselves from the rest of the crew. However, maintaining distance from the rest of the crew limits the information the leader is able to receive, and could burden the leader with the extreme stress of being an isolate in a separated group. Before either strategy is adopted, we must understand all potential consequences. It may be that a compromise strategy is required. For instance, in the early days of sailing, the ship's captain remained aloof from the activities of the crew, but he was allowed to bring aboard an individual of his choice who would act as his companion/confidant. If the demands of leadership are found to require the separation of the space commander from the rest of the crew, the role of companion to the leader may need to be revived.

The dilemma of leadership has a parallel in appropriate behavior for a follower. Just how far should an individual, not in a leadership role, press his or her views on others? When a crewmember follows a directive which he or she believes to be in error, and as a result a failure or accident occurs, he or she will probably feel guilty and be judged by others to have behaved inappropriately. On the other hand, when a crewmember for any reason fails to follow a directive, and as a result a failure or accident occurs, other people will surely brand that crewmember's behavior as inappropriate. Unfortunately, such *post facto* determinations do not help the crewmember to know when to speak up, when and how long to argue, when to refuse to comply, and when to accede. These issues constitute the dilemma of effective followership.

How to promote a healthy relationship between spacecrews and mission control represents another area of conflicting assumptions. Some argue that crews should be trained to include ground control in their concept of the spacecrew, thereby dispelling the expressions of hostility that are so often directed toward outsiders by members of confined groups. The opposing position is that antagonism is a natural outgrowth of the confinement experience, and that if hostility is deflected away from outsiders, crewmembers will turn their anger against each other or against themselves.

In each of these cases, competing or conflicting perspectives have led to differing beliefs about behavioral and interpersonal problems in space, and steps needed to be taken to rectify them. In such cases, research directed at the source of the conflict would be particularly important for spaceflight planning.

Neglected Research Areas

Research topics pertinent to extended spaceflight differ widely in the attention they have been given. In this section we wish to underscore a few of those topics which we believe are both important in themselves, and deserve special consideration because they have received relatively little attention in the past.

Multiple stressors— Combinations of stressors can aggrevate, cancel, or reverse the effects of single stressors. These varied effects have been observed across a wide range of stressors including physiological, psychological, environmental, etc. It is essential that the stressors of space be examined in combination to determine their cumulative and interactive effects on individual behavior and interpersonal relations.

Selecting in— The necessity of selecting those best qualified to deal with the conditions of space differs in kind from the more usual psychological research which emphasizes eliminating the unfit. Those charged with selecting top executives or successful candidates for prestigious professional schools face similar, but far more limited, problems. Selection for space must consider all aspects of the individual. This need presents a potentially large area of research in which the determination of factors involved in emotional balance will occupy a central role.

Groups as groups—Selection strategies generally emphasize the qualities of the individual. However, it is also necessary to understand how personal qualities intermesh with, and potentially complement, those of others. Studies are needed to address this latter issue. Also, the majority of studies of small groups are based on individuals who were brought together solely for the purpose of the study; that is, groups of aggregated individuals. This approach is helpful in understanding interactions that occur at the beginning of a relationship. What is also needed are studies to reveal how people who have known each other for long periods of time behave.

Mediated communications— The effect of mediated communications is a relatively new area of investigation. It is therefore not surprising that research in this area has concentrated on average Earth situations; that is, on situations in which mediated communication is an essentially peripheral activity and in which the preponderance of significant communication is conducted face to face. To understand the effects of mediated communication in space, research is needed in environments in which indirect communication, rather than face-to-face communication, forms the essential link.

Boundary roles— Individuals who form the focus of interaction between groups perform an essential yet demanding function. In the stressful environment of an extended space mission, the difficulties of people who fulfill such roles are likely to be exacerbated. We need to know more about the effects of various structural arrangements, the personal characteristics necessary for the boundary role, and the effect of the role on its occupants.

On-board training— In the future, extensive ground training regimens may not always be practical. In addition, some crews will eventually find themselves with little essential work to occupy their time. Therefore, we suggest a greater emphasis on on-board, rather than preflight, training. We suspect that a significant number of tasks, even those which are important to the success of the mission (but not to the safety of the crew), can be learned, and may be learned better, in flight. On-board training could also be used to train crew members for less essential work, such as training for multiple duties, or general educational training. The practicality of this approach to training needs to be carefully explored.

RESEARCH OPPORTUNITIES

It is our belief that future experimentation on human psychological and social adjustment to space must lean heavily on the use of operational or realistically simulated operational environments. Although the limited-variable laboratory experiment will continue to play a role in the quest for better understanding of life in space, there is a need to study the adjustment to complex sets of variables which approximate, if not duplicate, the space experience.

Situations and Environments

Situations requiring that groups be isolated and confined have provided, and will continue to provide, a fertile source of information relevant to space. Submarines and other naval vessels, oil rigs and supertankers, as well as manned weather stations, diving bells, and remote military bases all simulate certain aspects of isolation, confinement, and risk found in spaceflight. For instance, the benefits to be derived from in-group use of crisis-intervention techniques or other helping behaviors in space could be examined by studying similar techniques aboard ship. Or, questions regarding the changing use of leisure with time and experience in space could take advantage of repeat confinees such as submariners or Arctic explorers. Arctic oil drillers offer a special opportunity for studying large crews of 750 to 1,000 individuals. Simulation projects themselves should not be ignored. A systems test of hardware or hardware/human factors can be informative in regard to other aspects of performance or to the question of human interactions.

Some issues of interest to space can be addressed by studying ground populations in normal settings. For instance, we know little of what motivates individuals to persist in exercise programs. Acquisition of even such basic information could aid the design of exercise programs for space. Problem definition groups, such as those frequently convened in private industry or by NASA, constitute a mission-oriented group not unlike that of a spacecrew. A casual observation of the dynamics of such groups suggests that there is much to be learned from a careful assessment of how these groups move toward consensus.

Some research opportunities may come from novel sources. For instance, airbeds used to treat burn patients could be used to simulate sleep research for space. Bedrest studies, such as those conducted at Ames Research Center, have proven to be a good method of assessing many biomedical and, more recently, performance aspects of spaceflight. Bedrest studies could also constitute the setting for studying communications questions such as the use of mediated systems for affective links with family and friends, or for training experiments to improve communications between crew and ground control.

Most confined groups that have been the subject of research examination have been composed of highly similar individuals; as a result, we have only limited knowledge of how diverse crews would perform in space. Almost any situation providing the opportunity to study diverse crews working together in space-like environments could be highly informative. The U.S. Navy now has men and women serving side by side on certain vessels. These assignments provide a valuable opportunity to examine mixed-gender living and working arrangements over time. It has been proposed that the United Nations sponsor an International Sea Service to explore the ocean bottom. If a suggestion of this nature were to be implemented, it could provide an opportunity to study multinational crews. Special attention could be given to verbal and nonverbal communication as well as to the broader relational aspects of culturally diverse groups.

When dealing with very long-term effects, only a few environments offer simulation opportunities. Historically, prison populations have been rejected as sources of information relevant to space. Inmates of jails and crews on spaceflights probably have little in common. Yet both find themselves in long-term situations of high risk without the possibility of leaving. We would suggest that, rather than rejecting prison populations out of hand, one should examine carefully what might be learned in this environment. Similarities can also be found in institutions such as convents and seminaries, in religious or philosophical sects, and in kibbutzim. It should be remembered that information, even from seemingly unrelated sources, can provide important insights, especially when dealing with the long-duration effects so difficult for experimenters to study.

Spaceflight itself is an obvious but underused resource for conducting research on psychological and social adjustment to space. For some questions, such as the prolonged effects of 0-g, space is the only appropriate environment; for others, it is simply the preferred environment. In either case, it should not be overlooked. The development of a space laboratory or a space station would present the opportunity to observe "breakoff" or other transcendent experiences. The inclusion of payload specialists from the European Space Agency could provide useful material on mixed crews. James Beggs, the NASA Administrator, has announced intentions to fly private citizens aboard the Space Shuttle (NASA Activities, 13(10), Oct. 1982). This decision could open up a number of research opportunities; for example, a chance to examine the responses to weightlessness of individuals who are of average physical condition, and to explore social questions such as the acceptance of members who are not part of the traditional group.

Methods and Approaches

Concentrating on applied research means that researchers must contend with the difficulties and limitations of field research as described in our opening chapter. Along with identifying and developing opportunities in which useful parallels to space exist, researchers must also identify and develop appropriate methods and approaches. This may mean abandoning an experimental approach in favor of a quasi-experimental or even a nonexperimental orientation. Although some situations may provide an opportunity for simulated pre- and post-mission testing, others may require that the researcher rely on *post facto* or survey research techniques. A researcher should also consider the practicality of unobtrusive measures such as remote surveillance, or of participant observation, or even such unconventional approaches as role-playing.

For one interested in advancing the understanding of space, the important factor is to determine where and how information can be gathered. We believe that there are certain factors that will aid this process. The first is simply to think in terms of space questions. As alluded to above, many projects and institutional arrangements offer an opportunity to examine issues relevant to space; the opportunities suggested here are only a sampling of those that will occur to a watchful observer. An attitude of thinking in terms of the conditions of spaceflight can convert many environments into environments suitable for human spaceflight-related research.

Another factor involves looking for opportunities to "piggyback," or tag along, on ongoing programs. Some research, such as fear reactions, can, for ethical reasons, be done only by piggybacking on existing programs. But piggybacking has practical advantages, even when it is not essential. Most of the opportunities, such as those listed above, are conducted for purposes of their own — either in the public interest or to satisfy some private commercial endeavor. By piggybacking on such activities, needed space research can be done for a small fraction of the cost otherwise required.

Finally, research needs should be combined whenever possible. Again, the essential ingredient is to think in terms of human needs in spaceflight, rather than limiting one's view to narrow research interests. By attempting to include others whose research could proceed in parallel with one's own, a greater net benefit could be gained from the ongoing activity.

CONCLUSIONS

Human adjustment to space is a new field of study. As such, it lacks the coordination found in more mature research specialties. We have attempted in this book to advance the understanding of life in space by providing a framework in which the various behavioral and social issues could be considered and by attempting to organize and synthesize the relevant data.

We have also indicated topics that we see as fruitful areas of research. This in no way implies that the Space Agency stands ready to support such investigations, or even concurs in our judgments. In fact, no source could support the volume of research that is potentially important to extended spaceflight; the relevant issues are too diverse and the questions too many. Information must be gathered using whatever means are available. As stated in the introduction to this book, we are now at the point that marks the end of our sojourn as visitors in space and the beginning of our role as space inhabitants. The thoughtful contributions from many disciplines can ease this passage.

REFERENCES

Abbott, H. M.; and Duddy, J. H.: Weightless Simulation Using Water Immersion Techniques: An Annotated Bibliography. Lockheed Missiles and Space Co., Sunnyvale, Calif., 1965. (AD-649861)

Abe, K.; Amatomi, M.; and Kajiyama, S.: Genetical and Developmental Aspects of Susceptibility to Motion Sickness and Frostbite. Hum. Hered., vol. 20, 1970, pp. 507-516.

Adams, Carl R.; and Bulk, George K.: Zero Buoyancy: Simulation of Weightlessness to Evaluate Psychophysiological and Anthropomorphic Parameters that Affect Space Station Design. Aerosp. Med., vol. 38, no. 5, 1967, pp. 518-520.

Adams, J. S.: Inequity in Social Exchange. Advances in Experimental Social Psychology, vol. 2, Leonard Berkowitz, ed., Academic Press, 1965, pp. 287-299.

——: The Structure and Dynamics of Behavior in Organizational Boundary Roles. Handbook of Industrial and Organizational Psychology, Marvin D. Dunnette, ed., Rand-McNally College Pub., 1976, pp. 1175-1199.

Adams, O. S.; and Chiles, W. D.: Human Performance as a Function of the Work-Rest Cycle. USAF WADC TR 60-248, 1960. (AD-240654)

——: Human Performance as a Function of Work-Rest Ratio During Prolonged Confinement. USAF ASD TR 61-720, 1961.

——: Prolonged Human Performance as a Function of the Work-Rest Cycle. Aerosp. Med., vol. 34, no. 2, 1963a, pp. 132-138.

——: Human Performance and the Work-Rest Schedule. Human Factors in Technology, Edward Martin Bennett, J. Degan, and J. Spiegel, eds., McGraw-Hill Book Co., 1963b, pp. 38-64.

Adams, O. S.; Levine, R. B.; and Chiles, W. D.: Research to Investigate Factors Affecting Multiple-Task Psychomotor Performance. USAF WADC TR 59-120, 1959.

Adams, Virginia: Jane Crow in the Army: Obstacles to Sexual Integration. Psych. Today, vol. 14, no. 5, October 1980, pp. 50-59, 62, 64-65.

Adolfson, John: Human Performance and Behavior in Hyperbaric Environments. Acto Psychologica Gothoburgensia VI, John Elmgren, ed., Almquist and Wiksell (Stockholm), 1967.

Aiello, J. R.; Epstein, Y. M.; and Karlin, R. A.: Effects of Crowding on Electrodermal Activity. Sociological Symposium, vol. 14, 1975a, pp. 43-57.

———: Field Experimental Research on Human Crowding. Paper presented at the Western Psychological Assn. Meeting, Sacramento, Calif., April 1975b.

Aguilera, Donna C.; and Messick, Janice M.: Crisis Intervention: Theory and Methodology. Second ed., C. V. Mosby Co., St. Louis, Mo., 1974.

Akiskal, H. S.; and McKinney, W. T.: Overview of Recent Research in Depression — Integration of Ten Conceptual Models into a Comprehensive Clinical Frame. Arch. Gen. Psychiatry, vol. 32, no. 3, March 1975, pp. 285-305.

Alexander, M.; and Isaac, W.: Effect of Illumination and d-Amphetamine on the Activity of the Rhesus Macaque. Psychol. Rep., vol. 16, no. 1, 1965, pp. 311-313.

Alexander, S. J.; Cotzin, M.; Hill, C. H.; Ricciuti, E. A.; and Wendt, G. R.: Wesleyan University Studies on Motion Sickness: I. The Effects of Variation of Time Intervals Between Accelerations Upon Sickness Rates. J. Psychol., vol. 19, 1945a, pp. 49-62.

———: Wesleyan University Studies on Motion Sickness: II. A Second Approach to the Problem of the Effects of Variation of Time Intervals Between Accelerations upon Sickness Rates. J. Psychol., vol. 19, 1945b, pp. 63-68.

———: Wesleyan University Studies on Motion Sickness: III. The Effects of Various Accelerations upon Sickness Rates. J. Psychol., vol. 20, July 1945c, pp. 3-8.

———: Wesleyan University Studies on Motion Sickness: IV. The Effects of Waves Containing Two Acceleration Levels upon Sickness. J. Psychol., vol. 20, July 1945d, pp. 9-15.

———: Studies of Motion Sickness: Experimental Proof that Aviation Cadets Tell the Truth on Motion Sickness History Questionnaires. J. Psychol., vol. 30, 1955, pp. 403-409.

Allport, G. W.: The Nature of Prejudice. Addison-Wesley, 1954.

Alluisi, Earl A.: Methodology in the Use of Synthetic Tasks to Assess Complex Performance. Human Factors, vol. 9, no. 4, 1967, pp. 375–384.

———: Sustained Performance. Principles of Skill Acquisition, Edward Alfred Bilodeau and Ina McD. Bilodeau, eds., Academic Press, 1969, pp. 59–102.

———: Optimum Uses of Psychobiological, Sensorimotor, and Performance Measurement Strategies. Human Factors, vol. 17, Aug. 1975, pp. 309–320.

———: Performance Measurement Technology: Issues and Answers. Proceedings of the Symposium on Productivity Enhancement: Personnel Performance Assessment in Navy Systems, Oct. 12–14, 1977, Louis T. Pope and David Meister, eds., Navy Personnel Research and Development Center, 1977, pp. 343–359.

Alluisi, E. A.; Beisel, W. R.; Bartelioni, B. J.; and Coates, C. D.: Behavioral Effects of Tularemia and Sandfly Fever in Man. Infect. Dis., vol. 128, no. 6, 1973, pp. 710–717.

Alluisi, E. A.; and Chiles, W. D.: Sustained Performance, Work-Rest Scheduling, and Diurnal Rhythms in Man. Acta Psychol., vol. 27, 1967, pp. 436–442.

Alluisi, E. A.; Chiles, W. D.; and Hall T. J.: Combined Effects of Sleep Loss and Demanding Work-Rest Schedules on Crew Performance. USAF AMRL TR 64-63, 1964. (AD-606214)

Alluisi, Earl A.; Chiles, W. Dean; Hall, Thomas J.; and Hawkes, Glenn R.: Human Group Performance During Confinement. AMRL-TDR-63-87, Wright-Patterson AFB, Ohio, 1963.

Alluisi, E. A.; Coates, G. D.; and Morgan, B. B., Jr.: Effects of Temporal Stresses on Vigilance and Information Processing. Vigilance: Theory, Operational Performance, and Physiological Correlates, Robert H. Mackie, ed., Plenum Press, 1977, pp. 361–422. (NATO Conf. Ser. 3: Human Factors, vol. 3.)

Alluisi, E. A.; and Morgan, B. B., Jr.: Temporal Factors in Human Performance and Productivity. Human Performance and Productivity Series, vol. 3. Stress and Performance Effectiveness, Earl A. Alluisi and Edwin A. Fleishman, eds., Lawrence Erlbaum Assoc., 1982, pp. 165–247.

Alluisi, E. A.; and Thurmond, J. B.: Behavioral Effects of Infectious Diseases. Annual Progress Report. Performance Research Laboratory Rep. no. PR-65-3, Univ. of Louisville, Ky., 1965.

Altman, Irwin: Reciprocity of Interpersonal Exchange. J. Theory of Soc. Behav., vol. 3, no. 2, Oct. 1973, pp. 249–261.

———: The Environment and Social Behavior: Privacy, Personal Space, Territory and Crowding. Brooks-Cole, Monterey, Calif., 1975.

———: Privacy Regulation. Culturally Universal or Culturally Specific? J. Social Issues, vol. 33, no. 3, 1977, pp. 66–84.

Altman, Irwin; and Haythorn, William W.: Interpersonal Exchange in Isolation. Sociometry, vol. 28, no. 4, 1965, pp. 411–426.

———: The Ecology of Isolated Groups. Behav. Sci., vol. 12, no. 3, May 1967a, pp. 169–182.

———: The Effects of Social Isolation and Group Composition on Performance. Human Relations, vol. 20, no. 4, 1967b, pp. 313–339.

Altman, I.; and Taylor, D. A.: Social Penetration: The Development of Interpersonal Relationships. Holt, Rinehart and Winston, 1973.

Amir, Yehuda: Contact Hypothesis in Ethnic Relations. Psychol. Bull., vol. 71, no. 5, 1969, pp. 319–342.

Andrews, Gavin; Tennant, Christopher; Hewson, Daphne M.; and Vaillant, George E.: Life Event Stress, Social Support, Coping Style and Risk of Psychological Impairment. J. Nerv. and Men. Dis., vol. 166, no. 5, May 1978, pp. 307–316.

Andreyeva-Galanina, Ye. Ts.; Alekeseyev, S. V.; Kadyskin, A. V.; and Suvorov, G. A.: Noise and Noise Sickness. Meditsina Press, Leningrad, 1972. NASA TT F-748, July 1973.

Archea, John: The Place of Architectural Factors in Behavioral Theories of Privacy. J. Social Issues, vol. 33, no. 3, Summer 1977, pp. 116–137.

Argyle, Michael: Social Interaction. Methuen (London), 1969.

Aronson, Elliot; and Mills, Judson: The Effect of Severity of Initiation on Liking for a Group. J. Abnorm. and Soc. Psychol., vol. 59, Sept. 1959, pp. 177–181.

Asch, S. E.: Effects of Group Pressure on the Modification and Distortion of Judgements. Groups, Leadership and Men, H. Guetzkow, ed., Carnegie Press, Pittsburgh, 1951, pp. 177–190.

———: Studies of Independence and Conformity: I. A Minority of One Against a Unanimous Majority. Psychol. Monogr. 416, vol. 70, no. 9, 1956.

Aschoff, Jurgen: Features of Circadian Rhythms Relevant for the Design of Shift Schedules. Ergonomics, vol. 21, 1978, pp. 739–754.

Atkinson, J. W., ed.: Motives in Fantasy, Action, and Society: A Method of Assessment and Study. Van Nostrand, 1952.

Averill, James R.: Grief: Its Nature and Significance. Psychol. Bull., vol. 70, no. 6, 1968, pp. 721–748.

———: Personal Control Over Aversive Stimuli and Its Relationship to Stress. Psychol. Bull., vol. 80, no. 4, Oct. 1973, pp. 286–303.

Babin, B. A.; and Loiko, M. A.: Training Pilots for Action in Flight Emergencies. Voprosy Psikhologii, vol. 19, no. 1, Jan. 1973, pp. 134-136.

Bailey, Gerald C.; Nordlie, Peter G.; and Sistrunk, Frank: Teleconferencing: Literature Review, Field Studies, and Working Papers. Research Paper P-113, Institute for Defense Analyses, Washington, D.C., March 1964. (AD-480695)

Bair, James H.: Productivity Assessment of Office Information Systems Technology. Proceedings of the IEEE Symposium on Trends and Applications in Distributed Processing. National Bureau of Standards, May 18, 1978.

——: An Analysis of Organizational Productivity and the Use of Electronic Office Systems. American Society for Information Sciences, 43rd Annual Meeting, Anaheim, Calif., Oct. 5-10, 1980. Proceedings on Communicating Information, Alan R. Benenfeld and Edward John Kazlauskus, eds., Proc. ASIS, vol. 17, pp. 4-9, Knowledge Industries Pub., White Plains, New York, 1980.

Bakan, P.: Preliminary Tests of Vigilance for Verbal Material. Univ. of Illinois contract AF 33(058)-25726, issued by Human Resources Research Center, USAF. Research Tech. Note 52-7, Lackland AFB, Texas, Proj. no. 507-011-000, July 1952.

Bakeman, Roger; and Helmreich, Robert: Cohesiveness and Performance: Covariation and Causality in an Undersea Environment. J. Experim. Soc. Psychol., vol. 11, no. 5, Sept. 1975, pp. 472-489.

Bales, R. F.: Interaction Process Analysis. A Method for the Study of Small Groups. Addison-Wesley, 1950.

——: The Equilibrium Problem in Small Groups. Working Papers in the Theory of Action, Talcott Parsons, Robert F. Bales, and Edward A. Shils, eds., Free Press, Glencoe, Ill., 1953, pp. 111-162.

——: Task Roles and Social Roles in Problem Solving Groups. Readings in Social Psychology, third ed., Eleanor E. Maccoby, T. M. Newcomb, and E. Hartley, eds., Holt, Rinehart & Winston, 1958, pp. 437-446. (Prepared for the Committee on the Teaching of Social Psychology of the Society.)

——: Personality and Interpersonal Behavior. Holt, Rinehart & Winston, 1970.

Barker, Roger Garlock: Ecological Psychology: Concepts and Methods for Studying the Environment of Human Behavior. Stanford Univ. Press, Stanford, Calif., 1968.

Barnes, R.: Habitability Requirements for Multiman, Long-Duration Missions. Report prepared for Biotechnology and Human Research Division, OART, Washington, D.C., 1969.

Baron, Reuben M.; Mandel, David R.; Adams, Claire A.; and Griffen, Lynne M.: Effects of Social Density in University Residential Environments. J. Pers. and Soc. Psychol., vol. 4, no. 3, 1976, pp. 434-446.

Baron, Robert A.; and Bell, Paul A.: Aggression and Heat: The Influence of Ambient Temperature, Negative Effect, and a Cooling Drink on Physical Aggression. J. Pers. and Soc. Psychol., vol. 33, no. 3, 1976, pp. 245-255.

Barrett, G. V.; and Thornton, C. L.: Relationship Between Perceptual Style and Simulator Sickness. J. Appl. Psychol., vol. 52, no. 4, 1968, pp. 304-308.

Bart, Pauline B.: The Sociology of Depression. Explorations in Psychiatric Sociology, Paul M. Roman and Harrison M. Trice, eds., F. A. Davis Co., Philadelphia, 1974, pp. 139-155.

Basowitz, Harold; Korchin, Sheldon A.; and Grinker, Roy R.: Anxiety and Stress. McGraw-Hill Book Co., 1955.

Baum, Andrew; and Davis, Glen E.: Reducing the Stress of High Density Living: An Architectural Intervention. J. Pers. and Soc. Psychol., vol. 38, no. 3, 1980, pp. 471-481.

Baumel, I. P.; Robinson, S. M.; and Blatt, W. F.: Multi-Chamber System for Toxicity Studies in Mice at Simulated High Altitude. J. Pharm. Sci., vol. 56, no. 7, July 1967, pp. 918-919.

Bakwin, H.: Car-Sickness in Twins. Dev. Med. Child Neurol., vol. 13, no. 3, Jan. 1971, pp. 310-312.

Beck, A. T.: The Development of Depression: A Cognitive Model. The Psychology of Depression: Contemporary Theory and Research, Raymond J. Friedman and Martin M. Katz, eds., V. H. Winston, Washington, D.C., 1974, pp. 3-27.

Becker, Ernest: The Revolution in Psychiatry: The New Understanding of Man. Free Press of Glencoe, Collier-MacMillan Ltd. (London), 1964, pp. 108-135.

Becker, R. W.; Poza, F.; and Kryter, K. D.: A Study of Sensitivity to Noise. DOT-FA69WA-2211, SRI, Menlo Park, Calif., June 1971.

Behnke, A. R.; Thomson, R. M.; and Motley, E. P.: Psychologic Effects from Breathing Air at 4 Atmospheres Pressure. Amer. J. Physiol., vol. 112, July 1935, pp. 554-558.

Beisel, W. R.; Morgan, B. B., Jr.; Bartelloni, B. J.; Coates, G. D.; DeRubertis, F. R.; and Alluisi, E. A.: Symptomatic Therapy in Viral Illness: A Controlled Study of Effects on Work Performance. J. Amer. Med. Assoc., vol. 228, no. 5, 1974, pp. 581-584.

Beljan, J. R.: Desynchronosis Literature Analysis, Evaluation, and Data Summation. Part 1-A: Human Performance in the Aviation Environment. NASA CR-135934, Dec. 1972.

Bell, Paul A.: Effects of Noise and Heat Stress on Primary and Subsidiary Task Performance. Human Factors, vol. 20, no. 6, Dec. 1978, pp. 749-750.

Bem, Sandra L.: The Measurement of Psychological Androgyny. J. Consulting and Clin. Psychol., vol. 42, no. 2, April 1974, pp. 153-162.

——: Sex Role Adaptability: One Consequence of Psychological Androgyny. J. Pers. and Soc. Psychol., vol. 31, no. 4, April 1975, pp. 634-643.

Bem, S. L.; and Bem, D. J.: Case Study of a Nonconscious Ideology: Training the Woman to Know Her Place. Beliefs, Attitudes and Human Affairs, Daryl J. Bem, Brooks/Cole, Monterey, Calif., 1970, pp. 89-99.

Bennett, P. B.: Measurement and Mechanisms of Inert Gas Narcosis. J. Occupat. Med., vol. 11, no. 5, 1969, pp. 217-222.

Bennis, Warren: The Unconscious Conspiracy: Why Leaders Can't Lead. AMACOM, New York, 1976.

Bennis, Warren G.; and Shepard, Herbert A.: A Theory of Group Development. Human Relations, vol. 9, 1956, pp. 415-437.

Benson, A. J.: Spatial Disorientation and the "Break-off" Phenomenon. Aerosp. Med., vol. 44, no. 8, Aug. 1973, pp. 944-952.

Berglund, Birgitta; Berglund, Ulf; and Lindvall, Thomas: Psychological Processing of Odor Mixtures. Psychol. Rev., vol. 83, no. 6, 1976, pp. 432-441.

Bergum, Bruce O.; and Lehr, Donald J.: Vigilance Performance as a Function of Interpolated Rest. J. Appl. Psychol., vol. 46, no. 6, 1962, pp. 425-427.

——: Effects of Authoritarianism on Vigilance Performance. J. Appl. Psychol., vol. 47, no. 1, 1963, pp. 75-77.

Berkhout, J.: Simulated Time-Zone Shifts and Performance Ability: Behavioral, Electroencephalographic and Endocrine Effects of Transient Alterations in Environmental Phase. AGARD CP-74-70, AGARD Conference Proceedings on Rest and Activity Cycles for the Maintenance of Efficiency of Personnel Concerned with Military Flight Operations, June 1970.

Berkowitz, Leonard: Group Standards, Cohesiveness, and Productivity. Human Relations, vol. 7, 1954, pp. 509-519.

Berkun, Mitchell M.: Performance Decrement Under Psychological Stress. Human Factors, vol. 6, no. 1, Feb. 1964, pp. 21-30.

Berliner, David C.; Angell, David; and Shearer, James W.: Behaviours, Measures, and Instruments for Performance Evaluation in Simulated Environments. Paper presented at the Symposium and Workshop on the Quantification of Human Performance, Albuquerque, 1964. (Note: The report version was called "Study of Training Performance Evaluation Techniques." Am. Institute for Research, 1964.)

Berlo, David K.: The Process of Communication: An Introduction to Theory and Practice. Holt, Rinehart, and Winston, 1960.

Berry, Charles A.: Space Medicine in Perspective — A Critical Review of the Manned Space Program. J. Amer. Med. Assoc., vol. 201, no. 4, 1967, pp. 232–241.

——: Preliminary Clinical Report of the Medical Aspects of Apollos 7 and 8. NASA TM X-58027, 1969, pp. 197–218.

——: Summary of Medical Experience in the Apollo 7 Through 11 Manned Space Flights. Aerosp. Med., vol. 41, no. 5, 1970, pp. 500–519.

——: A View of Human Problems to be Addressed for Long Duration Space Flights. Aerosp. Med., vol. 44, no. 10, 1973a, pp. 1136–1146.

——: Weightlessness. Bioastronautics Data Book. Second ed., NASA SP-3006, 1973b, pp. 349–416.

Berry, C. A.; Coons, D. O.; Catterson, A. D.; and Kelly, G. F.: Man's Response to Long-Duration Flight in the Gemini Spacecraft. NASA SP-121, 1966, pp. 235–261.

Berry, Charles A.; and Catterson, Allen D.: Pre-Gemini Medical Predictions Versus Gemini Flight Results. Gemini Summary Conference. NASA SP-138, 1967.

Berry, Charles A.; and Homick, Gerry L.: Findings on American Astronauts Bearing on the Issue of Artificial Gravity for Future Manned Space Vehicles. Aerosp. Med., vol. 44, no. 2, Feb. 1973, pp. 163–168.

Berscheid, Ellen: Privacy: A Hidden Variable in Experimental Social Psychology. J. Social Issues, vol. 33, no. 3, 1977, pp. 85–101.

Berscheid, Ellen; and Walster, Elaine H.: Interpersonal Attraction. Second ed., Addison-Wesley, 1978.

Bevan, W.; Avant, L. L.; and Lankford, H. G.: Influence of Interpolated Periods of Activity and Inactivity Upon the Vigilance Decrement. J. Appl. Psychol., vol. 51, no. 4, Aug. 1967, pp. 352–356.

Bexton, W. H.; Heron, W.; and Scott, T. H.: Effects of Decreased Variation in the Sensory Environment. Canad. J. Psychol., vol. 8, 1954, pp. 70–76.

Billings, C. E.; and Cheaney, E. S.: Information Transfer Problems in the Aviation System. NASA TP-1875, 1981.

Billman, G. E.; Teoh, K.; Dickey, D. T.; and Stone, H. L.: Horizontal Body Casting and Baroreceptor Sensitivity: The Role of Central Blood Volume Shifts in the Rhesus Monkey. Aerosp. Med. Assoc. Preprint, 1981, pp. 82-83.

Bilodeau, J. W.: The Role of Simulation in Space Shuttle Training. Proceedings of the Summer Computer Simulation Conference, Toronto, Can., July 16-18, 1979. AFIPS Press, Montvale, N.J., 1980, pp. 813-818.

Birkhead, N. C.; Blizzard, J. J.; Issekutz, B., Jr.; and Rodahl, K.: Cardiodynamic and Metabolic Effects of Prolonged Bed Rest. WADD-AMRL-TDR-64-67, Wright-Patterson AFB, Ohio, Aug. 1964.

Birren, J. E.: Motion Sickness: Its Psychophysiological Aspects. A Survey Report on Human Factors in Undersea Warfare, National Research Council, Committee on Undersea Warfare, Panel on Psychology and Physiology, 1949, pp. 375-398.

Biryukov, Ye. N.; and Krasnykh, I. G.: The Change in the Optical Density of Bone Tissue and Calcium Metabolism in the Cosmonauts. Space Bio. Med., vol. 4, no. 9, 1970, pp. 58-62.

Blaney, Paul H.: Contemporary Theories of Depression: Critique and Comparison. J. Abnorm. Psychol., vol. 86, no. 3, June 1977, pp. 203-225.

Bluth, B. J.: The Truth About the Skylab Crew "Revolt." L-5 News, Sept. 1979, pp. 12-15.

———: Sociological Aspects of Permanent Manned Occupancy of Space. AIAA Student J., vol. 19, Fall 1981a, pp. 11-15, 44.

———: Soviet Space Stress. Science 81, vol. 2, no. 7, Sept. 1981b, pp. 30-35.

———: Staying Sane In Space. Mech. Eng., vol. 104, Jan. 1982, pp. 24-29.

Borsky, P. N.: A Comparison of a Laboratory and Field Study of Annoyance and Acceptability of Aircraft Noise Exposures. NASA CP-2772, Feb. 1977.

Bourne, G. H.; Nandy, K.; and Golarz De Bourne, M. N.: Muscle and the Weightless State. Hypodynamics and Hypogravics, Michael McCally, ed., Academic Press, 1968, pp. 187-212.

Boyle, C. P.: Manned and Womanned Space Flight. Space World, vol. 3, no. 171, March 1978, pp. 22-24.

Breed, Warren: Five Components of a Basic Suicide Syndrome. Life-Threatening Behavior, vol. 2, no. 1, Spring 1972, pp. 3-18.

Breeze, R. K.: Space Vehicle Environmental Control Requirement Based on Equipment and Physiological Criteria. ASD-TR-61-161. Pt. 1, Wright-Patterson AFB, Ohio, Nov. 1961.

Brehm, Jack Williams: A Theory of Psychological Reactance. Academic Press, 1966.

Brehmer, Berndt: Social Judgment Theory and the Analysis of Interpersonal Conflict. Psychol. Bull., vol. 83, Nov. 1976, pp. 985-1003.

Broadbent, D. E.: Perception and Communication. Pergamon Press (London), 1958.

——: Human Performance and Noise. Handbook of Noise Control. Second ed. C. M. Harris, ed., McGraw-Hill Book Co., 1979, pp. 17-1 to 17-20.

Brooks, Courtney G.; Grimwood, James M.; and Swenson, Loyd S.: Chariots for Apollo: A History of Manned Lunar Spacecraft. NASA SP-4205, 1979.

Broverman, Inge K.; Vogel, S. R.; Broverman, D. M.; Clarkson, F. E.; and Rosenkrantz, P. S.: Sex-Role Stereotypes: A Current Appraisal. J. Social Issues, vol. 28, no. 2, 1972, pp. 59-78.

Brown, Frederick M.; and Graeber, R. Curtis, eds.: Rhythmic Aspects of Behavior. Lawrence Erlbaum Asso., Inc., Hillsdale, N.J., 1982.

Brown, I. E.: Dual Task Methods of Assessing Work-Load. Ergonomics, vol. 21, no. 3, March 1978, pp. 221-224.

Bucher, Rue: Blame and Hostility in Disaster. Amer. J. Sociol., vol. 62, no. 5, March 1957, pp. 467-475.

Buck, Leslie: Reaction Time as a Measure of Perceptual Vigilance. Psychol. Bull., vol. 65, no. 5, 1966, pp. 291-304.

Burke, P. L.: Leadership Role Differentiation. Experimental Social Psychology, Charles Graham McClintock, ed., Holt, Rinehart & Winston, 1972, pp. 514-546.

Burns, N. M.; and Gifford, E. C.: Effects of Long-Term Confinement on Performance. Environmental Requirements of Sealed Cabins for Space and Orbital Flights — A Second Study. II. Rep. TED NAM AE-1403, Naval Air Material Center Air Crew Equipment Laboratory, Philadelphia, Pa., March 13, 1961.

Byrne, Donn Erwin: The Attraction Paradigm. Academic Press, 1971.

Byrne, D.; Gouaux, C.; Griffitt, W.; Lamberth, J.; Murakawa, N.; Prasad, M.; Prasad, A.; and Ramirez, M.: The Ubiquitous Relationship: Attitude Similarity and Attraction: A Cross-Cultural Study. Human Relations, vol. 24, no. 5, 1971, pp. 201-207.

Calhoun, J. B.: The Social Use of Space. Physiological Mammalogy, vol. 1, W. Mayer and Richard G. Van Gelder, eds., Academic Press, 1963, pp. 2-187.

Campbell, David P.: SVIB, Strong Vocational Inventory Blank, Manual for the Strong-Campbell Interest Inventory, T325 (merged form). Stanford University Press, Stanford, Calif., 1976.

Cantrell, G. K.; and Hartman, B. O.: Application of Time and Workload Analysis Technics to Transport Flyers, Sept. 1965-May 1966. SAM-TR-67-71, School of Aerospace Medicine, Brooks AFB, Texas, 1967.

Cantrell, R. W.: Effects of Noise Exposure. Effects of Long Duration Noise Exposure on Hearing and Health, Milton A. Whitcomb, ed., AGARD CP-171, Nov. 1975a, pp. C10-1 to C10-13.

——: Physiological Effects of Noise. Effects of Long Duration Noise Exposure on Hearing and Health, Milton A. Whitcomb, ed., AGARD CP-171, Nov. 1975b, pp. C11-1 to C11-11.

Capel, W. C.; Youngblood, David; and Stewart, G. T.: Note on Stress, Anxiety, and Related Defenses in a Controlled Situation. Psychol. Rep., vol. 27, no. 2, Oct. 1970, pp. 351-355.

Caplow, T.: How to Run Any Organization: A Manual of Practical Sociology. Dryden Press, New York, 1976.

Carkhuff, Robert R.: Helping and Human Relations: A Primer for Lay and Professional Helpers. Vol. I. Selection and Training; Vol. II, Practice and Research. Holt, Rinehart, and Winston, 1969-1970.

Carskadon, M. A.; and Dement, W. C.: Sleep Studies on a 90 Minute Day. EEG Clin. Neurol., vol. 39, no. 2, Aug. 1955, pp. 145-155.

Cartwright, D.: The Nature of Group Cohesiveness. Group Dynamics Research and Theory. Third ed. Dorwin Cartwright and Alvin Zander, eds., Harper & Row, 1968, pp. 91-109.

Casey-Stahmer, Anna E.; and Havron, M. Dean: Planning Research in Teleconference Systems. Rep. HSR-RR-73/10-ST-X, Human Science Research, Inc., McLean, Va., Sept. 1973.

Central Institute for the Deaf: Effects of Noise on People. EPA-68-01-05000, Dec. 31, 1971.

Chambers, Randall M.: Isolation and Disorientation. Physiological Problems in Space Exploration, James D. Hardy, ed., Chas. C. Thomas, 1964, pp. 231-297.

Champness, Brian G.: Bargaining at Bell Laboratories. Paper E/71270/CH, Communication Studies Group, London, England, 1971.

Chapanis, Alphonse; Ochsman, Robert B.; Parrish, Robert N.; and Weeks, Gerard D.: Studies in Interactive Communication. I — The Effects of Four Communication Modes on the Behavior of

Teams During Cooperative Problem-Solving. Human Factors, vol. 14, no. 6, Dec. 1972, pp. 487-509.

Chapman, L. F.; Winget, C. M.; Vernikos-Danellis, J.; and Evans, J. W.: Performance Capability and EEG Changes During 21 Days of Simulated Weightlessness (Bedrest). Paper presented at XIII Int. Cong. Aerosp. Med., Acapulco, Mexico, 1975.

Chapman, Loring F.; Winget, Charles M.; and Vernikos-Danellis, Joan: Flight Task Performance Capability During 21 Days Simulated Weightlessness and Intermittent Gravitational Stress. Aerosp. Med. Assoc. Preprint of 1976 Annual Scientific Meeting, Bal Harbour, Florida, May 10-13, 1976, p. 221.

Chase, G. A.; Grave, C.; and Rowell, L. B.: Independence of Changes in Functional and Performance Capacities Attending Prolonged Bed Rest. Aerosp. Med., vol. 37, no. 12, Dec. 1966, pp. 1232-1233.

Chernyshev, M: "Salyut-7" Station's Medical and Biological Research Equipment. Moskovskiy Komsomolets, Moscow, June 1982.

Chiles, W. Dean; and Adams, O. S.: Human Performance and the Work-Rest Schedule. ASD TR 61-270, Aeronautical Research Lab, Wright-Patterson AFB, Ohio, 1961.

Chiles, W. Dean; and Alluisi, Earl A.: On the Specification of Operator or Occupational Workload with Performance-Measurement Methods. Human Factors, vol. 21, no. 5, Oct. 1979, pp. 515-528.

Chiles, W. Dean; Alluisi, Earl A.; and Adams, Oscar S.: Work Schedules and Performance During Confinement. Human Factors, vol. 10, no. 2, 1968, pp. 143-196.

Chinn, H. I.: Evaluation of Drugs for Protection Against Motion Sickness Aboard Transport Ships: Report of Study by Army, Navy, Air Force Motion Sickness Team. J. Amer. Med. Assoc., vol. 160, no. 9, March 3, 1956, pp. 755-760.

Chinn, H. I.; and Smith, P. K.: Motion Sickness. Pharmacol. Rev., vol. 7, 1953, pp. 33-82.

Christian, J. J.: Endocrine Adaptive Mechanisms and the Physiological Regulation of Population Growth. Physiological Mammalogy, vol. 1, William V. Mayer and Richard G. Van Gelder, eds., Academic Press, 1963, pp. 189-353.

Christiansen, James E.; and Yoesting, Dean R.: The Substitutability Concept: A Need for Further Development. J. Leisure Res., vol. 9, 1977, pp. 188-207.

Christensen, Julien M.; and Simons, John C.: Human Performance in Space Systems. Lectures in Aerospace Medicine, USAF School of Aerospace Medicine, Brooks AFB, Texas, Feb. 9-12, 1970, pp. 400-457.

Clark, Brant; and Graybiel, Ashton: The Break-Off Phenomenon. J. Aviat. Med., vol. 28, no. 2, April 1957, pp. 121–126.

———: Human Performance During Adaptation to Stress in Pensacola Slow Rotation Room. Aerosp. Med., vol. 32, no. 2, Feb. 1961, pp. 93–106.

Cleveland, S. E.; Boyd, J.; Sheer, D.; and Reitman, E. E.: Effects of Fallout Shelter Confinement on Family Adjustment. Arch. Gen. Psychiatry, vol. 8, 1963, pp. 38–46.

Coburn, K. R.: A Report of the Physiological and Bacteriological Aspects of 20 Days in Full Pressure Suits, 20 Days at 27,000 Ft on 100% Oxygen, and 34 Days of Confinement. NASA CR-708, 1967.

Coch, Lester; and French, John R. P., Jr.: Overcoming Resistance to Change. Human Relations, vol. 1, 1948, pp. 512–532.

Cogoli, A.; Vallucki-Morf, M.; Mueller, M.; and Briegleb, W.: Effect of Hypogravity on Human Lymphocyte Activation. Aviat., Space, and Environ. Med., vol. 51, no. 1, Jan. 1980, pp. 29–34.

Cohen, Abraham I.: The Impact of the Death of a Group Member on a Therapy Group. Int. J. Group Psychotherapy, vol. 26, no. 2, April 1976, pp. 203–212.

Cohen, Harvey G.; Conrad, Donald W.; O'Brien, John F.; and Pearson, Richard G.: Effects of Noise Upon Human Information Processing. NASA CR-132469, 1974.

Cohen, S.; Evans, G. W.; Krantz, D. S.; and Stokols, D.: Physiological, Motivational and Cognitive Effects of Aircraft Noise on Children: Moving from the Laboratory to the Field. Amer. Psychol., vol. 35, no. 3, 1980, pp. 231–243.

Cohen, S.; Evans, G. W.; Krantz, D. S.; Stokols, D.; and Kelly, S.: Aircraft Noise and Children: Longitudinal and Cross-Sectional Evidence on Adaptation to Noise and the Effectiveness of Noise Abatement. J. Pers. and Soc. Psychol., vol. 40, no. 2, 1981, pp. 331–345.

Cohen, Sheldon: Aftereffects of Stress on Human Performance and Social Behavior: A Review of Research and Theory. Psychol. Bull., vol. 88, no. 1, 1980, pp. 82–108.

Cohen, Sheldon; Glass, David C.; and Phillips, Susan: Environment and Health. Handbook of Medical Sociology, Third ed., H. E. Freeman, S. Levine, and L. G. Reader, eds., Prentice Hall, 1979.

Collins, William E.; and Lentz, J. Michael: Some Psychological Correlates of Motion Sickness Susceptibility. Aviat., Space, and Environ. Med., vol. 48, no. 7, July 1977, pp. 587–594.

Colquhoun, W. P.: Circadian Variations in Mental Efficiency. Biological Rhythms and Human Performance, W. P. Colquhoun, ed., Academic Press, 1971, pp. 39-107.

Colquhoun, W. P.; Blake, M. J.; and Edwards, R. S.: Experimental Studies of Shift Work. II: Stabilized 8-Hour Shift Systems. Ergonomics, vol. 11, no. 6, 1968, pp. 527-546.

Communications Studies Group: Medium of Communication and Consensus. E/72210/SH, 1972.

Connell, S.: The 1973 Office Communications Survey. Communications Studies Group, P/74067/CN, 1974.

Conrath, David W.: Measuring the Computer's Impact on Organizational Structure. Proceedings of the 1st Int. Conf. on Computer Communication, Washington, D.C., Oct. 24-26, 1972. Stanley Winkler, ed. IEEE, New York, pp. 68-73.

Conroy, Richard Thomas Walter Lawrence; and Mills, J. N.: Human Circadian Rhythms. J. & A. Churchill (London), 1970.

Convertino, V.; Olsen, L.; Goldwater, D.; and Sandler, H.: Cardiovascular Responses to Exercise After Bed Rest (BR) in 45 to 55 Year Old Men and Women. Aerosp. Med. Assoc., Preprint of 1979 Annual Meeting, Washington, D.C., May 14-17, 1979, pp. 47-48.

Conyers, James H.: An Exploratory Study of Musical Tastes and Interests of College Students. Sociological Inquiry, vol. 33, 1963, pp. 8-66.

Cooper, Alec.: The Biochemistry of Affective Disorders. Brit. J. Psychiatry, vol. 113, Nov. 1967, pp. 1237-1264.

Cooper, G.: Cooper Reports on Details of MA-9 Flight. Aviat. Week and Space Tech., vol. 79, no. 16, Oct. 14, 1963, pp. 61-81.

Cooper, G. E.; and Harper, R. P., Jr.: The Use of Pilot Rating in the Evaluation of Aircraft Handling Qualities. NASA TN D-5153, 1969.

Cooper, Henry S. F., Jr.: A House In Space. Bantam Books, 1976.

———: Comments on B. J. Bluth's "The Truth About the Skylab Crew 'Revolt'." L-5 News, Sept. 1979, p. 13.

Cooper, Robert: Leader's Task Relevance and Subordinate Behavior In Industrial Work Groups. Human Relations, vol. 19, no. 1, 1966, pp. 57-84.

Covault, C.: Mission 6's EVA to Verify Capability. Aviat. Week and Space Tech., vol. 118, Jan. 3, 1983, pp. 47-49.

Cowan, Thomas Anthony; and Strickland, D. A., with the collaboration of Martin Stow, Suellen Lanstein, and John Bosley: The Legal Structure of a Confined Microsociety: A Report on the

Cases of Penthouse II and III. Space Sciences Lab., Internal Working Paper no. 34, Univ. of Calif., Berkeley, Calif., NASA CR-69513, Aug. 1965.

Cowings, P. S.: Combined Use of Autogenic Therapy and Biofeedback in Training Effective Control of Heart Rate by Humans. In Proceedings of International College of Psychosomatic Medicine, Third, Rome, Sept. 16-20, 1975, W. Luthe and F. Antonelli, eds. Vol. 4, Autogenic Therapy, Roma, L. Pozzi, 1977a, pp. 167-173.

———: Observed Differences in Learning Ability of Heart Rate Self-Regulation as a Function of Hypnotic Susceptibility. In Proceedings of International College of Psychosomatic Medicine, Third, Rome, Sept. 16-20, 1975. W. Luthe and F. Antonelli, eds. Vol. 4, Autogenic Therapy, Roma, L. Pozzi, 1977b, pp. 221-226.

Cowings, P. S.; Billingham, J.; and Toscano, B. W.: Toward the Solution of the Zero Gravity Sickness Syndrome: Simultaneous Control of Multiple Autonomic Responses During Coriolis Acceleration. Aerosp. Med. Assoc., Preprint of 1976 Annual Scientific Meeting, Bal Harbour, Florida, May 10-13, 1976, pp. 219-220.

———: Learned Control of Multiple Autonomic Responses to Compensate for the Debilitating Effects of Motion Sickness. Ther. Psychosomatic Med., vol. 4, 1977, pp. 318-323.

Cowings, Patricia S.; and Toscano, William B.: A Theory on the Evolutionary Significance of Psychosomatic Disease. Ther. Psychosomatic Med., vol. 4, 1977, pp. 404-409.

———: The Relationship of Motion Sickness Susceptibility to Learned Autonomic Control for Symptom Suppression. Aviat., Space, and Environ. Med., vol. 53, no. 6, June 1982, pp. 570-575.

Cramer, E. H.; and Flinn, D. E.: Psychiatric Aspects of the SAM Two-Man Space Cabin Simulator. SAM TDR 63-27, Sept. 1963.

Crandall, Rick: The Assimilation of Newcomers into Groups. Small Group Behavior, vol. 9, no. 3, 1978, pp. 331-336.

Crocq, L.; Rivolier, J.; and Cazes, G.: Selection and Psychological Adjustment of Individuals Living in Small Isolated Groups in the French Antarctic Stations. Polar Human Biology, O. G. Edholm and E. K. E. Gunderson, eds. Wm. Heineman Medical Books Ltd., Cambridge, England, 1973.

Culbert, Samuel A.: Trainer Self-Disclosure and Member Growth in Two T-Groups. J. Appl. Behav. Sci., vol. 4, no. 1, 1968, pp. 47-73.

Cunningham, Walter; and Herskowitz, Mickey: The All-American Boys. Macmillan, 1977.

Danesh, Hossain G.: Anger and Fear. Amer. J. Psychiatry, vol. 134, no. 10, Oct. 1977, pp. 1109-1112.

D'Atri, D. A.: Psychophysiological Responses to Crowding. Environ. and Behav., vol. 7, no. 2, June 1975, pp. 237-252.

D'Atri, D. A.; and Ostfeld, A. M.: Stress, Crowding and Blood Pressure in Man. Proceedings of the American Public Health Assoc., vol. 116, Nov. 4-8, 1973.

Davenport, F. W.; Congdon, S. P.; and Pierce, B. F.: The Minimum Volumetric Requirements of Man in Space. Paper 63-250, Paper presented at the AIAA Summer Meeting, Los Angeles, Calif., June 17-20, 1963.

Day, Robert C.; and Hamblin, Robert L.: Some Effects of Close and Punitive Styles of Supervision. Amer. J. Sociology, vol. 69, March 1964, pp. 499-510.

Day, R. M.: Ships Laws: Normative Structure for Isolated Groups. Texas Christian Univ., Institute of Behavioral Research, Fort Worth, Texas, 1979.

DeLamater, J.: Intimacy in a Coeducational Community. Explorations in Psychology, Albert A. Harrison, ed., Brooks/Cole, Monterey, Calif., 1974, pp. 278-290.

Dember, W. N.; and Earl, R. W.: Analysis of Exploratory, Manipulatory, and Curiosity Behaviors. Psychol. Rev., vol. 64, 1957, pp. 91-96.

Demling, R. H.; Perea, A.; Maly, J.; Moylan, J. A.; Jarrett, F.; and Balish, E.: The Use of a Laminar Airflow Isolation System for the Treatment of Major Burns. Amer. J. Surg., vol. 136, no. 3, Sept. 1978, pp. 375-378.

Derlega, V. J.; Wilson, M.; and Chaikin, A. L.: Friendship and Disclosure Reciprocity. J. Pers. and Soc. Psychol., vol. 34, no. 4, 1976, pp. 578-582.

Derlega, Valerian J.; and Chaiken, Alan L.: Privacy and Self-Disclosure in Social Relationships. J. Social Issues, vol. 33, no. 3, Summer 1977, pp. 102-115.

Dervin, Brenda: Mass Communications: Changing Conceptions of the Audience. Public Communication Campaigns, Ronald E. Rice and William J. Paisley, eds., Sage, Beverly Hills, Calif., 1981, pp. 71-87.

Desor, J. A.: Toward a Psychological Theory of Crowding. J. Personality and Soc. Psychol., vol. 21, no. 1, 1972, pp. 79-83.

Dickey, D. T.; Billman, G. E.; Teoh, K.; Sandler, H.; and Stone, H. L.: The Effects of Horizontal Body Casting on Blood Volume, Drug Responsiveness, and $+G_z$ Tolerance in the Rhesus Monkey. Aviat., Space, and Environ. Med., vol. 53, no. 2, 1982, pp. 142-146.

Dietlin, L. F.; and Johnston, R. S.: U.S. Manned Space Flight: The First Twenty Years — A Biomedical Status Report. Acta Astronautics, vol. 8, nos. 9-10, Sept.-Oct. 1981, pp. 893-906.

Dietrick, J. E.; Whedon, G. D.; and Shorr, E.: Effects of Immobilization upon Various Metabolic and Physiologic Functions of Normal Man. Amer. J. Med., vol. 4, Jan. 1948, pp. 3-36.

Dodge, Robin: Circadian Rhythms and Fatigue: A Discrimination of Their Effects on Performance. Aviat., Space, and Environ. Med., vol. 53, no. 11, Nov. 1982, pp. 1131-1137.

Doll, R. E.; Gunderson, E. K. E.: Group Size, Occupational Status, and Psychological Symptomatology in an Extreme Environment. J. Clin. Psychol., vol. 27, no. 2, April 1971, pp. 196-198.

Dorcus, R. M.: Hypnosis as a Tool for Investigating Some Problems of Space Flight. Symposium on Psychophysiological Aspects of Space Flight, Bernard E. Flaherty, ed., Columbia Univ. Press, New York, 1960.

Drabeck, Thomas E.; and Quarantelli, Enrico L.: Scapegoats, Villains, and Disasters. Trans-action, vol. 4, no. 4, March 1967, pp. 12-17.

Dubos, R.: The Social Environment. Environmental Psychology: Man and His Physical Setting, H. M. Proshansky, W. H. Ittelson, and L. G. Rivlin, eds., Holt, Rinehart and Winston, 1970, pp. 202-208.

Dunlap, R. D.: The Selection and Training of Crewmen for an Isolation and Confinement Study in the Douglas Space Cabin Simulator. Douglas Paper, Rept. 3446, Douglas Aircraft Co., Inc., Santa Monica, Calif., 1968.

Dunn, C. D. R.: Effect of Dehydration on Erythropoiesis in Mice: Relevance to the "Anemia" of Spaceflight. Aviat., Space, and Environ. Med., vol. 49, no. 8, Aug. 1978, pp. 990-993.

Dunn, C. D. R.; and Lange, R. D.: Erythropoietic Effects in Space Flight. Acta Astronautica, vol. 6, May-June 1979, pp. 725-732.

Dunn, C. D. R.; Leonard, J. I.; and Kimzey, S. L.: Interactions of Animal and Computer Models in Investigations of the "Anemia" of Space Flight. Aviat., Space, and Environ. Med., vol. 52, no. 11, Section 1, Nov. 1981, pp. 683-690.

Duntley, S. Q.; Austin, R. W.; Harris, J. L.; and Taylor, J. H.: Experiments on Visual Acuity and the Visibility of Markings on the Ground on Long-Duration Earth-Orbital Space Flight. NASA CR-1134, Nov. 1968.

Durkheim, Emile: Suicide: A Study in Sociology. The Free Press, Glencoe, Ill., 1951.

Dusek, E. Ralph: Effect of Temperature on Manual Performance. Protection and Functioning of the Hands in Cold Climates, Frank R. Fisher, ed. Proceedings of a conference held April 23-24, 1956, Natick, Mass. NAS/NRC Advisory Board on Quartermaster Research and Development, Committee on Environmental Protection, Washington, D.C., 1957, pp. 63-76.

Earls, J. H.: Human Adjustment to an Exotic Environment: The Nuclear Submarine. Arch. of Gen. Psychiatry, vol. 20, no. 1, 1969, pp. 117-123.
Eberhard, J. W.: The Problem of Off Duty Time in Long Duration Space Missions. Vol. II. NASA CR-96721, 1967.
Ebersole, J. H.: The New Dimensions of Submarine Medicine. New England J. of Med., vol. 262, 1960, pp. 599-610.
Edmonston, William E., Jr.: Familiarity and Musical Training in the Esthetic Evaluation of Music. J. Soc. Psychol., vol. 79, 1969, pp. 109-111.
Egan, Gerald: The Skilled Helper: A Model for Systematic Helping and Interpersonal Relating. Brooks/Cole, Monterey, Calif., 1975.
Ekman, P.; and Friesen, W. V.: Hand Movements. J. Communication, vol. 22, no. 4, 1972, pp. 353-374.
——: Unmasking the Face: A Guide to Recognizing Emotions From Facial Cues. Prentice Hall, 1975.
Ekman, Paul; Friesen, W. V.; and Ellsworth, P.: Emotions in the Human Face: Guidelines for Research and an Integration of Findings. Pergamon Press, 1972.
Elton, Martin C. J.; and Carey, John: Implementing Interactive Telecommunications Services. Final Report on Problems Which Arise During Implementation of Field Trials and Demonstration Projects. The Alternate Media Center, New York Univ., New York, NSF Grant no. APR-77-18697, 1980.
Engler, Nick; and Cheshire-Engler, Donna: Life After Liftoff. Omni, vol. 5, no. 8, May 1983, pp. 108-112, 133.
Environmental Protection Agency: Information on Levels of Environmental Noise Requisite to Protect Public Health and Welfare with an Adequate Margin of Safety. Rep. no. EPA 550/9-74-004, March 1974.
Epstein, M.: Renal Effects on Head-out Water Immersion in Man; Implications for Understanding of Volume Homeostasis. Physiol. Rev., vol. 58, no. 3, July 1978, pp. 529-581.
Epstein, S.: Toward a Unified Theory of Anxiety. Progress in Experimental Personality Research, Vol. 4, Brendan A. Mather, ed., Academic Press, 1968, pp. 1-89.

───: The Nature of Anxiety with Emphasis upon Its Relationship to Expectancy. Chapter 8. Anxiety: Current Trend in Theory and Research, Vol. II, Charles D. Spielberger, ed., Academic Press, 1972, pp. 291-337.

Epstein, Y. M.; and Karlin, R. A.: Effects of Acute Experimental Crowding. J. Appl. Soc. Psychol., vol. 5, no. 1, 1975, pp. 34-53.

Ertel, Ivan D.; and Morse M. L.: The Apollo Spacecraft: A Chronology. NASA SP-4009, 1969.

Esser, A. H.: Cottage Fourteen: Dominance and Territoriality in a Group of Institutionalized Boys. Small Group Behav., vol. 4, no. 3, 1973, pp. 131-146.

Esser, Aristide H.; Chamberlain, Amyaro S.; Chapple, Eliot D.; and Kline, Nathan S.: Territoriality of Patients on a Research Ward. Recent Advances in Biological Psychiatry, Joseph Wortis, ed., Plenum Press, 1964, pp. 37-44.

Evans, Gary W.: Environmental Cognition. Psychol. Bull., vol. 88, no. 2, 1980, pp. 259-287.

Farnsworth, Paul R.: Musical Taste: Its Measurement and Cultural Nature. Stanford Univ. Press, Stanford, Calif., 1950.

───: The Phenomenon of Musical Tastes. Psychol. Today, vol. 1, no. 4, Aug. 1967, pp. 14-19, 40-43.

Farrell, R. J.; and Smith, S.: Behavior of Five Men Confined for Thirty Days: Psychological Assessment During Project MESA. NASW-658, Rep. no. DZ-90586, The Boeing Company, Seattle, Wash., 1964.

Fawcett, R. E.; and Newman, P. P.: Operation Hideout. USN Medical Research Laboratory, Rep. no. 228, Bethesda, Md., 1953.

Fenz, Walter D.: Stress and Its Mastery: Predicting from Laboratory to Real Life. Canadian J. Behav. Sci., vol. 5, no. 4, Oct. 1973, pp. 332-346.

Fenz, Walter D.; and Epstein, Seymour: Gradients of Physiological Arousal in Parachutists as a Function of an Approaching Jump. Psychosomatic Med., vol. 29, no. 1, Jan.-Feb. 1967, pp. 33-51.

───: Stress: In the Air. Psychol. Today, vol. 3, no. 4, Sept. 1969, pp. 27-28, 58-59.

Fenz, Walter D.; and Jones, G. Brian: Individual Differences in Psychological Arousal and Performance in Sport Parachutists. Psychosomatic Med., vol. 34, no. 1, Jan. 1972, pp. 1-8.

Festinger, L.: A Theory of Social Comparison Processes. Human Relations, vol. 7, 1954, pp. 117-140.

Fidell, Sanford; Jones, Glenn; and Pearsons, Karl S.: Feasibility of a Novel Technique for Assessing Noise-Induced Annoyance. Contract DOT-OS-20103, Bolt, Beranek and Newman, 1973.

Fiedler, F. E.: A Theory of Leadership Effectiveness. McGraw-Hill Book Co., 1967.

——: Validation and Extension of the Contingency Model of Leadership Effectiveness: A Review of Empirical Findings. Psychol. Bull., vol. 76, no. 2, 1971, pp. 128-148.

——: Recent Developments in Research on the Contingency Model. Group Processes, Leonard Berkowitz, ed., Academic Press, 1978, pp. 209-226.

Fiedler, Fred Edward; Chemers, Martin M.; and Mahar, Linda: Improving Leadership Effectiveness: The Leader Match Concept. John Wiley & Sons, 1976.

Finkelman, Joy M.; and Glass, David C.: Reappraisal of the Relationship Between Noise and Human Performance by Means of a Subsidiary Task Measure. J. Appl. Psychol., vol. 54, 1970, pp. 211-213.

Fisher, B. Aubrey: Information Systems Theory and Research: An Overview. Communications Yearbook 2, Brent D. Ruben, ed., Transactions Books, New Brunswick, N.J., 1978, pp. 81-108.

Fleishman, E. A.: Psychomotor Tests in Drug Research. Drugs and Behavior. J. G. Miller and L. Uhr, eds., John Wiley & Sons, 1960, pp. 278-296.

——: The Description and Prediction of Perceptual-Motor Skill Learning. Training Research and Education, Robert Glaser, ed., Univ. of Pittsburgh, Pa., 1962, pp. 137-176.

——: The Structure and Measurement of Physical Fitness. Prentice-Hall, 1964.

——: Performance Assessment Based on an Empirically Derived Task Taxonomy. Human Factors, vol. 9, 1967, pp. 349-366.

Flinn, D. E.; Monroe, J. T., Jr.; Cramer, E. H.; and Hagan, D. H.: Observations in the SAM Two-Man Space Cabin Simulator. IV. Behavioral Factors in Selection and Performance. Aerosp. Med., vol. 32, no. 7, 1961, pp. 610-615.

Fort, A.: The Effects of Rapid Change in Time Zone on Circadian Variation in Psychological Functions. J. Physiol. (London), vol. 200, no. 2, 1969, p. 124.

Foushee, H. Clayton: The Role of Communications, Socio-Psychological, and Personality Factors in the Maintenance of Crew Coordination. Aviat., Space, and Environ. Med., vol. 53, no. 11, Nov. 1982, pp. 1062-1066.

Fox, J. G.: Background Music and Industrial Efficiency — A Review. Applied Ergonomics, vol. 2, no. 2, 1971, pp. 70-73.

Fraser, D. C.: The Relation of an Environmental Variable to Performance in a Prolonged Visual Task. Quart. J. Exp. Psychol., vol. 5, 1953, pp. 31-32.

Fraser, T. M.: The Effects of Confinement as a Factor in Manned Space Flight. NASA CR-511, 1966.

——: Confinement and Free Volume Requirements. Space Life Sciences, vol. 1, 1968a, pp. 428–466.

——: Leisure and Recreation in Long Duration Space Missions. Human Factors, vol. 10, no. 5, 1968b, pp. 453–488.

——: The Intangibles of Habitability During Long-Duration Space Missions. NASA CR-1084, 1968c.

Frazier, Shervert H.; and Carr, Arthur C.: Introduction to Psychopathology. Jack Aronson, New York, 1974.

Freedman, J.; Levy, A. S.; Buchanan, R. W.; and Price, J.: Crowding and Human Aggressiveness. J. Exp. Soc. Psychol., vol. 8, no. 6, Nov. 1972, pp. 528–548.

French, J. R. P., Jr.; and Raven, B. H.: The Bases of Social Power. Group Dynamics: Research and Theory. Second ed. Dorwin Cartwright and A. F. Zander, eds., Row & Peterson, New York, 1960, pp. 607–623.

French, John R. P., Jr.; Rodgers, Willard; and Cobb, Sidney: Adjustment as Person-Environment Fit. Coping and Adaptation, George V. Coelho, David A. Hamburg, and John E. Adams, eds., Basic Books, New York, 1974, pp. 316–333.

French, Wendell L.; and Bell, Cecil H., Jr.: Organization Development: Behavioral Science Interventions for Organization Improvement. Second ed. Prentice-Hall, 1978.

Freud, Sigmund: Mourning and Melancholia, 1917. Sigmund Freud Collected Papers, vol. 4, Hogarth Press (London) 1950, pp. 152–172.

Fried, Lena; and Berkowitz, Leonard: Music Hath Charms . . . and Can Influence Helpfulness. J. Appl. Soc. Psychol., vol. 9, no. 3, May-June 1979, pp. 199–208.

Frost, J. D., Jr.; Shumate, W. H.; Salamy, J. G.; and Booher, C. R.: Experiment M133. Sleep Monitoring on Skylab. Biomedical Results from Skylab, Richard S. Johnston and Lawrence F. Dietlein, eds., NASA SP-377, 1977, pp. 113–126.

Gagne, Robert Mills: Conditions of Learning. Holt, Rinehart, and Winston, 1964.

Garriott, O. K.; and Doerre, G. L.: Crew Efficiency on First Exposure to Zero-Gravity. Biomedical Results from Skylab, Richard S. Johnston and Lawrence F. Dietlein, eds., NASA SP-377, 1977, pp. 155-162.

Gartner, W. B.; and Murphy, M. R.: Pilot Workload and Fatigue: A Critical Survey of Concepts and Assessment Techniques. NASA TN D-8365, 1976.

Gauer, O. H.: Recent Advances in the Study of Whole Body Immersion. Acta Astronautica, vol. 2, 1975, pp. 39-49.

Gazenko, O. G., ed.: Aerospace Medicine Sixth All-Soviet Union Conference on Space Biology and Aerospace Medicine, Parts I and II. Kaluga, USSR, June 5-7, 1979.

Gazenko, O. G.; Genin, A. M.; and Yegorov, A. D.: Major Medical Results of the Salyut 6/Soyuz 185-Day Space Flight. Space: Mankind's Fourth Environment: Selected Papers from the XXXII Int. Astron. Cong., Rome, Italy, Sept. 6-12, 1981, L. G. Napolitano, ed., Pergamon Press (Oxford), 1982, pp. 275-293.

Gazenko, O. G.; Il'in, Yu. A.; Genin, A. M.; Kotovskaya, A. R.; Korolkov, V. I.; Tigranyan, R. A.; and Portugalov, V. V.: Principal Results of Physiological Experiments with Mammals Aboard the Cosmos-936 Biosatellite. Space Bio. and Aerosp. Med., vol. 14, no. 2, May 1980, pp. 33-37. Translated into English from Kosm. Biol. Avia. Kosm. Med. (USSR), vol. 14, no. 2, March-April 1980, pp. 22-25.

Gazenko, O. G.; Myasnikov, V. I.; Ioseliani, K. K.; Kozerenko, O. P.; and Uskov, F. N.: Important Problems of Space Psychology: As Evidenced by Salyut 6-Soyuz Manned Missions. Paper presented at the XXVII International Congress of Space Medicine, 1979.

Geen, R. C.; and O'Neal, E. C.: Activation of Cue-Elicited Aggression by General Arousal. J. Personality and Soc. Psychol., vol. 11, no. 3, 1969, pp. 289-292.

Gehlen, Frieda, L.: Toward a Revised Theory of Hysterical Contagion. J. Health and Soc. Behav., vol. 18, no. 1, March 1978, pp. 27-35.

General Electric Corp.: Space Station System Simulation: Results of a Four Man/30 Day Mission Simulation Program. Doc. no. 64SD679, Valley Forge Space Technology Center, Philadelphia, Pa., 1964.

George Washington Univ. Medical Center: Studies of Social Group Dynamics Under Isolated Conditions. Objective Summary of the Literature as It Relates to Potential Problems of Long Duration Space Flight. Sciences Communication Division, Washington, D.C., NASA CR-2496, 1974.

Gerard, Harold B.; and Mathewson, Grover C.: The Effect of Severity of Initiation on Liking for a Group: A Replication. J. Exp. Soc. Psychol., vol. 2, no. 3, 1966, pp. 278-287.

Gerard, Harold B.; and Rabbie, Jacob M.: Fear and Social Comparison. J. Abnorm. Soc. Psychol., vol. 62, no. 3, May 1961, pp. 586-592.

Gerathewohl, S. J.: Work Proficiency in the Space Cabin Simulator. Aerosp. Med., vol. 30, no. 10, Oct. 1959, pp. 722-735.

Gibson, Edward G.: Skylab 4 Crew Observations. Proceedings of the Skylab Life Science Symposium, Aug. 21-29, 1974, Lyndon B. Johnson Space Center, Houston. NASA TM X-58154, vol. 1, 1974, pp. 57-54.

Glass, David C.; and Singer, J. E.: Urban Stress: Experiments on Noise and Social Stressors. Academic Press, 1972.

Glazer, J. Henry: The Maltese Initiatives Within the United Nations — A Blue Planet Blueprint for Trans-National Space. Ecology Law Quarterly, vol. 4, 1974, pp. 279-318.

Goffman, Erving: The Presentation of Self in Everyday Life. Doubleday-Anchor, New York, 1959.

———: Asylums: Essays on the Social Situation of Mental Patients and Other Inmates. Doubleday-Anchor, New York, 1961.

Gogolev, K. I.; Aleksandrova, Ye. A.; and Shul'Zhenko, Ye. B.: Comparative Evaluation of Changes in the Human Body During Orthostatic (Headdown) Hypokinesia and Immersion. ND3111. Translated into English from Fisiologiya Cheloveka, vol. 6, no. 6, 1980, pp. 978-983.

Goldmark, Peter C.: The New Rural Society. University Lecture Series, Papers in Communication no. 5, Cornell Univ., Ithaca, New York, 1973.

Goldsmith, John R.: Health and Annoyance: Impact of Odor Pollution. EPA-650/1-75-001, Oct. 1973.

Goldwater, D.; Montgomery, L.; Hoffler, G. W.; Sandler, H.; and Poll, R.: Echocardiographic and Peripheral Vascular Responses of Men (Aged 46 to 55) to Lower Body Negative Pressure (LBNP) Following 10 Days of Bed Rest. Preprint of 1979 Annual Meeting, Aerospace Med. Assoc., May 14-17, 1979, Washington, D.C.

Gorokhov, A.; and Stapantsev, V.: Salyut-4: Stadium in Weightlessness. NASA TT F-16326, 1975.

Gouldner, A. W., Jr.: The Norm of Reciprocity: A Preliminary Statement. Amer. Sociol. Rev., vol. 25, 1960, pp. 161-179.

Graen, G.: Role Making Processes Within Complex Organizations. Handbook of Industrial and Organizational Psychology, Marvin Dunnette, ed., Rand McNally College Pub., 1976, pp. 1201-1245.

Graybiel, Ashton: Angular Velocities, Angular Accelerations, and Coriolis Accelerations. Foundations of Space Biology and Medicine, vol. II, book 1, NASA SP-374, 1975, pp. 247-304.

Graybiel, Ashton; Guedry, Fred E.; Johnson, Walter H.; and Kennedy, Robert S.: Adaptation to Bizarre Stimulation of the Semicircular Canals as Indicated by the Oculogyral Illusion. Aerosp. Med., vol. 32, no. 4, April 1961, pp. 321-327.

Graybiel, Ashton; and Lackner, James R.: Evaluation of the Relationship Between Motion Sickness Symptomatology and Blood Pressure, Heart Rate, and Body Temperature. Aviat., Space, and Environ. Med., vol. 51, no. 3, March 1980, pp. 211-214.

Graybiel, Ashton; Miller, Earl F., II; and Homick, J. L.: Experiment M-131, Human Vestibular Function. Proceedings of the Skylab Life Science Symposium, vol. 1, NASA TM X-58154, 1974, pp. 169-220.

———: Experiment M131. Human Vestibular Function. Susceptibility to Motion Sickness. Biomedical Results of Skylab. NASA SP-377, 1977, pp. 74-103.

Graybiel, Ashton; and Wood, Charles D.: Rapid Vestibular Adaptation in a Rotary Environment by Means of Controlled Head Movements. Aerosp. Med., vol. 40, no. 6, June 1969, pp. 638-643.

Greiff, Barrie S.; and Hunter, Preston K.: Tradeoffs: Executive, Family, and Organizational Life. Mentor Executive Library, New York, 1980.

Grether, Walter F.: Vibration and Human Performance. Human Factors, vol. 13, June 1971a, pp. 203-216.

———: Noise and Human Performance. AMRL-TR-70, June 1971b.

Griffitt, W.: Environmental Effects on Interpersonal Affective Behavior: Ambient Effective Temperature and Attraction. J. Pers. and Soc. Psychol., vol. 15, no. 3, 1970, pp. 240-244.

Griffitt, W.; and Veitch, R.: Hot and Crowded: Influence of Population Density and Temperature on Interpersonal Affective Behavior. J. Pers. and Soc. Psychol., vol. 17, no. 1, 1971, pp. 92-98.

Grodsky, Milton A.: The Use of Full Scale Mission Simulation for the Assessment of Complex Operator Performance. Human Factors, vol. 9, no. 4, 1967, pp. 341-348.

Grodsky, M. A.; and Bryant, J. P.: Integrated Mission Simulation for Long Term Space Flight. Paper presented at the AIAA Simulation for Aerospace Flight Conference, 1963. Also published as Report ND380289, Martin Co., Baltimore, Md.

Grodsky, M. A.; Glazer, D. L.; and Hopkins, A. H., Jr.: Analysis of Crew Performance in the Apollo Command Module, Phase I. ER-14262, Martin Co., Baltimore, Md., NASA CR-65755, 1966a.

Grodsky, M. A.; Moore, H. O.; and Flaherty, T. W.: Crew Reliability During Simulated Space Flight. J. Spacecraft Rockets, vol. 3, June 1966b, pp. 810-817.

Grodsky, M. A.; Warfield, J. T.; Flaherty, T. M.; Mandour, J. A.; and Hurley, R. W.: An Analysis of Pilot Reliability for Manned Space Flight. RM-178, Martin Co., Baltimore, Md. NASA CR-69032, 1964.

Grumman Aerospace Corp.: Use of the Ben Franklin Submersible as a Space Station Analog. Vol. II — Psychology and Physiology. OSR-70-5. Vol. III — Habitability, OSR-70-6. Bethpage, New York. NASA CR-102829, CR-102830, 1970.

Guedry, F. E., Jr.: Visual Control of Habituation to Complex Vestibular Stimulation in Man. Acta Otolaryng., vol. 58, Nov. 1964, pp. 377-389.

——: Comparison of Vestibular Effects in Several Rotating Environments. The Role of the Vestibular Organs in the Exploration of Space. NASA SP-77, 1965, pp. 243-255.

Guedry, F. E.; and Ambler, R. K.: Assessment of Reactions to Vestibular Disorientation Stress for Purposes of Aircrew Selection. AGARD Conference on Predictability of Motion Sickness in the Selection of Pilots, Paris, 1972. CP-109, 1973.

Gunderson, E. K. E.: Emotional Symptoms in Extremely Isolated Groups. Arch. Gen. Psychiat., vol. 9, no. 4, 1963, pp. 362-368.

——: Mental Health Problems in Antarctica. Arch. Environ. Health, vol. 17, no. 4, Oct. 1968, pp. 558-564.

Gunderson, E. K. E.; and Mahan, J. L.: Cultural and Psychological Differences Among Occupational Groups. J. Psychol., vol. 62, no. 2, 1966, pp. 287-304.

Gunderson, E. K. E.; and Nelson, P. D.: Adaptation of Small Groups to Extreme Environments. Aerosp. Med., vol. 34, no. 12, 1963, pp. 1111-1115.

Gunderson, E. K. Eric; and Nelson, Paul D.: Biographical Predictors of Performance in an Extreme Environment. J. Psychol., vol. 61, no. 1, Sept. 1965, pp. 59-67.

——: Criterion Measures of Extremely Isolated Groups. Personnel Psychol., vol. 19, no. 1, 1966, pp. 67-80.

Gunderson, E. K. E.; and Rahe, R. M.: Life, Stress and Illness. Chas. R. Thomas, Springfield, Ill., 1974.

Hackler, T.: Motion Malaise: It's as Old as Transportation. United Magazine, May 1982, pp. 115-116.

Hackman, J. R.: Group Influence on Individuals. Handbook of Industrial and Organizational Psychology. M. D. Dunnette, ed., Rand McNally College Pub., 1976, pp. 1455-1525.

Hackman, J. Richard; and Lawler, Edward E., III: Employee Reactions to Job Characteristics. J. Appl. Psychol. Monograph, vol. 55, no. 3, June 1971, pp. 259-286.

Hackman, J. Richard; and Morris, Charles C.: Group Tasks, Group Interaction Process, and Group Performance Effectiveness: A Review and Proposed Integration. Advances in Experimental Social Psychology, vol. 8, L. Berkowitz, ed., Academic Press, 1975, pp. 45-99.

——: Group Process and Group Effectiveness: A Reappraisal. Group Processes, Leonard Berkowitz, ed., Academic Press, 1978, pp. 57-66.

Hackman, J. Richard; and Oldham, Gres R.: Motivation Through the Design of Work: Test of a Theory. Tech. Rep. no. 6, School of Organization and Management, Department of Administrative Sciences, Yale University, New Haven, Conn., 1974.

Hackworth, J. R.: Relationship Between Spatial Density and Sensory Overload, Personal Space and Systolic and Diastolic Blood Pressure. Perceptual and Motor Skills, vol. 43, 1976, pp. 867-872.

Haggard, E. A.: Isolation and Personality. Personality Change, Philip Worchel and D. E. Byrne, eds., John Wiley and Sons, 1964, pp. 433-469.

Hall, Edward T.: The Silent Language. Doubleday, Garden City, New York, 1959.

Hammerton, M.; and Tickner, A. H.: An Investigation Into the Effects of Stress Upon Skilled Performance. Ergonomics, vol. 12, no. 6, Nov. 1969, pp. 851-855.

Hammes, J. A. and Ahearn, T. R.: Shelter Occupancy Studies at the University of Georgia. Final Report for 1967, AD-673-778; Final Report for 1966, AD-653-881.

Hammes, J. A.; and Osborne, R. T.: Survival Research in Group Isolation Studies. J. Appl. Psychol., vol. 19, no. 6, 1965, pp. 418-421.

Hammes, J. A.; and Watson, James A.: Behavior Patterns of Groups Experimentally Confined. Perceptual and Motor Skills, vol. 20, no. 3, pt. 2, June 1965, pp. 1269-1272.

Hammes, John A.: Shelter Occupancy Studies at the University of Georgia. Final Report for 1965, AD-635-501; Final Report for 1964, AD-615-004.

Hammes, John A.; Ahearn, Thomas R.; and Foughner, James W.: Shelter Occupancy Studies at the University of Georgia. Final Report for 1968, AD-688-099.

Hammes, John; Ahearn, T. R.; and Keith, J. F., Jr.: A Chronology of Two Weeks Fallout Shelter Confinement. J. Clin. Psychol., vol. 21, no. 4, 1965, pp. 452-456.

Hammond, K. R.; and Brehmer, B.: Quasi-rationality and Distrust: Implications for International Conflict. Human Judgment and Social Interaction, Leon Rappaport and D. A. Summers, eds., Holt, Rinehart, & Winston, 1973, pp. 338-392.

Hammond, K. R.; Stewart, T. R.; Brehmer, B.; and Steinmann, D. O.: Social Judgment Theory. Human Judgment and Decision Processes, Martin F. Kaplan and Steven Schwartz, eds., Academic Press, 1975.

Hanna, Thomas D.; and Gaito, John: Performance and Habitability Aspects of Extended Confinement in Sealed Cabinets. Aerosp. Med., vol. 31, no. 5, May 1960, pp. 399-406.

Harburg, E.; Erfurt, J. C.; Hauenstein, L. S.; Chape, D.; Schull, W. J.; and Schork, M. A.: Socio-ecological Stress, Suppressed Hostility, Skin Color, and Black-White Blood Pressure: Detroit, Psychosomatic Med., vol. 35, no. 4, July and Aug. 1973, pp. 276-296.

Harburg, E.; Schull, W. J.; Erfurt, J. C.; and Schork, M. A.: A Family Set Method for Estimating Heredity and Stress: A Pilot Survey of Blood Pressure Among Negroes in High and Low Stress Areas, Detroit, 1966-1967. J. Chronic Dis., vol. 23, no. 2, 1970, pp. 69-81.

Harris, C. Stanley: Effects of Acoustic Stimuli on the Vestibular System. The Disorientation Incident, AGARD CP-95, pt. 1, A7-1-A7-11, 1972. Also AMRL-TR-71-58, 1972.

Harrison, R. Van: Person-Environment Fit and Job Stress. Stress at Work, Cary L. Cooper and Roy Payne, eds., John Wiley and Sons, 1978.

Hartman, B. O.; and Flinn, D. E.: Crew Structure in Future Space Missions. Lectures in Aerospace Medicine, Brooks AFB, Texas, Feb. 3-7, 1964, pp. 50-72.

Hartman, B. O.; and McNee, R. C.: Psychometric Characteristics of Astronauts. USAF School of Aviation Medicine, Rep. SAM-TR-77-5, SAM Review 1-77167930, 1977. (AD-AO39228)

Hattner, R. S.; and McMillan, D. E.: Influence of Weightlessness Upon the Skeleton: A Review. Aerosp. Med., vol. 39, no. 3, Aug. 1968, pp. 849-855.

Hauty, G. T.; and Adams, T.: Phase Shifts of the Human Circadian System and Performance Deficit During the Periods of Transition: I. East-West Flight. Aerosp. Med., vol. 37, no. 7, July 1966a, pp. 668-674.

———: Phase Shifts of the Human Circadian System and Performance Deficit During the Periods of Transition: II. West-East Flight. Aerosp. Med., vol. 37, no. 10, Oct. 1966b, pp. 1027-1033.

———: Phase Shifts of the Human Circadian System and Performance Deficit During the Periods of Transition: III. North-South Flight. Aerosp. Med., vol. 37, no. 12, Dec. 1966c, pp. 1257-1262.

Havron, M. D.; and Butler, L. F.: Evaluation of Training Effectiveness of the 2-PH-2 Flight Trainer Research Tool. TR NAVTRADEVCEN 1915-00-1, U.S. Naval Training Device Center, Port Washington, New York, 1957, p. 150.

Haythorn, W. W.: The Composition of Groups: A Review of the Literature. Acta Psychologica, vol. 28, no. 2, 1968, pp. 97-128.

———: Interpersonal Stress in Isolated Groups. Social and Psychological Factors in Stress, Joseph E. McGrath, ed., Holt, Rinehart and Winston, 1970.

———: The Miniworld of Isolation: Laboratory Studies. Man in Isolation and Confinement, J. E. Rassmussen, ed., Aldine Pub., New York, 1973, pp. 219-240.

Haythorn, William W.; and Altman, Irwin: Together in Isolation. Transaction, vol. 4, no. 3, Jan./Feb. 1967, pp. 18-23.

Haythorn, William W.; Altman, Irwin; and Myers, Thomas U.: Emotional Symptomatology and Subjective Stress in Isolated Pairs of Men. J. Exper. Res. in Pers., vol. 1, no. 4, 1966, pp. 290-305.

Haythorn, W. W.; McGrath, J. E.; Hollander, E. P.; Latané, B., Helmreich, R.; and Radloff, R.: Group Processes and Interpersonal Interaction. Human Factors in Long-Duration Spaceflight. Space Science Board, National Research Council, Washington, D.C., 1972, pp. 160-178.

Helmreich, R.: Psychological Research in Tektite II. Man-Environment Systems, vol. 3, no. 2, 1973, pp. 125-127.

———: Prolonged Stress in Sealab II — A Field Study of Individual and Group Reactions. Ph.D. Thesis, RPT TR-1, Yale Univ., New Haven, Conn., 1967. (AD-651974)

Helmreich, R.; Wilhelm, J.; Tanner, T. A.; Sieber, J. E.; and Burgenbach, S.: A Critical Review of the Life Sciences Project Management at Ames Research Center for the Spacelab Mission Development Test III, NASA TP-1364, Jan. 1979a.

Helmreich, R. L.; Wilhelm, J. A.; and Runge, T. E.: Psychological Considerations in Future Space Missions. The Human Factors in Outer Space Production, Stephen T. Cheston and David L. Winter, eds., American Association for the Advancement of Science, Selected Symposium no. 50, Washington, D.C., 1980, pp. 1-23.

Helmreich, Robert L.; Spence, J. T.; and Holahan, C. K.: Psychological Androgyny and Sex Role Flexibility: A Test of Two Hypotheses. J. Pers. and Soc. Psychol., vol. 37, no. 10, Oct. 1979b, pp. 1631-1644.

Heron, Woodburn: The Pathology of Boredom. Scien. Amer., vol. 196, no. 1, Jan. 1957, pp. 52-56.

Hodge, James R.: They That Mourn. J. Religion and Health, vol. 11, no. 3, July 1972, pp. 229-240.

Hodges, W. F.; and Spielberger, C. D.: The Effects of Threat of Shock on Heart Rate for Subjects Who Differ in Manifest Anxiety and Fear of Shock. Psychophysiology, vol. 2, no. 4, 1966, pp. 287-294.

Holland, John L.: A Theory of Vocational Choice. J. Counseling Psychol., vol. 6, 1959, pp. 35-45.

——: Making Vocational Choices: A Theory of Careers. Prentice-Hall, 1973.

——: Vocational Preferences. Handbook of Industrial and Organizational Psychology, Marvin Dunnette, ed., Rand McNally College Pub., 1976, pp. 522-570.

Hollander, Edwin P.: Leadership Dynamics: A Practical Guide to Effective Relationships. Free Press, Glencoe, Ill., 1978.

Hollander, Edwin P.; and Julian, James W.: Studies in Leader Legitimacy, Influence, and Innovation. Advances in Experimental Social Psychology, vol. 5, Leonard Berkowitz, ed., Academic Press, 1970, pp. 33-69.

——: A Further Look at Leader Legitimacy, Influence, and Innovation. Group Processes, Leonard Berkowitz, ed., Academic Press, 1978, pp. 153-175.

Holmes, Thomas H.; and Rahe, Richard H.: The Social Readjustment Rating Scale. J. Psychosomatic Res., vol. 11, no. 2, 1967, pp. 213-218.

Holt, A. C.; and da Silva, A. J.: Skylab Simulator Visual Displays and Training for Joint Observing Programs. Appl. Opt., vol. 16, April 1977, pp. 950-960.

Holton, E. M.; Turner, R. T.; and Baylink, D. J.: Quantitative Analysis of Selected Bone Parameters. U.S. Experiments Flown on the Soviet Satellite Cosmos 936. S. N. Rosenzweig and K. A. Souza, eds., NASA TM-78526, 1979, pp. 135-183.

Homick, J. L.; Degioanni, J.; Reschke, M. F.; Leach, L. S.; and Kohl, R. L.: An Evaluation of the Time Course of Efficacy of Transdermally Administered Scopolamine in the Prevention of Motion Sickness. Preprints of the 1982 Annual Scientific Meeting of the Aerosp. Med. Assoc., Bal Harbour, Fla., May 10-13, 1982.

Homick, J. L.; and Miller, E. F., II: Apollo Flight Crew Vestibular Assessment. Biomedical Results of Apollo. NASA SP-368, 1975, pp. 323–340.

Homick, J. L.; Reschke, M. F.; and Miller, E. F., II: Effects of Prolonged Exposure to Weightlessness on Postural Equilibrium. Biomedical Results from Skylab, R. S. Johnston and L. F. Dietlein, eds., NASA SP-377, 1977, pp. 104–112.

Hough, R. W.; and Panko, R. R.: Teleconferencing Systems: A State-of-the-Art Survey and Preliminary Analysis. SRI Project 3735, April 1977. PB-268455/3, NSI/RA-770163.

Hull, D. H.; Wolthuis, R. A.; Gillingham, K. K.; and Triebwasser, J. H.: Relaxed +G Tolerance in Healthy Men: Effect of Age. J. Appl. Physiol., vol. 45, no. 4, 1978, pp. 626–629.

Isay, R. A.: The Submariners' Wives' Syndrome. Psychiat. Quart., vol. 42, 1968, pp. 647–652.

Izard, Carrol E.: Comments on paper "Intervention in Relation to Anxiety in School," by Beeman N. Phillips, Roy P. Martin, and Joel Meyers. Anxiety — Current Trends in Theory and Research, vol. II, ch. 12, Chas. D. Spielberger, ed., pp. 410–464. Academic Press, 1972, pp. 465–468.

Jackson, J. K.; Wamsley, J. R.; Bonura, M. S.; and Seeman, J. S.: Program Operational Summary — Operational 90-Day Manned Test of the Regenerative Life Support System. NASA CR-1835, 1972.

Jacobs, Selby; and Ostfeld, Adrian: An Epidemiological Review of the Mortality of Bereavement. Psychosom. Med., vol. 39, no. 5, Sept.-Oct. 1977, pp. 344–357.

Jacobs, Thomas Owen: Leadership and Exchange in Formal Organizations. Human Resources Research Organization, Alexandria, Va., 1970 [i.e., 1971].

Jacobson, Gerald: Crisis Theory and Treatment Strategy: Some Socio-Cultural and Psychodynamic Considerations. J. Nerv. Ment. Dis., vol. 141, Aug. 1965, pp. 209–218.

Janis, Irving Lester: Groupthink. Psychol. Today, vol. 5, no. 6, June 1971, pp. 43–46.

———: Victims of Groupthink: Psychological Study of Foreign-Policy Decisions and Fiascoes. Houghton-Mifflin, 1972.

———: Decision Making: A Psychological Analysis of Conflict, Choice, and Commitment. The Free Press, New York, 1977.

Jenney, L. L.; Older, H. J.; and Cameron, B. J.: Measurement of Operator Workload in an Information Processing Task. NASA CR-2150, 1972.

Jex, H. R.; Peters, R. A.; DiMarco, R. J.; and Allen, R. W.: The Effects of Bedrest on Crew Performance During Simulated Shuttle Reentry. Vol. II: Control Task Performance. NASA CR-2367, 1974.

Johansen, Robert; Miller, Richard H.; and Vallee, Jacques: Group Communication Through Electronic Media: Fundamental Choices and Social Effects. The Delphi Method: Techniques and Applications, Harold A. Linstone and Murray Turnoff, eds., Addison-Wesley, 1975, pp. 517-534. (Also in: Educational Technology, Aug. 1974, pp. 7-20.)

Johansen, Robert; Vallee, Jacques; and Collins, Kent: Learning the Limits of Teleconferencing: Design of a Teleconference Tutorial. Evaluating New Telecommunication Services, Martin C. J. Elton, William A. Lucas, and David W. Conrath, eds., Plenum Press, 1977, pp. 385-398.

Johansen, Robert; Vallee, Jacques; and Spangler, Kathleen: Electronic Meetings: Technical Alternatives and Social Choices. Addison-Wesley, 1979.

Johnson, C. C.: Skylab Experiment M487: Habitability/Crew Quarters. Advances in the Astronautical Sciences, William C. Schneider and Thomas E. Hanes, eds., Amer. Astron. Soc., Tarzana, Calif., vol. 31, pt. 1, 1974, pp. 313-333.

Johnson, R. L.; Hoffler, G. W.; Nicogossian, A. E.; Bergman, S. A.; and Jackson, M. M.: Lower Body Negative Pressure: Third Manned Skylab Mission. Vol. 2. Proceedings of the Skylab Life Science Symposium, NASA TM X-58154, 1974, pp. 545-595.

Johnston, Richard S.; and Dietlein, Lawrence F., eds.: Biomedical Results From Skylab. NASA SP-377, 1977.

Jones, Edward E.; and Archer, Richard L.: Are There Special Effects of Personalistic Self-Disclosure? J. Exper. Soc. Psychol., vol. 12, no. 2, 1976, pp. 180-193.

Jones, Edward E.; and Gordon, Eric M.: Timing of Self-Disclosure and Its Effect on Personal Attraction. J. Pers. and Soc. Psychol., vol. 24, no. 3, 1972, pp. 358-365.

Jones, F. Nowell; and Woskow, Morris H.: On the Intensity of Odor Mixture. Ann. New York Acad. Sci., vol. 116, art. 2, 1964, pp. 484-494.

Jonsson, Anders; and Harsson, Lennart: Prolonged Exposure to a Stressful Stimulus (Noise) as a Cause of Raised Blood-Pressure in Man. The Lancet, No. 8002, Jan. 8, 1977, pp. 86-87.

Jordan, J. P.; Sykes, H. A.; Crownover, J. C.; Schaffe, C. L.; Simmons, J. B., II; and Jordon, D. P.: Simulated Weightlessness: Effects of Bioenergetic Balance. Aviat., Space, and Environ. Med., 51, no. 2, Feb. 1980, pp. 132-136.

Jourard, S.: Self-Disclosure and Other Cathexis. J. Abnorm. and Soc. Psychol., vol. 59, 1959, pp. 428–431.

———: Self-Disclosure: An Experimental Analysis of the Transparent Self. Van Nostrand-Reinhold, 1971.

Jourard, Sidney M.; and Landsman, Murry J.: Cognition, Cathexis, and the "Dyadic Effect" in Men's Self-Disclosing Behavior. Merrill-Palmer Quart., vol. 6, 1960, pp. 178–185.

Kahn, R. L.: The Work Module — A Tonic for Lunchpail Lassitude. Psychol. Today, vol. 6, Feb. 1973, pp. 35–40.

Kanas, N. A.; and Fedderson, W. E.: Behavioral, Psychiatric, and Sociological Problems of Long Duration Missions. NASA TM X-58067, 1971.

Kaplan, Rachel: Some Physiological Benefits of Gardening. Environ. Behav., vol. 50, no. 2, June 1973, pp. 145–162.

Katlin, Edward S.; and Silver Hoffman, Linda: Sex Differences and Self Report of Fear: A Psychosociological Assessment. J. Abnorm. and Soc. Psychol., vol. 85, no. 6, Dec. 1976, pp. 607–610.

Katz, Daniel; and Kahn, Robert L.: The Social Psychology of Organizations. Second ed. John Wiley and Sons, 1966.

Kelly, G. F.; and Coons, D. O.: Medical Aspects of Gemini Extra-Vehicular Activities. Gemini Summary Conference, NASA SP-138, 1967, pp. 107–126.

Kelman, Herbert C.: Privacy and Research with Human Beings. J. Social Issues, vol. 33, no. 3, Summer 1977, pp. 169–195.

Kendall, A. David; and Lindvall, Thomas, eds.: Evaluation of Community Odor Exposure. Report of a Symposium sponsored by the Environmental Protection Agency at Arthur D. Little, Inc., Cambridge, Mass., April 26–29, 1971, PB-204989.

Kennedy, Robert S.; and Graybiel, Ashton: Validity of Tests of Canal Sickness in Predicting Susceptibility to Airsickness and Seasickness. Aerosp. Med., vol. 33, no. 8, Aug. 1962, pp. 935–938.

Kerckhoff, Alan Chester; and Back, Kurt W.: The June Bug. A Study of Hysterical Contagion. Appleton-Century-Crofts, New York, 1968.

Kerwin, J. P.: Skylab 2 Crew Observations and Summary. Proceedings of the Skylab Life Sciences Symposium. NASA TM X-58154, vol. 1, 1974, pp. 55–59.

Kiesler, C. A.; Kiesler, S.; and Pallak, M.: The Effects of Commitment to Future Interaction on Reactions to Norm Violations. J. Personality, vol. 35, no. 4, 1967, pp. 585–599.

Kimmel, D. C.: Adulthood and Aging: An Interdisciplinary View. Second ed. John Wiley and Sons, 1980.

Kinsey, J. L.: Psychological Aspects of Nautilus Transpolar Cruise. U.S. Armed Forces Medical J., vol. 10, no. 4, April 1959, pp. 451-462.

Kiritz, Stewart; and Moos, Ralph H.: Physiological Effects of Social Environments. Psychosom. Med., vol. 36, no. 2, March-April 1974, pp. 96-114.

Klapp, Orrin Edgar: Currents of Unrest: An Introduction to Collective Behavior. Holt, Rinehart and Winston, 1972.

Kleiber, Douglas A.: Free-time Activity and Psycho-Social Adjustment in College Students: A Preliminary Analysis. J. Leisure Res., vol. 12, no. 3, 1980, pp. 205-212.

Klein, K. E.; Backhausen, F.; Bruner, H.; Eichhorn, J.; Jovy, D.; Schotte, J.; Vogt, L.; and Wegmann, H. M.: Die Abhängigkeit der Orthostase und Beschleunigungstoleranz von Körperbau and Leistungsfahigkeit. Internationale Zeitschrift Fuer Angewandte Physiologie Einschleisslich Arbeitsphysiologie, vol. 26, no. 3, 1968, pp. 205-226.

Klein, K. E.; Bruner, H.; Holtmann, H.; Rehme, H.; Stolze, H.; Steinhoff, W. D.; and Wegmann, H. J.: Circadian Rhythms of Pilots Efficiency and Effects of Multiple Time Zone Change. Aerosp. Med., vol. 41, no. 2, Feb. 1970, pp. 125-138.

Klein, K. E.; Bruner, H.; Jovy, D.; Vogt, L.; and Wegmann, H. M.: Influence of Stature and Physical Fitness on Tilt-Table and Acceleration Tolerance. Aerosp. Med., vol. 40, no. 3, March 1969, pp. 293-297.

Klein, K. E.; Wegmann, H. M.; and Hunt, Bonnie I.: Desynchronization of Body Temperature and Performance Circadian Rhythm as a Result of Outgoing and Homegoing Transmeridian Flights. Aerosp. Med., vol. 43, no. 2, Feb. 1972, pp. 119-132.

Klein, K. E.; Wegmann, H. M.; and Kuklinkski, P.: Athletic Endurance Training — Advantage for Space Flight: The Significance of Physical Fitness for Selection and Training of Spacelab Crews. Recent Advances in Space Medicine, J. Collin, ed., Technical Editing and Reproduction Ltd., 1977. Also Aviat., Space, and Environ. Med., vol. 48, March 1977, pp. 215-222.

Klein, T. J.: A Workload Simulation Model for Predicting Human Performance Requirements in the Pilot-Aircraft Environment. Paper presented at the 14th Annual Convention, Human Factors Society, San Francisco, 1970.

Kleinhans, B.; and Taylor, D. A.: Group Process, Productivity, and Leadership. Social Psychology: An Introduction, Bernard Seidenberg and Alvin Snadowski, eds., Free Press, New York, 1976, pp. 377–406.

Kleitman, N.: The Sleep-Wakefulness Cycle in Submarine Personnel. A Survey Report: Human Factors in Undersea Warfare, National Research Council, Committee on Undersea Warfare. Panel on Psychology and Physiology. Waverly Press, Baltimore, Md., 1949, pp. 329–341.

——: Sleep and Wakefulness. Univ. Chicago Press, 1963.

Knapp, Ronald J.; and Capel, William C.: Stress in the Deep: A Study of Undersea Divers in Controlled Dangerous Situations. J. Appl. Psychol., vol. 61, no. 4, Aug. 1976, pp. 507–512.

Kobasa, Suzanne C.; Hilker, Robert R. J.; and Maddi, Salvatore R.: Who Stays Healthy Under Stress? J. Occupat. Med., vol. 21, no. 9, Sept. 1979, pp. 595–598.

Kornilova, L. N.; Syrykh, G. D.; Tarasov, I. N.; and Yakovleva, I. Ya.: Results of the Investigation of the Otolith Function in Manned Space Flights. NASA TM-76103, June 1980. Translated from Vestnik Otorino-laringologii, vol. 6, 1979, pp. 21–24.

Kottenhoff, H.; and Lindahl, L. E. H.: Laboratory Studies on the Psychology of Motion Sickness. Acta Psychol., vol. 17, 1960, pp. 89–112.

Kubis, J. F.: Isolation, Confinement and Group Dynamics in Long Duration Spaceflight. Astronautica Acta, vol. 17, Apr. 1972, pp. 45–72.

Kubis, J. F.; Elrod, J. T.; Rusnak, R.; and Barnes, J. E.: Apollo 15 Time and Motion Study. NASA CR-128695, 1972.

Kubis, J. F.; Elrod, J. T.; Rusnak, R.; Barnes, J. E.; and Saxon, S. C.: Apollo 16 Time and Motion Study, NASA CR-128696, 1972.

Kubis, J. F.; and McLaughlin, E. J.: Psychological Aspects of Space Flight. Trans. New York Acad. Sci., Ser. 2, vol. 30, no. 2, Dec. 1967, pp. 320–330.

Kubis, J. F.; McLaughlin, E. J.; Jackson, J. M.; Rusnak, R.; McBride, G. H.; and Saxon, S. V.: Task and Work Performance on Skylab Missions 2, 3, and 4: Time and Motion Study — Experiment M151. Biomedical Results from Skylab, R. S. Johnston and L. F. Dietlein, eds. NASA SP-377, 1977, pp. 136–154.

Kurash, S.; Andzheyevska, A.; and Gurski, Ya.: Morphological Changes in Different Types of Rat Muscle Fibers During Long Term Hypokinesia. Space Biol. and Aerosp. Med., vol. 14, no. 16, 1980, pp. 45–52.

Lamb, L. E.: Aeromedical Evaluation for Space Pilots. Lectures in Aerospace Medicine, Feb. 3-7, 1964. USAF School of Medicine, Brooks AFB, Texas, 1964, pp. 119-142. (AD-445253)

Lamb, Lawrence E.; Johnson, Robert L.; and Stevens, Paul M.: Cardiovascular Deconditioning During Chair Rest. Aerosp. Med., vol. 35, no. 7, 1964a, pp. 646-649.

Lamb, Lawrence E.; Johnson, Robert L.; Stevens, P. M.; and Welch, B. E.: Cardiovascular Deconditioning From a Space Cabin Simulator Confinement. Aerosp. Med., vol. 35, no. 5, 1964b, pp. 420-428.

Lamb, Lawrence E.; Stevens, Paul M.; and Johnson, Robert L.: Hypokinesia Secondary to Chair Rest from 4 to 10 Days. Aerosp. Med., vol. 36, no. 8, Aug. 1965, pp. 755-763.

Lang, G.: Hypertension Disease (Gipertonicheskaia bolezn'), Medgiz: Moscow, 1950.

Langdon, David E.; and Hartman, Bryce: Performance Upon Sudden Awakening. SAM TR-62-17, School of Aerospace Medicine, Brooks AFB, Texas, 1961.

Langdon, Laurence E.; Gabriel, Richard F.; and Abell, Paul A.: Behavioral Acoustics: The Impact of Space Simulator Noise on Crew Members. Preliminary Results from an Operational 90-Day Manned Test of a Regenerative Life Support System. A. O. Pearson and D. C. Grana, eds., NASA SP-261, 1971, pp. 471-478.

Langer, E. J.; and Rodins, J.: The Effects of Choice and Enhanced Personal Responsibility for the Aged. A Field Experiment in an Institutional Setting. J. Pers. and Soc. Psychol., vol. 34, no. 2, Aug. 1976, pp. 191-196.

Lauber, J. K.: Resource Management of the Flight Deck: Background and Statement of the Problem. Resource Management on the Flight Deck, G. E. Cooper, M. D. White, and J. K. Lauber, eds., NASA CP-2120, 1980, pp. 3-16.

Laufer, Robert S.; Proshansky, Harold M.; and Wolfe, Maxine: Some Analytic Dimensions of Privacy. Environmental Psychology: People and Their Physical Settings. Second ed. Harold M. Proshansky, William H. Ittelson, and Leanne G. Rivlin, eds. Holt, Rinehart and Winston, 1976, pp. 206-217.

Law, P.: Personality Problems in Antarctica. Med. J. Australia, 47th year, vol. I, no. 8, Feb. 1960, pp. 273-282.

Law, Phillip: The Mental Strain of Antarctic Solitude. UNESCO Courier, vol. 15, no. 6, June 1963, pp. 26-30.

Lawler, Edward Emmet: Motivation in Work Organizations. Brooks/Cole, Monterey, Calif., 1973.

―――: Control Systems in Organizations. Handbook of Industrial and Organizational Psychology, M. Dunnette, ed., Rand McNally College Pub., 1976, pp. 1247-1291.

Lazarus, Arnold A.: Learning Theory and the Treatment of Depression. Behav. Res. and Therapy, vol. 6, no. 1, 1968, pp. 83-89.

Lederer, L. G.; and Kidera, G. J.: Passenger Comfort in Commercial Air Travel with Reference to Motion Sickness. Internat. Rec. Med., vol. 167, no. 12, Dec. 1954, pp. 661-668.

Lefcourt, H. M.: The Function of the Illusions of Control and Freedom. Amer. Psychol., vol. 28, no. 5, May 1973, pp. 417-425.

Lentz, J. Michael: Nystagmus, Turning Sensations and Illusory Movement in Motion Sickness Susceptibility. Aviat., Space, and Environ. Med., vol. 47, no. 9, Sept. 1976, pp. 932-936.

Lentz, J. Michael; and Collins, William E.: Motion Sickness Susceptibility and Related Behavioral Characteristics in Men and Women. Aviat., Space, and Environ. Med., vol. 48, no. 4, April 1977, pp. 316-322.

Leon, Arthur S.; and Blackburn, H.: The Relationship of Physical Activity to Coronary Heart Disease and Life Expectancy. Ann. New York Acad. of Sci., vol. 301, 1977, pp. 561-578.

Leon, H. A.; Serova, L. V.; and Landaw, S. A.: Effect of Weightlessness and Centrifugation (1 X g) on Erthrocyte Survival in Rats Subjected to Prolonged Space Flight. Preprint, Aerosp. Med. Assoc., 1978, pp. 210-211. Also published in U.S. Experiments Flown on the Soviet Satellite Cosmos 936, 1978. NASA TM-78526, 1978, pp. 60-76.

Leonov, A. A.; and Lebedev, V. I.: Psychological Compatibility in Interplanetary Flight. JPRS Report 57878, 1973; translated from Russian Language Journal Voprosy Filosofii, Moscow, no. 9, 1972, pp. 14-27.

―――: Psychological Problems of Interplanetary Flight. NASA TT F-16536, Translation of article from Russian Language Journal, Izdatel'stvo Nauka, B. F. Lomov, ed., Moscow, Oct. 1975, pp. 4-248.

Lester, David: The Unique Qualities of Telephone Therapy. Psychotherapy: Theory, Research, and Practice, vol. 11, no. 3, Fall 1974, pp. 219-221.

Lester, James T.: Stress: On Mount Everest. Psychol. Today, vol. 3, no. 4, Sept. 1969, pp. 30-32; 62.

Leventhal, Gerald S.: Fairness in Social Relationships. Contemporary Topics in Social Psychology, John W. Thibaut, J. T. Spence, and R. C. Carson, eds., Learning Press, Morristown, N.J., 1975, pp. 211-240.

Levanthal, H.; and Lindsley, D. B.: Subjective States. Human Factors in Long Duration Spaceflight. Space Science Board, National Research Council, Washington, D. C., 1972, pp. 144-159.

Levanthal, Howard: Fear: For Your Health. Psychol. Today, vol. 1, no. 5, Sept. 1967, pp. 52-58.

——: Experimental Studies of Anti-Smoking Communications. Smoking, Health, and Behavior, Edgar F. Borgatta and Robert R. Evans, eds., Aldine, Chicago, Ill., 1968.

——: Findings and Theory in the Study of Fear Communication. Advances in Experimental Social Psychology, vol. 5. Leonard Berkowitz, ed. Academic Press, 1970.

Levanthal, Howard; Jones, Susan; and Trembly, Grevilda: Sex Differences in Attitude and Behavior Change Under Conditions of Fear and Specific Instructions. J. Exp. Soc. Psychol., vol. 2, no. 4, 1966, pp. 387-399.

Levin, Hannah A.; and Askin, Frank: Privacy in the Courts: Law and Social Reality. J. Social Issues, vol. 38, no. 3, Summer 1977, pp. 138-153.

Levy, R. A.; Jones, D. R.; and Carlson, E. H.: Biofeedback Rehabilitation of Airsick Aircrew. Aviat., Space, and Environ. Med., vol. 52, no. 2, Feb. 1981, pp. 118-121.

Lewis, Kurt; Lippitt, R.; and White, R. K.: Patterns of Aggressive Behavior in Experimentally Created Social Climates. J. Soc. Psychol., vol. 10, 1939, pp. 271-299.

Lewinsohn, P. M.: A Behavioral Approach to Depression. The Psychology of Depression: Contemporary Theory and Research. Raymond J. Friedman and Martin M. Katz, eds., V. H. Winston, Washington, D.C., 1974a, pp. 157-185.

——: Clinical and Theoretical Aspects of Depression. Innovative Treatment Methods in Psychopathology, Karen S. Calhoun, H. E. Adams, and K. M. Mitchell, eds. John Wiley and Sons, 1974b, pp. 63-120.

Lidvall, H. F.: Mechanisms of Motion Sickness as Reflected in the Vertigo and Nystagmus Responses to Repeated Caloric Stimuli. Acta Oto-Laryngol., vol. 55, Nov.-Dec. 1962, pp. 527-536.

Likers, Rensis: New Patterns of Management. McGraw-Hill Book Co., 1961.

Lindemann, Erich: Symptomatology and Management of Acute Grief. Stress and Coping, an Anthology, Alan Monat and Richard S. Lazarus, eds. Columbia Univ. Press, 1977, pp. 334-348.

——: Beyond Grief: Studies in Crisis Intervention. Jason Aronson, New York, 1979.

Link, Mae Mills: Space Medicine in Project Mercury. NASA SP-4003, 1965.

Link, M. M.; and Gurovskiy, N. N.: Training of Cosmonauts and Astronauts. Foundations of Space Biology and Medicine: Space Medicine and Biotechnology, M. Calvin and O. G. Gazenko, eds., NASA SP-374, vol. 3, 1975, pp. 438–452.

Llano, G. A.: Airmen Against the Sea: An Analysis of Sea Survival Experiences. ADTIC Publication G104, Research Studies Institute, Maxwell AFB, Ala., Oct. 1955, p. 114.

Lockhart, J. M.: Extreme Body Cooling and Psychomotor Performance. Ergonomics, vol. 11, 1968, pp. 249–269.

Loo, C.: Important Issues in Researching the Effects of Crowding on Humans. Representative Res. in Soc. Psychol., vol. 4, no. 1, Jan. 1973, pp. 219–226.

Lott, Albert J.; and Lott, Bernice E.: Group Cohesiveness and Interpersonal Attraction: A Review of the Relationships With Antecedent and Consequent Variables. Psychol. Bull, vol. 64, no. 4, 1965, pp. 259–305.

Lovelace, W. Randolph, II; Schwichtenberg, A. H.; Luft, Ulrich C.; and Secrest, Robert R.: Selection and Maintenance Program for Astronauts for the National Aeronautics and Space Administration. Aerosp. Med., vol. 33, no. 6, June 1962, pp. 667–684.

Luft, U. C.; Myhre, L. G.; Leoppky, J. A.; and Venters, M. D.: A Study of Factors Affecting Tolerance of Gravitational Stress Stimulated by Lower Body Negative Pressure. Specialized Physiological Studies in Support of Manned Space. NASA CR-147498, 1976, pp. 1–60.

Lugg, D. J.: The Adaptation of a Small Group to Life on an Isolated Antarctic Station. Polar Human Biology, O. G. Edholm and E. K. E. Gunderson, eds., Wm. Heinemann, London, 1973, pp. 401–409.

Lynch, Kevin: The Image of the City. MIT Press, Cambridge, Mass., 1960.

Lynch, Theodore N.; Jenson, Robert L.; Stevens, P. M.; Johnson, R. L.; and Lamb, L. E.: Metabolic Effects of Prolonged Bed Rest. Their Modification by Simulated Altitude. Aerosp. Med., vol. 38, no. 1, Jan. 1967, pp. 10–20.

McCarthy, John Joseph: Why Managers Fail, and What to Do About It. Second ed., McGraw-Hill Book Co., 1979.

McClintock, C. G.: Game Behavior and Social Motivation in Interpersonal Settings. Experimental Social Psychology, Charles Graham McClintock, ed., Holt, Rinehart & Winston, 1972, pp. 271–297.

―――: Social Values: Their Definition, Measurement and Development. J. Res. and Dev. in Education, vol. 12, no. 1, Fall 1978, pp. 121-137.

McCormack, P. D.: A 2-Factor Theory of Vigilance. Br. J. Psychol., vol. 53, no. 4, 1962, pp. 357-363.

McDonnell Douglas Astronautics Company: 60-Day Manned Test of a Regenerative Life Support System with Oxygen and Water Recovery. Pt. II: Aerospace Medicine and Man-Machine Results. NASA CR-98501, 1968.

McGuire, Frederick; and Tolchin, Sidney: Group Adjustment at the South Pole. J. Mental Science, vol. 107, no. 450, Sept. 1961, pp. 954-960.

Machell, R. M.; Bell, L. E.; Prim, J. W., III; and Shyken, N. P.: Summary of Gemini Extravehicular Activity. Gemini Summary Conference, NASA SP-138, 1967, pp. 127-146.

Mack, Pauline Berry: Bone Density Changes in a Macaca Nemestrina Monkey During the Biosatellite III Project. Aerosp. Med., vol. 42, no. 8, Aug. 1971, pp. 828-833.

McKenzie, Richard E.; Hartman, Bryce O.; and Welch, B. E.: Observations in the SAM Two-Man Space Cabin Simulator. III. System Operator Performance Factors. Aerosp. Med., vol. 32, no. 7, July 1961, pp. 603-609.

Mackie, Robert R.; O'Hanlon, James F.; and McCauley, Michael: A Study of Heat, Noise and Vibration in Relation to Driver Performance and Physiological Status. Rep. 1735, DOT Bureau of Motor Carrier Safety, Federal Highway Admin. and National Highway Traffic Safety Admin., Washington, D.C., Dec. 1974. PB-238829/6.

Mackworth, N. H.: Researches on the Measurement of Human Performance. Medical Research Council Rep. no. 268, 1950.

Mackworth, N. H.; Kaplan, I. T.; and Metlay, W.: Eye Movements During Vigilance. Percept. Mot. Skills, vol. 18, April 1964, pp. 397-402.

Maki, Judith E.; Thorngate, Warren B.; and McClintock, Charles C.: Prediction and Perception of Social Motives. J. Pers. and Soc. Psychol., vol. 37, no. 2, Feb. 1974, pp. 203-220.

Mandler, G.; Mandler, J. M.; and Uviller, E. T.: Autonomic Feedback: The Perception of Autonomic Activity. J. Abnorm. Soc. Psychol., vol. 56, no. 3, May 1958, pp. 367-375.

Mann, Richard D.: A Review of the Relationship Between Personality and Performance in Small Groups. Psychol. Bull., vol. 56, July 1959, pp. 241-270.

Mann, Richard Dewey; Gibbard, Graham S.; and Hartman, John Jacob: Interpersonal Styles and Group Development. John Wiley and Sons, 1972.

Margolis, George Frederick; Bernheim, F.; and Hurteau, William W., Jr.: Antagonistic Effects of Certain Drugs at Low Barometric Pressures. J. Aviat. Med., vol. 22, no. 3, 1951, pp. 190-193, 234.

Margulis, Stephen T.: Preface to Privacy. Man-Environment Interactions: Evaluations and Applications, Part II, Vol. 6, S. T. Margulis, vol. ed. Environmental Design Research Assoc., Washington, D.C., 1974, and Dowden, Hutchinson, and Rose, Inc., Stroudsberg, Pa., 1975.

——: Conceptions of Privacy: Current Status and Next Steps. J. Social Issues, vol. 33, no. 3, Summer 1977, pp. 5-51.

Maris, Ronald W.: Social Forces in Urban Suicide. Dorsey Press, Homewood, Ill., 1969.

Marshall, Eliot: Psychotherapy Works, But for Whom? Science, vol. 207, no. 4430, Feb. 1980, pp. 506-508.

Marshall, Nancy J.: Environmental Components of Orientation Toward Privacy. Proceedings of the Second Annual Design Research Association Conference, October 1970, Pittsburgh, Pa., John Archea and Charles Eastman, eds., Dowden, Hutchinson, and Ross, Inc., Stroudsburg, Pa. Distributed by Halsted Press, a Division of John Wiley and Sons, Inc., 1970, pp. 246-251.

Mason, Robert M.; and Garden, John L., eds.: Guiding the Development of a Controlled Ecological Life Support System. Report on the NASA Ames Workshop, January 8-12, 1979. NASA CR-162452, 1979.

Mathews, Kenneth F., Jr.; and Canon, Lance Kirkpatrick: Environmental Noise Level Determinant of Helping Behavior. J. Pers. and Soc. Psychol., vol. 32, no. 4, Oct. 1975, pp. 571-577.

Matsnev, E. I.; Iakovleva, I. I.; Tarasov, I. K.; Alekseev, V.; Kornilova, L. N.; Mateev, A. D.; and Gorgiladze, G. I.: Space Motion Sickness: Phenomenology, Countermeasures, and Mechanisms. Aviat., Space, and Environ. Med., vol. 54, April 1983, pp. 312-317.

Medina, Miguel A.: Alteration of Drug Action and Metabolism Under Reduced Pressure. Lectures in Aerospace Medicine, 7th Series, Feb. 9-12, 1970, Brooks AFB, Texas, 1970, pp. 304-327.

Mehrabian, Albert: Silent Messages. Wadsworth Publ. Co., Belmont, Calif., 1970.

Miles, Stanley: Underwater Medicine. Third ed., Lippincott, Philadelphia, 1969.

Miller, Dorothy Hillyer: Suicidal Careers: Toward a Symbolic Interaction Theory of Suicide. Dissertation, Dept. of Social Welfare, Univ. of California, Berkeley, 1967.

Miller, D. H.; and Goleman, D. J.: Predicting Post-Release Risk Among Hospitalized Suicide Attempters. Omega, the Journal of Death and Dying, vol. 1, 1970, pp. 71-84.

Miller, D. R.: The Study of Social Relationships: Situation, Identity, and Social Interaction. Psychology: A Study of a Science, Vol. 5, S. Koch, ed. McGraw-Hill Book Co., 1963, pp. 639-737.

Miller, Earl F.; Graybiel, Ashton; Kellogg, Robert S.; and O'Donnel, Robert D.: Motion Sickness Susceptibility Under Weightless and Hypergravity Conditions Generated by Parabolic Flight. Aerosp. Med., vol. 40, no. 8, Aug. 1969, pp. 862-868.

Miller, James G.: Toward a General Theory for the Behavioral Sciences. Amer. Psychologist, vol. 10, 1955, pp. 513-531.

——: Information Input, Overload and Psychopathology. Amer. J. Psychiatry, vol. 116, Feb. 1960, pp. 695-704.

——: Living Systems. McGraw-Hill Book Co., 1978.

Miller, Lawrence M.: Behavior Management: The New Science of Managing People at Work. John Wiley & Sons, 1978.

Miller, Norman; and Zimbardo, Philip G.: Motives for Fear-Induced Affiliation: Emotional Comparison or Interpersonal Similarity? J. Personality, vol. 34, no. 4, 1966, pp. 481-503.

Miller, Perry B.; Johnson, Robert L.; and Lamb, Lawrence E.: Effects of Four Weeks of Absolute Bed Rest on Circulatory Functions in Man. Aerosp. Med., vol. 35, no. 12, Dec. 1964, pp. 1194-1200.

——: Effects of Moderate Physical Exercise During Four Weeks of Bed Rest on Circulatory Functions in Man. Aerosp. Med., vol. 36, no. 11, Nov. 1965, pp. 1077-1082.

Miller, R. B.: Task Analysis and Task Taxonomy: Inventive Approach. Paper presented at the American Psychological Association Convention, Chicago, Ill., Sept. 3-7, 1965.

Mintz, Alexander: Non-Adaptive Group Behavior. J. Abnorm. and Soc. Psychol., vol. 46, 1951, pp. 150-159.

Misumi, Jyuji; and Shilakashi, Sanshiro: An Experimental Study of the Effects of Supervisory Behavior on Productivity and Morale in a Hierarchical Organization. Human Relations, vol. 19, no. 3, 1966, pp. 297-308.

Mitchell, Janet Lee: Out-of-Body Experiences: A Handbook. McFarland and Co., Inc. (London), 1981.

Money, K. E.: Motion Sickness. Physiol. Rev., vol. 50, no. 1, Jan. 1970, pp. 1-39.

Monk, T. H.; Knauth, P.; Folkard, S.; and Rutenpranz, J.: Memory Based on Performance Measures in Studies of Shiftwork. Ergonomics, vol. 21, no. 10, Oct. 1978, pp. 819-826.

Morey, Emily R.: Spaceflight and Bone Turnover. Correlation with a New Rat Model of Weightlessness. BioScience, vol. 29, no. 3, March 1979, pp. 168-172.

Morgan, Ben B., Jr.; and Alluisi, E. A.: Synthetic Work: Methodology for Assessment of Human Performance. Percept. Mot. Skills, vol. 35, no. 3, Dec. 1972, pp. 835-845.

Morgan, Thomas E.; Ulvedal, Frode; and Welch, B. E.: Observations in the SAM Two-Man Space Cabin Simulator. II. Biomedical Aspects. Aerosp. Med., vol. 32, no. 7, July 1961, pp. 591-602.

Morgan, William P.: The Mind of the Marathoner. Psych. Today, vol. 11, no. 11, April 1978, pp. 38-49.

Morley, Ian E.; and Stephenson, Geoffrey M.: Interpersonal and Inter-party Exchange: A Laboratory Simulation of an Industrial Negotiation at the Plant Level. Brit. J. Psychol., vol. 60, no. 4, 1969, pp. 543-545.

———: Formality in Experimental Negotiations: A Validation Study. Brit. J. Psychol., vol. 61, no. 3, 1970, pp. 383-384.

Morley, Wilbur E.; Messick, Janice M.; and Aquilera, Donna C.: Crisis: Paradigms of Intervention. J. Psychiatr. Nurs. and Ment. Health Serv., vol. 5, no. 6, Nov.-Dec. 1967, pp. 537-538.

Morris, R. L.: Spacelab Cost Reduction Alternatives Study: Vol. III, Crew Training Task Analysis. Final Report. NASA CR-147702, 1975.

Morway, D. A.; Lathrop, R. G.; Chambers, R. M.; and Hitchcock, L., Jr.: The Effects of Prolonged Water Immersion on the Ability of Human Subjects to Make Position and Force Estimations. U.S. Naval Air Development Center Report NADC-MA-6115-5, 1963.

Moss, Sidney Z.; and Moss, Miriam S.: Separation as a Death Experience. Child Psychiatry and Human Development, vol. 3, no. 3, Spring 1973, pp. 187-194.

Mostert, Noel: Supership. Alfred A. Knopf, Inc., 1974.

Mullin, C. S.: Some Psychological Aspects of Isolated Antarctic Living. Amer. J. Psychiatry, vol. 117, 1960, pp. 323-325.

Murphy, M. R.: Analysis of Eighty-four Commercial Aviation Incidents: Implications for a Resource Management Approach to Crew Training. Proceedings of the Annual Reliability and Maintainability Symposium, 1980, pp. 298-307.

Murray, Raymond H.; and McCally, Michael: Combined Environmental Stresses. Bioastronautics Data Book, Second ed. James F.

Parker and Vita R. West, eds. NASA SP-3006, 1973, pp. 881-914.
Myasnikov, V. I.; Panchenkova, E. F.; and Uskov, F. N.: Prospect of Using Radio and TV Communication Data in the Medical Supervision of Cosmonauts in Flight. Preprint of paper presented at the 25th Int. Cong. of Aviat. and Space Med., Helsinki, Finland, Sept. 4-9, 1977.

NASA: Medical Evaluation and Standards for Astronaut Selection: NASA Class I — Pilot Astronaut (JSC-11569). Prepared by Space and Life Sciences Directorate, NASA, Lyndon B. Johnson Space Center, Houston, Texas, Jan., 1977a.

———: Medical Evaluation and Standards for Astronaut Selection: NASA Class II — Mission Specialist (JSC-11570). Prepared by Space and Life Sciences Directorate, NASA, Lyndon B. Johnson Space Center, Houston, Texas, Jan., 1977b.

———: Medical Evaluation and Standards for Astronaut Selection: NASA Class III — Payload Specialist (JSC-11571). Prepared by Space and Life Sciences Directorate, NASA, Lyndon B. Johnson Space Center, Houston, Texas, April 1977c.

NASA, Office of Manned Spaceflight: Flammability, Odor and Offgassing Requirements and Test Procedures for Materials in Environments that Support Combustion. NHB 8060.1A, Feb. 1974.

Nagatsuka, Yasuhiro; and Suzuki, Yukio: Studies on Sensory Deprivation. II. Part 2. Effects of Sensory Deprivation upon Perceptual and Motor Functions. Tohoyu Psychological Folia, vol. 22, nos. 3-4, 1964, pp. 64-68.

Nardini, J. E.; Herrmann, R. S.; and Rasmussen, J. E.: Navy Psychiatric Assessment Program in the Antarctic. Amer. J. Psychiatry, vol. 119, 1962, pp. 97-105.

Natani, Kirmach: Future Directions for Selecting Personnel. Human Factors of Outer Space Production, T. Stephen Cheston and David L. Winter, eds., Westview Press, Boulder, Colorado, 1980, pp. 25-63.

Natani, K.; and Shurley, J. T.: Sociopsychological Aspects of a Winter Vigil at a South Pole Station. Human Adaptability to Antarctic Conditions. Antarctic Research Series, E. K. E. Gunderson, ed., American Geophysical Union, Washington, D.C., vol. 22, 1974, pp. 89-114.

Natani, K.; Shurley, J. T.; and Joern, A. T.: Interpersonal Relationships, Job Satisfaction, and Subjective Feelings of Competence: Their Influence Upon Adaptation to Antarctic Isolation. Polar Human Biology, O. G. Edholm and E. K. E. Gunderson, eds., Wm. Heinemann (London), 1973.

Natani, K.; Shurley, J. T.; Pierce, C. M.; and Brooks, R. E.: Long Term Changes in Sleep Patterns in Men on the South Polar Plateau. Arch. Intern. Med., vol. 125, no. 4, April 1970, pp. 655–659.

National Research Council: Noise Abatement: Policy Alternatives for Transportation. Rep. ISBN-0-309-02648-2, Analytical Studies for the U.S. Environmental Protection Agency, Washington, D.C., Oct. 1977.

Nedzel, A. J.: Influence of Negative Pressure on the Toxicity of Morphone Sulfate in White Mice. J. Aviat. Med., vol. 26, no. 2, April 1955, pp. 164–165.

Nelson, Paul D.: Psychological Aspects of Antarctic Living. Military Med., vol. 130, May 1965, pp. 485–489.

Neuringer, Charles: Rigid Thinking in Suicidal Individuals. J. Consulting Psych., vol. 28, 1964, pp. 54–58.

Newsom, Bernard D.; Brady, James F.; and Goble, Guy J.: Equilibrium and Walking Changes Observed at 5, 7½, 10, and 12 rmp in the Revolving Space Station Simulator. Aerosp. Med., vol. 36, no. 4, April 1965, pp. 322–326.

Newsom, B. D.; Brady, J. F.; Shafer, W. A.; and French, R. S.: Adaptation to Prolonged Exposures in the Revolving Space Station Simulator. Aerosp. Med., vol. 37, no. 8, Aug. 1966, pp. 778–783.

Nicogossian, A. E.; and Parker, J. F., Jr.: Space Physiology and Medicine. NASA SP-447, 1982.

Novikov, V. E.; and Il'in, E. A.: Age-related Reactions of Rat Bones to Their Unloading. Aviat., Space, and Environ. Med., vol. 52, no. 9, Sept. 1981, pp. 551–553.

Noyes, R., Jr.; and Kletti, R.: Depersonalization in the Face of Life-Threatening Danger: A Description. Psychiatry, vol. 39, no. 1, Feb. 1976, pp. 19–27.

Oberg, James E.: Red Star in Orbit. Random House Inc., 1981.

O'Leary, Brian: The Making of an Ex-Astronaut. Houghton-Mifflin Co., 1970.

O'Lone, Richard G.: New Roles Seen for Human Eyes in Space. Aviat. Week, vol. 83, no. 9, Aug. 30, 1965, pp. 51–53.

O'Neal, H. A.; Bond, G. F.; Lanphear, R. E.; and Odum, T.: Project Sealab Summary Report: An Experimental Eleven-Day Under Sea Saturation Dive at 193 Feet. Sealab I Project Group, ONR Report ACR-108, Washington, D.C., 1965.

Ordiway, V.; and Raymon, R. S.: Case Report of an In-Flight Incident Involving an Aircraft Commander with a Psychiatric Illness. Aerosp. Med., vol. 45, no. 5, March 1974, pp. 316–317.

Orne, Martin T.: The Nature of Hypnosis: Artifact and Essence. J. Abnorm. Soc. Psychol., vol. 58, May 1959, pp. 277-299.

Osgood, Charles Egerton: An Alternative to War or Surrender. Univ. of Illinois Press, Urbana, Ill., 1962.

Page, R. N.; Dagley, C.; and Smith, S.: Manned Environmental System Assessment. MESA Program, Final Report Contract NASW-658, No. 02-00487-5, The Boeing Company, Seattle, Wa., June 1964.

Palamarek, Daniel L.; and Rule, Brendon G.: The Effects of Ambient Temperature and Insult on the Motivation to Retaliate or Escape. Motivation and Emotion, vol. 3, no. 1, March 1979, pp. 82-92.

Palmai, G.: Psychological Observations on an Isolated Group in Antarctica. Brit. J. Psychiatry, vol. 109, 1963, pp. 364-370.

Parin, V. V.; Kosmalinskiy, F. P.; and Dushkov, B. A.: Kosmicheskaya Biologiya i Meditsina, Moscow, Prosveshcheniye, 1970.

Parin, V. V.; Vinogradov, V. M.; and Razumeev, A. N.: Problems in Space Pharmacology. Kosmicheskaya Biologiya i Meditisina, vol. 3, no. 1, 1969, pp. 20-32.

Parker, David M.; and Wilsoncroft, W. E.: Intensity of Motion Sickness Symptoms as a Function of Apparent Autonomic Balance. J. Gen. Psychol., vol. 98, April 1978, pp. 253-257.

Parker, J. F., Jr.: The Identification of Performance Dimensions through Factor Analysis. Human Factors, vol. 9, no. 4, Aug. 1967, pp. 367-373.

Parker, J. F., Jr.; and Every M. G.: Habitability Issues in Long Duration Undersea and Space Missions. NASA CR-130537, 1972.

Parker, J. F., Jr.; Reilly, R. E.; Dillon, R. F.; Andrews, T. G.; and Fleishman, E. A.: Development of a Battery of Tests for Measurement of Primary Perceptual-Motor Performance. BioTechnology Inc. Contract Rep., 1965.

Patterson, Cecil Holden: Theories of Counseling and Psychotherapy. Third ed., Harper and Row Publishers, Inc., 1980.

Patterson, Peter H.: The Military Man. New York State J. Med., vol. 75, no. 12, Oct. 1975, pp. 2132-2135.

Paykel, E. S.: Recent Life Events and Clinical Depression. Life Stress and Illness, Ellsworth K. Eric Gunderson and Richard H. Rahe, eds., Chas. C. Thomas, Springfield, Ill., 1974, pp. 134-163.

Pearlman, C. A., Jr.: Separation Reactions of Married Women. Amer. J. of Psychiatry, vol. 126, 1970, pp. 946-950.

Pearson, David W.; and Thackray, Richard I.: Consistency of Performance Change and Automatic Response as a Function of Expressed Attitude Toward a Specific Stress Situation. Psychophysiology, vol. 6, no. 5, 1970, pp. 561-568.

Pearson, R. G.; Hart, F. D.; and O'Brien, J. F.: Individual Differences in Human Annoyance Response to Noise. N. Carolina State Univ., Raleigh, N.C., NASA CR-144921, 1975.

Pepler, R. D.: Warmth and Performance: An Investigation in the Tropics. Ergonomics, vol. 2, 1958, pp. 63–88.

———: Extreme Warmth and Sensorimotor Coordination. J. Appl. Physiol., vol. 14, no. 3, May 1959, pp. 383–386.

Perrow, Charles: Complex Organizations: A Critical Essay. Second ed., Scott, Foresman and Company, Glenview, Ill., 1979.

Perry, C. J. C.: Psychiatric Selection of Candidates for Space Missions. J. Amer. Med. Assn., vol. 194, Nov. 22, 1965, pp. 841–844.

———: A Psychiatric "Back Up System" for Selection of Space Crews. Amer. J. Psychiatry, vol. 123, no. 7, Jan. 1967, pp. 821–825.

Perry, Samuel: Acute Psychotic States. Psychiatric Emergencies, Robert A. Glick, Arthur T. Mayerson, Edwin Robbins, and John A. Talbott, eds. Grune and Stratton, New York, 1976, pp. 51–58.

Petrov, Y. A.: Habitability of Spacecraft. Foundations of Space Biology and Medicine, vol. 3, M. Calvin and O. G. Gazenko, eds., NASA SP-374, 1975, pp. 157–192.

Pinsky, L. S.; Osborne, W. Z.; Bailey, J. V.; Benson, R. E.; and Thompson, L. F.: Light Flashes Observed by Astronauts on Apollo 11 Through Apollo 17. Science, vol. 183, no. 4128, March 8, 1974, pp. 957–959.

Pishchik, V.: Goals and Equipment of Soviet-French Space Medicine Experiment. Meditsinskaya Gazeta, July 2, 1982, p. 4.

Pope, Frederick E.; and Rogers, Terrance A.: Some Psychiatric Aspects of an Arctic Survival Experiment: A Clinical Note on Thirteen Scientists. J. Nerv. and Ment. Dis., vol. 146, no. 6, 1968, pp. 433–445.

Popov, V. A.; and Boyko, N. I.: Vision During Space Flight. NASA TM X-60574, 1967.

Poulton, E. C.: Environment and Human Efficiency, ch. 12. Chas. C. Thomas, Springfield, Ill., 1970.

Preber, L.: Vegetative Reactions in Caloric and Rotatory Tests: A Clinical Study With Special Reference to Motion Sickness. Acta Oto-laryngol., vol. 144, suppl., 1958, pp. 1–119.

Quarantelli, E. L., ed.: Disasters: Theory and Research. Sage Studies in Sociol., vol. 13, Int. Sociol. Assoc., 1978.

Radloff, R.; and Helmreich, R.: Groups Under Stress: Psychological Research in Sealab II. Appleton-Century-Crofts, New York, 1968.

———: Stress: Under the Sea. Psychol. Today, vol. 3, no. 4, Sept. 1969, pp. 28-29, 59-60.

Rahe, Richard H.; and Arthur, Ronsom J.: Life Change and Illness Studies: Past History and Future Directions. J. Human Stress, vol. 4, no. 1, March 1978, pp. 3-15.

Ramsay, R. W.: Behavioral Approaches to Bereavement. Behav. Res. and Therapy, vol. 15, no. 1, 1977, pp. 131-135.

Rasmussen, E.: NASA Scientist Uses Biofeedback to Stop Motion Sickness. NASA News Release no. 80-32, May 6, 1980, pp. 1-5.

Rasmussen, J. E.: Psychologic Discomforts in 1962 Navy Protective Shelter Tests. J. Amer. Dietetic Assn., vol. 42, 1963, pp. 109-116.

Rasmussen, J. E.; and Wagner, C. M.: Psychological Studies. Studies of the Bureau of Yards and Docks Protective Shelter: I. Winter Trials. NRL Rept. 5882, Dec. 1962, pp. 81-102.

Rawls, J. R.; Hopper, A. E.; and Rawls, D. J.: Variables Thought to Determine Personal Space: An Opinion Sample. Texas Christian Univ. Inst. of Behavioral Research, Fort Worth, Texas, 1969.

Rawls, J. R.; McGaffey, C. N.; Trego, R. E.; and Sells, S. B.: Some Determiners of Interpersonal Climate in Relation to Long Distance Space Missions. Texas Christian Univ. Inst. of Behavioral Research, Fort Worth, Texas, 1968.

Rayman, Russell B.; and McNaughton, Grant B.: Decompression Sickness: USAF Experience 1970-1980. Aviat., Space, and Environ. Med., vol. 54, no. 3, March 1983, pp. 258-260.

Reason, J. T.: An Investigation of Some Factors Contributing to Individual Variation in Motion Sickness Susceptibility. FPRC Rep. No. 1277, Ministry of Defense, Flying Personnel Research Committee, London, 1968.

Reason, J. T.; and Brand, J. J.: Motion Sickness. Academic Press, 1975.

Reason, J. T.; and Diaz, E.: Simulator Sickness in Passive Observers. FPRC Rep. No. 1310, Ministry of Defense, Flying Personnel Research Committee, London, 1971.

Reason, J. T.; and Graybiel, A.: Factors Contributing to Motion Sickness Susceptibility: Adaptability and Receptivity. AGARD Conference Proceedings CP-109, Prediction of Motion Sickness in the Selection of Pilots, Paris, 1972.

Redgrove, June A.: Menstrual Cycles. Biological Rhythms and Human Performance. William Peter Colquhoun, ed., Academic Press, 1971, pp. 211-240.

Reinberg, A.; Vieux, N.; Ghata, J.; Chaumont, A. J.; and LaPorte, A.: Circadian Rhythms Amplitude and Individual Ability to Adjust to Shift Work. Ergonomics, vol. 21, no. 10, 1978, pp. 763-766.

Remek, V.: Communication Problems of International Crews. 30th Cong. Int. Astron. Fed., Munich, Germany, Sept. 17-29, 1977.

Richman, J.; and Rosenbaum, M. A.: A Clinical Study of Role Relationships in Suicidal and Non-Suicidal Psychiatric Patients. Paper presented at the Fifth Conf. on Suicide Prevention, London, England, Sept. 24-27, 1969. Proceedings edited by Richard Fox, The ASN, Vienna, 1970.

Rickarby, Geoffrey A.: Four Cases of Mania Associated with Bereavement. J. Nerv. and Ment. Dis., vol. 165, no. 4, Oct. 1977, pp. 255-262.

Rimpler, A.: Der Einfluss des Trainingszustandes auf die köperliche Leislungsfähigkeit in der Höhe. DLR-FB 70-08, ZLDI, 1970.

Rippere, Vicky: What's the Thing to Do When You're Feeling Depressed: A Pilot Study. Behav. Res. and Therapy, vol. 15, no. 2, 1977, pp. 185-191.

Robbins, Edwin; Angus, Bertrand; and Stern, Marvin: Drug Abuse. Psychiatric Emergencies, Robert A. Glick, Arthur T. Mayerson, Edwin Robbins, and John A. Talbott, eds., Grune and Stratton, New York, 1976, pp. 161-180.

Rogers, Everett M.; and Kincaid, D. Lawrence: Communication Networks: Towards a New Paradigm for Research. The Free Press, New York, 1981.

Rogers, Everett M.; and Rogers, Rekha Agarwala: Communication in Organizations. The Free Press, New York, 1976.

Rogers, Jon: Environmental Needs of Individuals and Groups. Proceedings of Annual Conf. of the California Council of the American Inst. of Architecture, Monterey, Calif., Oct. 1973.

———: Environmental Preference of Mixed Gender Crew in Isolation. Univ. of Alabama, Huntsville, Ala., 1978. Paper presented at American Psychol. Assn., Toronto, Canada, 1977.

Rohrer, J. H.: Interpersonal Relationships in Isolated Small Groups. Psychophysiological Aspects of Space Flight, Bernard E. Flaherty, ed., Columbia Univ. Press, New York, 1961, pp. 263-271.

Rose, C. L.; and Cohen, M. L.: Relative Importance of Physical Activity for Longevity. Annals of the New York Acad. Sci., vol. 301, 1977, pp. 671-702.

Rosengren, Karl Erik; Arvidson, Peter; and Sturesson, Dahn: The Barseback Panic — Radio Program as a Negative Summary Event. Acta Sociologica, vol. 18, no. 4, 1975, pp. 303-321.

Roth, E. M., ed.: Compendium of Human Responses to the Aerospace Environment. NASA CR-1205, vols. 1-4, 1968.

Rothstein, L. D.; and Kikoshima, A.: Human Information Processing Under Stress. Final Report of NASA Contract NAS2-9614, 1978.

Rowes, Barbara: Housewives in Space. Omni, vol. 4, no. 9, June 1982, pp. 64-67, 128.

Rubin, Robert T.; and Rahe, Richard H.: U.S. Navy Demolition Team Training. Life Stress and Illness, E. K. Eric Gunderson and Richard H. Rahe, eds., Chas. C. Thomas, Springfield, Ill., 1974, pp. 208-226.

Ruff, G. E.: Psychological and Psychophysiological Indices of Stress. Unusual Environments and Human Behavior: Physiological and Psychological Problems of Man in Space, Neal Murray Burns, R. M. Chambers, and E. Hendler, eds. Free Press of Glencoe (London), 1963, pp. 33-60.

Ruff, G. E.; Levy, E. Z.; and Thaler, V. H.: Studies of Isolation and Confinement. Aerosp. Med., vol. 30, 1959, pp. 599-604.

Ruff, George E.; and Levy, Edwin Z.: Psychiatric Evaluation of Candidates for Spaceflight. Amer. J. Psychiatry, vol. 116, no. 5, Nov. 1959, pp. 385-391.

Ruffell Smith, H. P. R.: A Simulator Study of the Interaction of Pilot Workload with Errors, Vigilance, and Decisions. NASA TM-78482, 1979.

Rushing, William A.: Individual Behavior and Suicide. Suicide, Jack Porter Gibbs, ed., Harper and Row Publishers, Inc., 1968, pp. 96-121.

Ryan, Michael G.; and Craig, James G.: Intergroup Telecommunication: The Influence of Communications Medium and Role Inducted Status Level on Mood, and Attitudes Toward the Medium and Discussion. Communication Research Center, Dept. of Communications, Ottawa, Ontario, Canada, 1975.

Ryback, Ralph S.; Lewis, Oliver F.; and Lessard, Charles S.: Psychobiologic Effects of Prolonged Bed Rest (Weightless) in Young Healthy Volunteers (Study II). Aerosp. Med., vol. 42, no. 5, May 1971b, pp. 529-535.

Ryback, Ralph S.; Trimble, Ralph W.; Lewis, Oliver F.; and Jennings, Charles L.: Psychobiologic Effects of Prolonged Weightlessness (Bed Rest) in Young Healthy Volunteers. Aerosp. Med., vol. 42, no. 4, April 1971a, pp. 408-415.

Sanders, Mark; Gustanski, Jerry; and Lawton, Mike: Effects of Ambient Illumination on Noise Level of Groups. J. Appl. Psych., vol. 59, no. 4, Aug. 1974, pp. 527-528.

Sandler, H.; Goldwater, D.; Rositano, S. A.; Sawin, C. F.; and Booher, C. R.: Physiologic Response of Male Subjects Ages 46 to 55 Years to Shuttle Flight Simulation. Preprint of Aerosp. Med. Assoc., Annual Scientific Meeting, May 14-17, 1979, Washington, D.C., pp. 43-49.

Sandler, Harold; and Winter, David L.: Physiological Responses of Women to Simulated Weightlessness: A Review of the First Female Bed-Rest Study. NASA SP-430, 1979.

Sarnoff, Irving; and Zimbardo, Philip G.: Anxiety, Fear and Social Affiliation. J. Abnorm. and Soc. Psychol., vol. 62, no. 2, March 1961, pp. 356-363.

Sawin, C. F.; Nicogossian, A. E.; Rummel, J. A.; and Michel, E. L.: Pulmonary Function Evaluation during the Skylab and Apollo-Soyuz Missions. Aviat., Space, and Environ. Med., vol. 47, no. 2, Feb. 1976, pp. 168-172.

Scales, J. T.; Winter, G. D.; and Bloch, M.: Levitation, A Possible Means of Treating Burns. Research in Burns, A. B. Wallace and A. W. Wilkinson, eds., E & S Livingstone Ltd. (London), 1965, pp. 266-282.

Scarr, S.: Social Introversion-Extroversion as a Heritable Response. Child Development, vol. 40, no. 3, 1969, pp. 823-832.

Schachter, Stanley: Deviation, Projection, and Communication. J. Abnorm. and Soc. Psychol., vol. 46, 1951, pp. 190-207.

——: The Psychology of Affiliation; Experimental Studies of the Sources of Gregariousness. Stanford Univ. Press, Stanford, Calif., 1959.

Schalling, D.; Cronholm, B.; and Asberg, M.: Components of State and Trait Anxiety as Related to Personality and Arousal. Emotions: Their Parameter and Measurements, Lennart Levi, ed., Raven Press, New York, 1975, pp. 603-617.

Schopler, J.; and Walton, M.: The Effects of Expected Structure, Expected Enjoyment and Participants' Internality-Externality Upon Feelings of Being Crowded. Dept. of Psychology, Univ. of North Carolina, Chapel Hill, N.C., 1974.

Schreuder, Otis B.: Medical Aspects of Aircraft Pilot Fatigue with Special Reference to the Commercial Jet Pilot. Aerosp. Med., vol. 17, no. 4, sec. II, April 1966, pp. 1-44.

Schroder, Harold M.; Driver, Michael J.; and Streufert, Siegfried: Human Information Processing; Individual and Group Functioning in Complex Social Situations. Holt, Rinehart and Winston, 1967.

Schwartz, Gary E.; Davidson, Richard J.; and Goleman, Daniel J.: Patterning of Cognitive and Somatic Processes in the Self-Regulation of Anxiety: Effects of Meditation vs. Exercise. Psychosomatic Med., vol. 40, no. 4, June 1978, pp. 321-328.

Schwartz, Shalom; and David, Amit B.: Responsibility and Helping in an Emergency: Effects of Blame, Ability and Denial of Responsibility. Sociometry, vol. 39, no. 4, Dec. 1976, pp. 406-415.

Scott, T. H.; Bexton, W. H.; Heron, W.; and Doane, B. K.: Cognitive Effects of Perceptual Isolation. Can. J. Psychol., vol. 13, Sept. 1959, pp. 200-209.

Seeman, J. S.; Singer, R. V.; and McLean, M. V.: Habitability. Preliminary Results from an Operational 90-Day Manned Test of a Regenerative Life Support System, A. O. Pearson and D. C. Grana, eds., NASA SP-261, 1971, pp. 393-414.

Seligman, M. E. P.: Depression and Learned Helplessness. The Psychology of Depression: Contemporary Theory and Research, Raymond J. Friedman and Martin M. Katz, eds., V. H. Winston, 1974, pp. 83-125.

Seligman, Martin E. P.; Klein, David C.; and Miller, William R.: Depression. Handbook of Behavior Modification and Behavior Theory, Harold Leitenberg, ed., Prentice-Hall, 1976, pp. 168-210.

Sells, S. B.: A Model for the Social System for the Multiman Extended-Duration Space Ship. Aerosp. Med., vol. 37, Nov. 1966, pp. 1130-1135.

Sells, S. B.; and Gunderson, E. K. E.: A Social System Approach to Long-Duration Missions. Human Factors in Long Duration Spaceflight, Space Science Board, National Research Council, Washington, D.C., 1972, pp. 179-208.

Selye, Hans: Stress Without Distress. J. P. Lippincott Co., 1974.
———: The Stress of Life. Revised Edition. McGraw-Hill Book Co., 1976.

Serxner, J. L.: An Experience in Submarine Psychiatry. Amer. J. Psychiatry, vol. 125, no. 1, July 1968, pp. 25-30.

Shackel, B.: A Note on Panel Layout for Number of Identical Items. Ergonomics, vol. 2, May 1959, pp. 247-253.

Sharpe, Mitchell R.: Living in Space: The Astronaut and His Environment. Doubleday and Co., Inc., 1969.

Shashkov, V. W.; and Yegorov, B. B.: Problems of Pharmacology in Space Medicine. Farmakologiya i Toksikologiya, vol. 42, no. 4, 1979, pp. 325-330.

Shaw, M. E.: Group Dynamics: The Psychology of Small Group Behavior. Second ed. McGraw-Hill Book Co., 1976.

Shears, Loyda M.; and Gunderson, E. K. E.: Stable Attitude Factors in Natural Isolated Groups. J. Soc. Psychol., vol. 7, no. 2, 1966, pp. 199-204.

Sherif, Muzafer; Harvey, O.; White, B.; Hood, W.; and Sherif, C.: Intergroup Conflict and Cooperation: The Robbers' Cave Experiment. Revised edition. University Book Exchange (Oklahoma Univ. Inst. of Group Relations Publication), 1961.

Sherif, Muzafer; and Sherif, Carolyn Wood: Social Psychology. Harper & Row Publishers, Inc., 1969.

Sherrod, Drury R.: Crowding, Perceived Control and Behavioral Aftereffects. J. Appl. Soc. Psychol., vol. 4, April-June 1974, pp. 171-186.

Short, J. A.: Bargaining and Negotiation — An Exploratory Study. Paper E/71065/SH, Communications Studies Group, London, England, 1971.

———: Conflicts of Opinion and Medium of Communication. Paper E/72001/SH, Communications Studies Group, London, England, 1972a.

———: Medium of Communication, Opinion Change, and Solution of a Problem of Priorities. Paper E/72245/SH, Communications Studies Group, London, England, 1972b.

———: The Effects of Medium of Communication on Persuasion, Bargaining, and Perception of the Other. Paper E/73100/SH, Communications Studies Group, London, England, 1973.

———: Effect of Medium of Communication on Experimental Negotiations. Human Relations, vol. 27, 1974, pp. 225-234.

Short, John; Williams, Ederlyn; and Christie, Bruce: The Social Psychology of Telecommunications. John Wiley and Sons, Inc., London, 1976.

Shurley, J. T.: Antarctica Is Also A Prime Natural Laboratory for the Behavioral Sciences. Polar Human Biology, O. G. Edholm and E. K. E. Gunderson, eds., Wm. Heinemann, London, 1973, pp. 430-435.

Shurley, Jay T.; Natani, Kirmach; and Sengel, Randal: Ecopsychiatric Aspects of a First Human Space Colony. Paper 77-550, presented at Third Princeton/AIAA Conference on Space Manufacturing Facilities, Princeton, N.J., May 9-12, 1977.

Simons, John C.: An Introduction to Surface-Free Behavior. Ergonomics, vol. 7, no. 1, 1964, pp. 23-36.

Sinaiko, H. Wallace: Teleconferencing, Preliminary Experiments. Institute for Defense Analyses, Washington, D.C. Research Paper P-108, Nov. 1963. (AD-601932)

Singer, S. Fred, ed.: Manned Laboratories in Space. Springer-Verlag, 1968.

Smelser, Neil J.: Theory of Collective Behavior. Routledge, 1962.
Smith, Carroll W.; Corbin, Bob G.; and Olree, Harry D.: Developing and Maintaining Physical Fitness Suitable for Prolonged Spaceflight. Summary Report for January 1974–March 1976, Vol. 1, NASA CR-14778, June 1976.
Smith, D.: Communication and Negotiation Outcome. J. Communication, vol. 19, no. 3, Sept. 1969, pp. 248–256.
Smith, Seward: Studies of Small Groups in Confinement. Sensory Deprivation: Fifteen Years of Research. J. P. Zubek, ed., Appleton-Century-Crofts, 1969, pp. 374–401.
Smith, Seward; and Haythorn, W. W.: Effects of Compatibility, Crowding, Group Size and Leadership Seniority on Stress, Anxiety, Hostility and Annoyance in Isolated Groups. J. Pers. and Soc. Psychol., vol. 22, no. 1, April 1972, pp. 67–79.
Smith, W. M.: Observations of the Lifetime of a Small Isolated Group: Structure, Danger, Boredom and Vision. Psychol. Rep., vol. 19, no. 2, 1966, pp. 475–514.
Solomon, Richard L.; and Corbit, John D.: An Opponent-Process Theory of Motivation. Psychol. Rev., vol. 81, no. 2, March 1974, pp. 119–145.
Sours, J. A.: The "Break-off Phenomenon": A Precipitant of Anxiety in Jet Aviators. Arch. Gen. Psychiat., vol. 13, Nov. 1965, pp. 447–456.
Space Sciences Board: Human Factors in Long Duration Space Flight. National Research Council, Washington, D.C., 1972.
Sparvieri, F.: Effetti Della Depressione Barometrica e di Alcuni Formaci Tranquillanti Sulla Attivita Motoria Spontanea del Topino. Riv. Med. Aero., vol. 23, no. 3, July-Sept. 1960, pp. 397–402.
Spence, Janet T.; and Helmreich, Robert L.: Masculinity and Femininity: Their Psychological Dimensions, Correlates, and Antecedents. Univ. of Texas Press, Austin, Texas, 1978.
Spence, Janet T.; Helmreich, Robert L.; and Holahan, Carole K.: Negative and Positive Components of Psychological Masculinity and Femininity and Their Relationships to Self-Reports of Neurotic and Acting-out Behaviors. J. Pers. and Soc. Psychol., vol. 37, no. 10, Oct. 1979, pp. 1673–1682.
Spielberger, Charles D.: Conceptual and Methodological Issues in Anxiety Research. Anxiety: Current Trends in Theory and Research, vol. I, ch. 14, Chas. D. Spielberger, ed., Academic Press, 1972, pp. 481–493.
Stager, P.; Proulx, P.; Walsh, B.; and Fudakowski, T.: Bilingual Air-Traffic Control in Canada. Human Factors, vol. 22, Dec. 1980, pp. 655–670.

Stapley, B.: Collected Papers on the Remote Meeting Table. Paper W/73298/ST, Communications Studies Group, London, England, 1973.

Stapp, J. P., ed.: Orbital International Laboratory and Space Sciences. Conference, Air Force Missile Development Center, Holloman AFB, Cloudcraft, New Mexico, 1969.

Stegemann, J.; Meier, O.; Shipka, W.; Hartlieb, W.; Hemmer, B.; and Tiebes, U.: Effects of a Multi-Hour Immersion With Intermittent Exercise on Urinary Excretion and Tilt Table Tolerance in Athletes and Nonathletes. Aviat., Space, and Environ. Med., vol. 46, no. 1, Jan. 1975, pp. 26-29.

Stein, David D.; Hardyick, Jane Allyn; and Smith, M. Brewster: Race and Belief: An Open and Shut Case. J. Pers. and Soc. Psychol., vol. 1, no. 4, 1965, pp. 281-289.

Steiner, Ivan D.: Group Process and Productivity. Academic Press, 1972.

———: Task Performing Groups. Contemporary Topics in Social Psychology, John W. Thibaut, J. T. Spence, and R. C. Carson, eds., General Learning Press, Morristown, N.J., 1976, pp. 393-422.

Stewart, John D.; Clark, Brant; Cowings, Patricia; and Toscano, William B.: Learned Regulation of Autonomic Responses to Control Coriolis Motion Sickness: Its Effects on Other Vestibular Functions. Preprint of the Aerosp. Med. Assoc. 1978 Annual Meeting, May 8-11, 1978, New Orleans, La., pp. 133-135.

Sticha, Paul J.; Hunter, Gregory M.; and Randall, L. Scott: Research Into Teleconferencing. TR-80-9-314, Decisions and Designs, Inc., McLean, Va., Feb. 1981.

Stockton, William S.; and Wilford, John N.: Spaceliner: The New York Times Report on the Columbia's Voyage. Times Books, 1981.

Stokols, Daniel A.: On the Distinction Between Density and Crowding: Some Implication for Future Research. Psychol. Rev., vol. 79, no. 3, May 1972, pp. 275-277.

———: The Experience of Crowding in Primary and Secondary Environments. Environ. and Behav., vol. 8, no. 1, March 1976, pp. 49-86.

———: Environmental Psychology. Ann. Rev. Psychol., vol. 29, 1978a, pp. 253-295.

———: In Defense of the Crowding Concept. Advances in Environmental Psychology. Vol. 1, The Urban Environment, Andrew Baum, Jerome E. Singer, and Stuart Valins, eds. Lawrence Associates, Hillsdale, N.J., 1978b, pp. 111-130.

———: A Congruence Analysis of Human Stress. Stress and Anxiety, Vol. 6, Irwin G. Sarason and Charles D. Spielberger, eds. Halsted Press, 1979, pp. 27-53.

Stokols, Daniel; Ohlig, Walter; and Resnick, Susan M.: Perception of Residential Crowding, Classroom Experience and Student Health. Human Ecology, vol. 6, no. 3, Sept. 1978, pp. 233-252.

Stokols, Daniel; Rall, Marilyn; Pinner, Berna; and Schopler, John: Physical, Social, and Personal Determinants of the Perception of Crowding. Environ. and Behav., vol. 5, no. 1, 1973, pp. 87-115.

Storm, W. F.; and Giannetta, C. L.: Effects of Hypercapnia and Bedrest on Psychomotor Performance. Aerosp. Med., vol. 45, no. 4, 1974, pp. 431-433.

Strange, R. E.; and Klein, W. J.: Emotional and Social Adjustment of Recent Winter-Over Parties in Isolated Antarctic Stations. Polar Human Biology, O. G. Edholm and E. K. E. Gunderson, eds., Wm. Heinemann Medical Books Ltd., London, England, 1973, pp. 410-416.

Streimer, I.; Turner, D. P. W.; and Volkmer, K.: Human Manual Force Production Capabilities in the Underwater Environment. SID 66-1562, North American Rockwell, 1966.

Streimer, I.; Turner, D. P. W.; Volkmer, K.; and Guerin, D.: Effects of the Underwater Environment upon Work Efficiency of Divers. Rep. no. 70-117, Man Factors, Inc., 1970.

Strope, W. E.; Etter, H. S.; Goldbeck, R. A.; Heiskell, R. H.; and Sheard, J. S.: Preliminary Report on the Shelter Occupancy Test of 3-17 December, 1959. USN RDL-TR-418, Naval Radiological Defense Laboratory, San Francisco, Calif., 1960. (AD-237130)

Strope, W. E.; Schultze, D. P.; and Pond, J. I.: Preliminary Report of the Shelter Occupancy Test of 25-29 July 1960. USN RDL-TR-502, Naval Radiological Defense Laboratory, San Francisco, Calif., 1961. (AD-253856)

Strughold, H.: Physiological Day-Night Cycle in Global Flights. J. Aviat. Med., vol. 23, no. 5, Oct. 1952, pp. 464-473.

Strughold, H.; and Hale, H. B.: Biological and Physiological Rhythms. Foundations of Space Biology and Medicine, vol. II, book 2, M. Calvin and O. G. Gazenko, eds. NASA SP-374, 1975, pp. 535-548.

Stupakov, G. P.: Artificial Gravity as a Means of Preventing Atrophic Skeletal Changes. Space Biol. and Aerosp. Med., vol. 115, no. 4, July-Aug. 1981, pp. 88-90.

Suedfeld, Peter: Restricted Environmental Stimulation: Research and Clinical Application. Wiley Interscience, New York, 1980.

Sullins, W. R.; and Rogers, J. G.: Report of Preliminary Results, NASA Grant NSG-2058, Univ. of Alabama, Huntsville, Ala., Oct. 1975.
Sundstrom, E.: Interpersonal Behavior and the Physical Environment. Social Psychology, Second ed., Lawrence S. Wrightman, ed., Brooks-Cole, Monterey, Calif., 1977.
Sundstrom, Eric; and Altman, Irwin: Field Study of Territorial Behavior and Dominance. J. Pers. and Soc. Psychol., vol. 30, no. 1, 1974, pp. 115-124.
Sundstrom, Eric; Burt, Robert E.; and Kamp, Douglas: Privacy at Work: Architectural Correlates of Job Satisfaction and Job Performance. Acad. Mgmt. J., vol. 23, no. 1, 1980, pp. 101-117.
Surwillo, Walter W.; and Quilter, Reginald W.: Vigilance, Age, and Response-Time. Amer. J. Psychol., vol. 77, 1964, pp. 614-620.

Talmon, Y.: Mate Selections in Collective Settlements. Amer. Sociol. Review, vol. 29, 1964, pp. 491-508.
Taylor, D. A.; Altman, I.; Wheeler, L.; and Kushner, E. N.: Personality Factors Related to Response to Social Isolation and Confinement. J. Cons. and Clin. Psychol., vol. 33, no. 4, 1969, pp. 411-419.
Teichner, Warren H.: Manual Dexterity in the Cold. J. Appl. Physiol., vol. 11, no. 3, Nov. 1957, pp. 333-338.
Teichner, W. H.; Arees, E.; and Reilly, R.: Noise and Human Performance: A Psychophysiological Approach. Ergonomics, vol. 6, no. 1, 1963, pp. 83-97.
Thackray, Richard I.; and Pearson, David W.: Effects of Cognitive Appraisal of Stress on Heart Rate and Task Performance. Perceptual and Motor Skills, vol. 27, no. 2, 1968, pp. 651-658.
Thomas, K.: Conflict and Conflict Management. Handbook of Industrial and Organizational Psychology, M. Dunnette, ed., Rand McNally College Pub., 1976, pp. 889-937.
Thompson, Teresa L.; and Seibold, David R.: Stigma Management in Normal-Stigmatized Interactions: Test of the Disclosure Hypothesis and a Model of Stigma Acceptance. Human Communication Res., vol. 4, no. 3, Spring 1978, pp. 231-242.
Thornton, W. E.; Hoffler, G. W.; and Rummel, J. A.: Anthropometric Changes and Fluid Shifts. Biomedical Results from Skylab, R. S. Johnston and L. F. Dietlein, eds., NASA SP-377, 1977, pp. 330-338.
Thornton, W. E.; and Rummel, J. A.: Muscular Deconditioning and Its Prevention in Space Flight. Biomedical Results from Skylab.

R. S. Johnston and L. F. Dietlein, eds., NASA SP-377, 1977, pp. 191-197.

Tobias, Cornelius A.; and Grigor'yev, Yu. G.: Ionizing Radiation. Foundations of Space Biology and Medicine, vol. II, book 2, M. Calvin and O. G. Gazenko, eds., NASA SP-374, 1975, pp. 473-531.

Toscano, William B.; and Cowings, Patricia S.: Transference of Learned Autonomic Control for Symptom Suppression Across Opposite Directions of Coriolis Acceleration. Preprint of the Aerosp. Med. Assoc. 1978 Annual Meeting, New Orleans, La., May 8-11, 1978, pp. 132-133.

——: Reducing Motion Sickness: A Comparison of Autogenic-Feedback Training and an Alternative Cognitive Task. Aviat., Space, and Environ. Med., vol. 53, no. 5, May 1982, pp. 449-453.

Toufexis, Anastasia: The Hazards of Orbital Flight. Time, Feb. 28, 1983, p. 48.

Truchaud, M.: Influence de l'attitude sur le Seuil de la douleur. I. Etude Experimentale Generale. Ann. Pharm. Franc, vol. 24, 1966, pp. 281-294.

Tuckman, Bruce: Developmental Sequence in Small Groups. Psychol. Bull., vol. 63, no. 6, 1965, pp. 384-399.

Tyler, D. B.; and Bard, P.: Motion Sickness. Physiol. Rev., vol. 29, Oct. 1949, pp. 311-369.

Ushakov, A. S.; Myasnikov, V. I.; Shestkov, B. P.; Agureev, A. N.; Belakovsky, M. S.; and Runyantseva, M. P.: Effect of Vitamin and Amino Acid Supplements on Human Performance During Heavy Mental and Physical Work. Aviat., Space, and Environ. Med., vol. 49, no. 10, Oct. 1978, pp. 1184-1187.

Vachon, Mary L. S.: Grief and Bereavement Following the Death of a Spouse. Can. Psychiat. Assn. J., vol. 21, no. 1, Feb. 1976, pp. 35-44.

Vaillant, G. E.: Natural History of Male Psychological Health. V. The Relation of Choice of Ego Mechanisms of Defense to Adult Adjustment. Arch. Gen. Psychiat., vol. 33, no. 5, May 1976, pp. 535-545.

Van Huss, Wayne D.; and Heusner, William W.: Spaceflight Research Relevant to Health, Physical Education, and Recreation, With Particular Reference to Skylab's Life Science Experiments. Final Rep., Michigan State Univ., E. Lansing, Mich., June 1979.

Veltfore, Helene Rank; and Lee, George F.: The Coconut Grove Fire: A Study in Scapegoating. J. Abnorm. Psychol., vol. 38, no. 2, Supp., April 1943, pp. 138-154.

Vernikos-Danellis, Joan; Winget, C. M.; Leach, Carolyn S.; and Rambaut, Paul C.: Circadian, Endocrine and Metabolic Effects of Prolonged Bedrest: Two 56-Day Bedrest Studies. NASA TM X-3051, 1974.

Vernikos-Danellis, J.; Winget, C. M.; Leach, C. S.; Rosenblatt, L. S.; Lymen, J.; and Beljan, J. R.: Space Motion Sickness Medications: Interference with Biomedical Parameters. Acta Astronautica, vol. 4, Nov.-Dec. 1977, pp. 1159-1169.

Vinsel, Anne; Brown, Barbara B.; Altman, Irwin; and Foss, Carolyn: Privacy Regulation, Territorial Displays and Effectiveness of Individual Functioning. J. Pers. and Soc. Psychol., vol. 39, no. 6, 1980, pp. 1104-1115.

Voas, R. B.: Project Mercury Astronaut Training Programs. Symposium on Psychophysiological Aspects of Space Flight, Bernard E. Flaherty, ed., Columbia Univ. Press, 1961, pp. 99-161.

Vogt, Fred B.; Mack, Pauline B.; and Johnson, Philip C.: Tilt Table Response and Blood Volume Changes Associated with Thirty Days of Recumbency. Aerosp. Med., vol. 37, no. 8, Aug. 1966, pp. 771-777.

——: Tilt Table Response and Blood Volume Changes Associated with Fourteen Days of Recumbency. Aerosp. Med., vol. 38, no. 1, Jan. 1967, pp. 43-48.

Volynkin, U. M.; and Vasil'yev, P. V.: Some Results of Medical Studies Conducted During the Flight of the "Voskhod." The Problems of Space Biology, N. M. Sisakyan, ed., NASA TT F-528, 1969, pp. 52-66.

Von Gierke, Henning E.; Nixon, Charles W.; and Guignard, J. C.: Noise and Vibration. Foundations of Space Biology and Medicine, vol. II, book 1, M. Calvin and O. G. Gazenko, eds., NASA SP-374, 1975, pp. 355-405.

Vroom, V. H.: Handbook of Industrial and Organizational Psychology. Marvin D. Dunnette, ed., Rand McNally, 1976, pp. 1527-1551.

Vroom, Victor H.; and Yetton, Philip W.: Leadership and Decision Making. Univ. of Pittsburgh Press, Pittsburgh, Pa., 1976.

Waligora, J. M.: Physical Forces Generating Acceleration, Vibration and Impact. The Physiological Basis for Spacecraft Environmental Limits, J. M. Waligora (coor.), NASA RP-1045, 1979.

Wallis, D.; and Samuel, J. A.: Some Experimental Studies of Radar Operating. Ergonomics, vol. 4, 1961, pp. 155-168.

Walster, Elaine; Berscheid, Ellen; and Walster, G. William: New Directions in Equity Research. J. Pers. and Soc. Psychol., vol. 25, Feb. 1973, pp. 151-176.

Watters, H.; and Steadman, J.: Development and Analysis of a Modular Approach to Payload Specialist Training. Training of Spaceruns for Spacelab. NASA TM X-64980, 1976.

Webb, W. B.; and Agnew, H. W., Jr.: Reaction Time and Serial Response Efficiency on Arousal from Sleep. Percept. Mot. Skills, vol. 18, no. 3, June 1964, pp. 783–784.

———: Sleep and Waking in a Time-Free Environment. Aerosp. Med., vol. 45, no. 6, June 1974, pp. 617–622.

———: Sleep Efficiency for Sleep-Wake Cycles of Varied Length. Psychophysiology, vol. 12, no. 6, Nov. 1975, pp. 637–645.

Webb, Wilse B.; and Cartwright, Rosalind D.: Sleep and Dreams. Ann. Rev. Psychol., vol. 29, 1978, pp. 223–252.

Weber, Max: The Theory of Social and Economic Organization. A. M. Henderson and T. Parsons, trans. and eds., Oxford Univ. Press, New York, 1947.

Weick, Karl E.: Organization Design: Organizations as Self-Designing Systems. Organizational Dynamics, Autumn 1977, pp. 31–46.

Weigel, Richard G.; Dinges, Norman; Dyer, R.; and Straumfjord, A.: A Perceived Self-Disclosure, Mental Health, and Who is Liked in Group Treatment. J. Couns. Psychol., vol. 19, Jan. 1972, pp. 47–52.

Weinstein, Neil D.: Individual Differences in Reactions to Noise: A Longitudinal Study in a College Dormitory. J. Appl. Psychol., vol. 63, no. 4, 1978, pp. 458–466.

Weitzman, E. D.; Nogeire, D.; Perlow, M.; Fukushima, D.; Sassin, J.; McGregor, P.; Gallagher, T. F.; and Hellman, L.: Effects of a Prolonged 3-Hr Sleep-Wake Cycle on Sleep States, Plasma Cortisol, Growth Hormone and Body Temperature. Man. J. Clin. Endocrinol. Metab., vol. 38, June 1974, pp. 1018–1030.

Welford, Alan Traviss: Fundamentals of Skill. Methuen and Co., London, 1968.

Wenger, Dennis E.: Community Response to Disaster, Functional and Structural Alterations. Disasters: Theory and Research, E. L. Quarantelli, ed., Sage Publishing, Beverly Hills, Calif., 1978.

Wenger, M. A.: The Measurement of Individual Differences in Autonomic Balance. Psychosom. Med., vol. 3, 1941, pp. 427–434.

Westin, Alan I.: Privacy and Freedom. Atheneum, New York, 1967.

Weston, J. R.; and Kristen, C.: Teleconferencing: A Comparison of Attitudes, Uncertainty and Interpersonal Atmospheres in Mediated and Face-to-Face Group Interaction. The Social Policy and Programs Branch, Dept. of Communications, Ottawa, Ontario, Canada, December 1973.

Wexley, Kenneth N.; and Yuki, G. A.: Organizational Behavior and Personnel Psychology. Richard D. Irwin, Inc., Homewood, Ill., 1977.

Weybrew, Benjamin B.: Psychological Problems of Prolonged Submarine Submergence. Unusual Environment and Human Behavior: Physiological and Psychological Problems of Man in Space, Neal M. Burns, Randall M. Chambers, and Edwin Hendler, eds., Free Press, London, 1963.

Wheeler, Lawrence R.; and Grotz, Janis: Conceptualization and Measurement of Reported Self-Disclosure. Human Communication Research, vol. 2, no. 4, Summer 1976, pp. 338–346.

Whitaker, L. A.; Graham, W. P., III; Parsi, R.; and Olson, R. M.: Laminar Airflow in Major Burn Treatment. Pennsylvania Med., vol. 79, no. 6, June 1976, pp. 50–51.

White, P. D.; Nyberg, J. W.; Finney, L. M.; and White, W. J.: Influence of Periodic Centrifugation on Cardiovascular Functions of Man During Bed Rest. NASA CR-65422, 1966.

White, W. J.: Effects of Transient Weightlessness on Brightness Discrimination. Aerosp. Med., vol. 36, no. 4, 1965a, pp. 327–331.

———: A Space Laboratory. Role of the Vestibular Organs in the Exploration of Space. NASA SP-77, 1965b, pp. 209–213.

White, W. J.; Nyberg, J. W.; White, P. D.; Grimes, R. H.; and Finney, L. M.: Biomedical Potential of a Centrifuge in an Orbiting Laboratory. DAC-SM-48703, Douglas Aircraft Company, Santa Monica, Calif., 1965. SSD-TR-64-209-Suppl. (AD-472550).

Whittenburg, J. A.; Ross, S.; and Andrews, T. G.: Sustained Perceptual Efficiency as Measured by the Mackworth "Clock" Test. Percept. Mot. Skills, vol. 6, 1956, pp. 109–116.

Whyte, W. F.: Models for Building and Changing Social Organizations. Human Organization, vol. 26, no. 1, 1967, pp. 22–31.

Wicker, Allen W.: An Introduction to Ecological Psychology. Brooks/Cole, Monterey, Calif., 1979.

Wilding, J. M.; and Meddis, R.: A Note on Personality Correlates of Motion Sickness. Brit. J. Psychol., vol. 63, no. 4, Nov. 1972, pp. 619–620.

Wilkinson, Robert T.: Comparison of Paced, Unpaced, Irregular, and Continuous Display in Watch-Keeping. Ergonomics, vol. 4, 1961, pp. 259–267.

———: After-effects of Sleep Deprivation. J. Exp. Psychol., vol. 66, no. 5, 1963, pp. 439–442.

──: Sleep Deprivation: Performance Test for Partial and Selective Sleep Deprivation. Progress in Clinical Psychology, vol. 7, Lawrence Edwin Abt and Bernard F. Riess, eds., Grune and Stratton, New York, 1969, pp. 28-43.

Wilkinson, Robert T.; and Stretton, M.: Performance After Awakening at Different Times of Night. Psychonomic Sci., vol. 23, no. 4, May 1971, pp. 283-285.

Williams, Harold L.; Kearney, Ometta F.; and Lubin, Ardie: Signal Uncertainty and Sleep Loss. J. Exp. Psychol., vol. 69, no. 4, 1965, pp. 401-407.

Wilson, C. L., ed.: Project Mercury Candidate Evaluation Program. WADC-TR-59-505, Wright-Patterson AFB, Dayton, Ohio, Dec. 1959, (AD-234-739).

Winget, C. M.; DeRoshia, C. W.; and Sandler, H.: Influences of Horizontal Hypokinesia on Performance and Circadian Physiological Rhythms in Female Humans. Physiologist, vol. 22, no. 6, 1979, pp. S-79-S-80.

Winget, C. M.; Vernikos-Danellis, J.; DeRoshia, C. W.; and Cronin, S. E.: Rhythms During Hypokinesis. Biorhythms and Human Reproduction, Michael Ferin, F. Halberg, R. M. Richart, and R. L. Vande Wiele, eds., John Wiley and Sons, Inc., 1974, pp. 575-587.

Witkin, H. A.: The Nature and Importance of Individual Differences in Perception. J. Pers., vol. 18, 1949a, pp. 145-170.

──: Perception of Body Position and of the Position of the Visual Field. Psychol. Man., vol. 63, no. 7, 1949b, pp. 1-46.

Witkin, H. A.; Lewis, H. B.; Hertzman, M.; Machover, K.; Meissner, P.; Brehnall, P.; and Wopner, S.: Personality Through Perception. Harper and Row, 1954.

Wolfe, M.; and Golan, M. B.: Privacy and Institutionalization. Paper presented at the meeting of the Design Research Assn., Vancouver, B.C., May 1976.

Wolfe, T.: The Right Stuff. Farrar, Straus, Giroux, New York, 1979.

Wood, Charles D.; Kennedy, Robert S.; and Graybiel, Ashton: Review of Antimotion Sickness Drugs from 1954-1964. Aerosp. Med., vol. 36, no. 1, Jan. 1965, pp. 1-4.

Wood, Charles D.; Manno, Joseph E.; Manno, Barbara R.; Redetzki, Helmut M.; Wood, Mary; and Vekovius, W. Allen: Side Effects of Antimotion Sickness Drugs. Aviat., Space, and Environ. Med., vol. 55, no. 2, 1984, pp. 113-116.

Wood, D. T.: The Relationship Between Static Anxiety and Acute Physical Activity. American Correctional Therapy J., vol. 31, no. 3, May-June 1977, pp. 67-69.

Woodhead, Muriel M.: Effect of Brief Loud Noise on Decision Making. J. Acoust. Soc. Amer., vol. 31, no. 10, 1959, pp. 1329-1331.
——: Searching a Visual Display in Intermittent Noise. J. Sound and Vibration, vol. 1, no. 2, 1964a, pp. 157-161.
——: The Effect of Bursts of Noise on an Arithmetic Task. Amer. J. Psychol., vol. 77, no. 4, Dec. 1964b, pp. 627-633.
Woodling, C. H.; Faber, S.; vanBockel, J. J.; Olasky, C. C.; Williams, W. K.; Mire, J. L. C.; and Homer, J. R.: Apollo Experience Report: Simulation of Manned Space Flight for Crew Training. NASA TN D-7112, 1973.
Woodson, Wesley E.; and Conover, D. W.: Human Engineering Guide for Equipment Designers. Univ. of California Press, Berkeley, Calif., 1965.
Worthy, Morgan; Garry, Albert L.; and Kahn, Gay M.: Self Disclosure as an Exchange Process. J. Pers. and Soc. Psychol., vol. 13, no. 1, 1969, pp. 59-63.
Wright, J. E.; Vogel, J. A.; Sampson, J. B.; Knapik, J. J.; Patton, J. F.; and Daniels, W. L.: Effects of Travel Across Time Zones (Jetlag) on Exercise Capacity and Performance. Aviat., Space, and Environ. Med., vol. 54, no. 2, Feb. 1983, pp. 132-137.
Wrightsman, L. S., Jr.: Effects of Waiting with Others on Changes in Level of Felt Anxiety. J. Abnorm. Soc. Psychol., vol. 61, 1960, pp. 216-222.
Wurster, W. H.; Burchard, E. C.; and von Restorff, W.: Comparison of Oral and TTS-scopolamine with Respect to Anti-motion Sickness Potency and Psychomotor Performance. Preprints of the 1981 Annual Scientific Meeting of the Aerosp. Med. Assn. ISSN 0065-3764, San Antonio, Tex., May 4-7, 1981.

Yakovleva, Ya. I.; Kornilova, L. N.; Tarasov, I. K.; and Alekseyev, V. N.: Results of Studies of Cosmonauts' Vestibular Function and Spatial Perception. Space Biol. and Aerosp. Med., vol. 16, no. 1, Jan.-Feb. 1982, pp. 26-33.
Yanowitch, R. E.: Crew Behavior in Accident Causation. Aviat., Space, and Environ. Med., vol. 48, no. 10, Oct. 1977, pp. 918-921.
Yeremin, A. V.; Bogdashevskiy, R. M.; and Baburin, Ye. F.: Preservation of Human Performance Capacity under Prolonged Space Flight Conditions. Weightlessness: Medical and Biological Research, NASA TT F-16105, 1975, pp. 365-383.
Yingling, Robert W.: Classification of Reaction Patterns in Listening to Music. J. Res. in Music Education, vol. 10, no. 2, 1962, pp. 105-120.

Yuganov, E. M.; Gorshkov, A. I.; Kasian, I. I.; Brianov, I. I.; Kolosov, J. A.; Kopanev, V. I.; Lebedev, V. I.; Papov, N. I.; and Solodovnik, F. A.: Vestibular Reactions of Cosmonauts During the Flight in the Voskhod Spaceship. Aerosp. Med., vol. 37, no. 7, July 1966, pp. 691-694.

Ziller, R. C.: Group Dialectics: The Dynamics of Groups Over Time. Human Development, vol. 20, 1977, pp. 293-308.

Zimbardo, P. G.; and Formica, Robert: Emotional Comparison and Self Esteem as Determinants of Affiliation. J. Personality, vol. 31, no. 2, 1963, pp. 141-162.

Zimberg, Shelton: Alcohol Abuse. Psychiatric Emergencies, Robert A. Glick, Arthur T. Mayerson, Edwin Robbins, and John A. Talbott, eds., Grune and Stratton, New York, 1976, pp. 147-160.

Zubek, John P., ed.: Sensory Deprivation. Appleton-Century-Crofts, New York, 1969.

——: Sensory Isolation: Fifteen Years of Research at the University of Manitoba. Studia Psychologica, vol. 16, no. 4, 1974, pp. 265-274.

Zubek, J. P.; Aftanas, M.; Hasek, J.; Sansom, W.; Schludermann, E.; Wilgosh, L.; and Winocur, G.: Intellectual and Perceptual Changes During Prolonged Perceptual Deprivation: Low Ilumination and Noise Level. Percept. Mot. Skills, vol. 15, no. 1, 1962, pp. 171-198.

Zubek, John P.; Pushkar, Dolores; Sansom, Wilma; and Gowing, J.: Perceptual Changes after Prolonged Sensory Isolation (Darkness and Silence). Can. J. Psychol., vol. 15, 1961, pp. 83-101.

AUTHOR INDEX

Abbott, H. M., 40
Abe, K., 43
Abell, P. A., 73
Adams, C. A., 89
Adams, C. R., 40
Adams, J. S., 275,288
Adams, O. S., 114,123,124,127, 128
Adams, T., 132,133,134
Adams, V., 148,149
Adolfson, J., 244
Aftanas, M., 130
Agarwala-Rogers, R., 192,205, 208
Agnew, H. W., Jr., 123,133,139
Aguilera, D. C., 252
Agureev, A. N., 30
Ahearn, T. R., 65,76,78,161,164
Aiello, J. R., 87,89
Akiskal, H. S., 236
Alekeseyev, S. V., 71
Aleksandrova, Ye. A., 20
Alekseev, V., 48
Alekseyev, V. N., 35
Alexander, M., 69
Alexander, S. J., 36
Allen, R. W., 23
Allport, G. W., 151,152
Alluisi, E. A., 20,109,114,120, 123,124,125,129,133,134, 136,139
Altman, I., 10,11,13,15,84,85, 88,89,92,93,128,130,146,147, 159,178,193
Amatomi, M., 43

Ambler, R. K., 44
Amir, Y., 151,152
Andrews, G., 238,255
Andrews, T. G., 110,128
Andreyeva-Galanina, Ye. Ts., 71
Andzheyevska, A., 20
Angell, D., 110
Angus, B., 251
Archea, J., 86
Archer, R. L., 91,93
Arees, E., 71
Argyle, M., 199
Aronson, E., 169
Arthur, R. J., 237
Arvidson, P., 229
Asberg, M., 256
Asch, S. E., 93
Aschoff, J., 133,135
Askin, F., 84
Atkinson, J. W., 155
Austin, R. W., 52
Avant, L. L., 129
Averill, J. R., 72,247

Babin, B. A., 226
Baburin, Ye. F., 108
Back, K. 232
Backhausen, F., 26
Bailey, G. C., 197
Bailey, J. V., 53
Bair, J. H., 204,206

Bakan, P., 128
Bakeman, R., 173
Bakwin, H., 43
Bales, R. F., 13,164,177
Balish, E., 140
Bard, P., 42
Barker, R. G., 171
Barnes, J. E., 118
Barnes, R., 107
Baron, R. A., 95
Baron, R. M., 89
Barrett, G. V., 40,46
Bart, P., 236
Bartelloni, B. J., 114
Basowitz, H., 223
Baum, A., 87
Baumel, I. P., 31
Baylink, D. J., 34
Beck, A. T., 236
Becker, E., 236
Becker, R. W., 72
Behnke, A. R., 244
Beisel, W. R., 114
Belakovsky, M. S., 40
Beljan, J. R., 47,126
Bell, C. H., Jr., 267
Bell, L. E., 116
Bell, P. A., 95,113
Bem, D. J., 156
Bem, S. L., 156
Benjamin, B., 248
Bennett, P. B., 244
Bennis, W. G., 174,177
Benson, A. J., 244
Benson, R. E., 53
Berglund, B., 69
Berglund, U., 69
Bergman, S. A., 27
Bergum, B. O., 129
Berkhout, J., 134
Berkowitz, L., 74,173
Berkun, M. M., 222
Berlinger, D. C., 110
Bernheim, F., 31

Berry, C. A., 2,9,20,21,35,36,51,
 62,63,64,68,70,115,132,137,
 145,189,211,271
Berscheid, E., 85,149,171,275
Bevan, W., 129
Bexton, W. H., 129
Billingham, J., 44,49
Billings, C. E., 191
Billman, G. E., 20
Bilodeau, J. W., 116
Birkhead, N. C., 21
Birren, J. E., 37
Biryukov, Ye. N., 20
Blackburn, H., 81
Blake, M. J. F., 135
Blaney, P. H., 236
Blatt, W. F., 31
Blizzard, J. J., 21
Bloch, M., 140
Bluth, B. J., 11,64,92,227,275,
 289,290
Bogdashevskiy, R. M., 108
Bonura, M. S., 62
Booher, C. R., 137
Borsky, P. N., 72,74
Bourne, G. H., 22
Boyd, I., 11,12,13
Boyko, N. I., 52
Boyle, C. P., 29
Brady, J. F., 34
Brand, J. J., 35,38,43,48,50
Breed, W., 250,251,253
Breeze, R. K., 60
Brehm, J. W., 271
Brehmer, B., 291,292
Brettnall, P., 46
Brianov, I. I., 36
Briegleb, W., 21
Broadbent, D. E., 71,129
Brooks, C. G., 116
Brooks, R. E., 136
Broverman, D. M., 156
Broverman, I. K., 156
Brown, B. B., 92

Brown, F. M., 32
Brown, I. E., 113
Bruner, H., 26,133
Bryant, J. P., 116
Buchanan, R. W., 89
Bucher, R., 231
Buck, L., 128
Bulk, G. K., 40
Burchard, E. C., 33
Burgenbauch, S. F., 70,195,210, 271,274,289
Burke, P. J., 164,165
Burns, N. M., 128
Burt, R. B., 87
Butler, L. F., 40
Byrne, D., 158,265

Calhoun, J. B., 87
Cameron, B. J., 125
Campbell, D. P., 283
Canon, L. K., 71
Cantrell, G. K., 125
Cantrell, R. W., 70
Capel, W. C., 223
Caplow, T., 297
Carden, J. L., 63
Carey, J., 203
Carkhuff, R. R., 241
Carlson, E. H., 49,50
Carr, A. C., 232
Carskadon, M. A., 123
Cartwright, D., 169,173
Cartwright, R. D., 123
Casey-Stahmer, A. E., 197
Catterson, A. D., 20,137
Cazes, G., 95
Central Institute for the Deaf, 72
Chaikin, A. L., 84,93
Chamberlain, A. S., 90
Chambers, R. M., 123,138
Champness, B. G., 197
Chapanis, A., 197

Chape, C., 88
Chapman, L. F., 22,213
Chapple, E., 90
Chase, G. A., 21
Chaumont, A. J., 133
Cheaney, E. S., 191
Chemers, M. M., 167
Chernyshov, M., 66
Cheshire-Engler, D., 49
Chiles, W. D., 20,114,123,124, 125,127,128
Chinn, H. I., 42
Christensen, J. M., 110,122
Christian, J. J., 87
Christiansen, J. E., 77
Christie, B., 197,198,201
Clark, B., 37,44,243,244
Clarkson, F. E., 156
Cleveland, S. E., 11,12
Coates, G. D., 114,129,139
Cobb, S., 60
Coburn, K. R., 64,65,67
Coch, L., 168
Cogoli, A., 21
Cohen, A. I., 250
Cohen, H. C., 71
Cohen, M. L., 81
Cohen, S. 59,72,74,96,103
Collins, K., 202
Collins, W. E., 43,44,45
Colquhoun, W. P., 132,135,136
Communications Studies Group, 198
Congdon, S. P., 61
Connell, S., 201
Conover, D. W., 107
Conrad, D. W., 71
Conrath, D. W., 187
Conroy, R. T. W. L., 132,134
Convertino, V., 27
Conyers, J. E., 74
Coons, D. O., 107,116,137
Cooper, G., 51
Cooper, G. E., 125

399

Cooper, H. S. F., Jr., 8,92,127,
  132,233,279,289,290
Cooper, R., 165
Coppen, A., 236
Corbin, B. G., 80
Corbit, J. D., 226
Cotzin, M., 36
Covault, C., 122
Cowan, T. A., 90
Cowings, P. S., 44,45,49,50,51
Craig, J. G., 202
Cramer, E. H., 9,13,130,138,250
Crandall, R., 178
Crocq, L., 95
Cronholm, B., 256
Cronin, S. E., 133
Crownover, J. C., 20
Culbert, S. A., 92
Cunningham, W., 74,148,155,
  275

Dagley, C., 70
Danesh, H. G., 220
Daniels, W. L., 134
daSilva, A. J., 116
D'Atri, D. A., 88
Davenport, E. W., 61
David, A. B., 231
Davidson, R. J., 242,256
Davis, G. E., 87
Day, R. C., 210
Day, R. M., 154,282,294
Degioanni, J., 33
DeLamater, J., 149
Dember, W. N., 68
Dement, W. C., 123
Demling, R. H., 140
Derlega, V. J., 84,93
DeRoshia, C. W., 22,133,213
Dervin, B., 196
DeRubertis, F. R., 114
Desor, J. A., 87
DeWaard, J., 239
Diaz, E., 40

Dickey, D. T., 20
Dietlein, L. F., 20,31
Dietrick, J. E., 20
Dillon, R. F., 110
DiMarco, R. J., 23
Dinges, N., 92
Doane, B. K., 129
Dodge, R., 134
Doerre, G. L., 119
Doll, R. E., 153,160,225
Donnerstein, E., 71
Dorcus, R. M., 242
Drabeck, T. E., 231
Driver, M. J., 191
Dubos, R., 87
Duddy, J. H., 40
Dunlap, R. D., 241
Dunn, C. D. R., 20
Duntley, S. Q., 52
Durkheim, E., 248
Dusek, E. R., 66
Dushkov, B. A., 30,126
Dyer, R., 92

Earl, R. W., 68
Earls, J. H., 9,12,90,211,236
Eberhard, J. W., 65,67,77
Ebersole, J. H., 79,250
Eddowes, E. E., 77
Edmonston, W. E., Jr., 74
Edwards, R. S., 135
Egan, G., 241
Eichhorn, J., 26
Ekman, P., 193
Ellsworth, P., 193
Elrod, J. T., 118
Elton, M., 203
Engler, N., 49
Environmental Protection
  Agency, 72
Epstein, M., 20
Epstein, S., 220,221,223,224,
  226,254
Epstein, Y. M., 87,89

Erfunt, L., 88
Ertel, I. D., 116
Esser, A. H., 90
Etter, H. S., 65,163-164
Evans, G. W., 62,74,96
Evans, J. W., 22,213
Every, M. G., 75

Faber, S., 107
Farnsworth, P. R., 74
Farrell, R. J., 70
Fawcett, R. E., 130
Fedderson, W. E., 5,9,13,14,15, 145,146,151,269,280,289
Fenz, W. D., 223,224,225,226 254,282
Festinger, L., 10,84
Fidell, S., 72
Fiedler, F. E., 166,167
Finkelman, J. M., 113
Finney, L. M., 34
Fisher, B. A., 5
Fitzgerald, R. G., 248
Flaherty, T. M., 116,117
Fleishman, E. A., 109,110,113
Flinn, D. E., 9,13,130,138,228, 231,235,250
Folkard, S., 133
Formica, R., 230
Fort, A., 134
Foss, C., 92
Foughner, J. W., 65,76,78
Foushee, H. C., 190,230
Fox, J. G., 74
Fraser, D. C., 129
Fraser, T. M., 61,79,107
Frazier, S. H., 232
Freedman, J., 89
French, J. R. P., Jr., 60,168,262, 300
French, R. S., 34
French, W. L., 267

Freud, S., 236
Fried, Lena, 74
Friesen, W. V., 193
Frost, J. D., Jr., 137
Fudakowski, T., 191
Fukushima, D., 123

Gabriel, R. F., 73
Gagne, R. M., 109
Gaito, J., 20
Gallagher, T. F., 123
Garriott, O. K., 119
Garry, A. L., 91
Gartner, W. B., 125,126
Gauer, O. H., 20
Gazenko, O. G., 25,34,41,79, 132,210,211
Geen, R. G., 71
Gehlen, F. L., 232
General Electric Corp., 64
Genin, A. M., 25,34
George Washington University, 13,14,15,154,157,163,171, 208
Gerard, H. B., 169,226,230
Gerathewohl, S. J., 145
Ghata, J., 133
Giannetta, C. L., 23
Gibbard, G. S., 177
Gibson, E. G., 38
Gifford, E. C., 128
Gillingham, K. K., 27
Glass, D. C., 59,72,88,96,113
Glazer, D. L., 116
Glazer, J. H., 213
Goble, G. J., 34
Goffman, E., 83
Gogolev, K. 20
Golan, M. B., 83,92
Golarz De Bourne, M. N., 22
Goldbeck, R. A., 65,163-164
Goldmark, P., 200,201

401

Goldsmith, J. R., 69
Goldwater, D., 27,28
Goleman, D. J., 242,250,256
Gordon, E. M., 91
Gorgiladze, G. I., 48
Gorokhov, A., 80
Gorshkov, A. I., 36
Gouaux, C., 158
Gouldner, A. W., 297
Gowing, J., 130
Graeber, R. C., 32
Graen, G., 261
Graham, W. P., III, 140
Grave, C., 21
Graybiel, A., 33,35,36,37,42,44,
    46,47,49,50,243,244
Greiff, B. S., 298,299
Grether, W. F., 53,71,74
Griffen, L. M., 89
Griffitt, W., 66,158
Grigor'yev, Yu. G., 19
Grimes, R. H., 34
Grimwood, J. M., 116
Grinker, R. R., 223
Grodsky, M. A., 116,117
Grotz, J., 92
Group for the Advancement of
    Psychiatry, 77
Grumman Aerospace Corp.,
    15,64,70,93
Guedry, F. E., 33,34,44,48,50
Guerin, D., 123
Guignard, J. C., 73
Gunderson, E. K. E., 5,12,13,14,
    16,130,145,146,147,153,154,
    161,170,234,275
Gurovskiy, N. 115
Gurski, Ya., 20
Gustanski, J., 69

Hackler, T., 32
Hackman, J. R., 175,176,184,
    277,283

Hackworth, J. R., 89
Hagen, D. H., 9,13,130
Haggard, E. A., 174
Hale, H. B., 137
Hall, E. T., 193
Hall, T. J., 20,114,124
Hamblin, R., 210
Hammerton, M., 223,226
Hammes, J. A., 12,65,76,78,130,
    161,164
Hammond, K. R., 291
Hanna, T. D., 20
Hansson, L., 73
Harburg, E., 88
Hardyck, J. A., 151
Harper, R. P., Jr., 125
Harris, C. S., 70
Harris, J. L., 52
Harrison, R. V., 60
Hart, F. D., 72
Hartlieb, W., 26
Hartman, B. O., 125,130,139,
    228,231,235,239
Hartman, J., 177
Harvey, O., 170,295
Hasek, J., 130
Hattner, R. S., 20
Hauenstein, L. S., 88
Hauty, G. T., 132,133,134
Havron, M. D., 40,197
Hawkes, G. R., 20,114
Haythorn, W. W., 5,10,13,15,61,
    67,93,95,128,130,145,146,
    147,158,159,161,176,263
Heiskell, R. H., 65,163-164
Hellman, L., 123
Helmreich, R. L., 5,7,10,13,15,
    64,70,145,146,147,150,155,
    156,157,158,165,170,173,
    176,195,210,211,222,225,
    270,271,274,281,289,301
Hemmer, B., 26
Heron, W., 9,129
Herrmann, R. S., 16

Herskowitz, M., 148,156,275
Hertzman, K., 46
Heusner, W. W., 65,80
Hewson, D. M., 238,255
Hilker, R. R. J., 238
Hill, C. J., 36
Hitchcock, L., Jr., 123
Hodge, J. R., 248
Hodges, W. F., 221
Hoffler, G. W., 27,28,121
Hoffman, L. S., 225
Holahan, C. K., 156
Holland, J. L., 283
Hollander, E. P., 5,10,13,145,
    146,147,158,162,163,165,
    166,168,176,263
Holmes, T. H., 237
Holt, A. C., 116
Holtman, H., 133
Holton, E. M., 34
Homer, J. R., 107
Homick, J. L., 33,35,36,41,42,
    46,47
Hood, W., 170,295
Hooper, F. A., Jr., 77
Hopkins, A. H., Jr., 116
Hopper, A. E., 152
Hough, R. W., 197,202
Hull, D. H., 27
Hunt, B. I., 133,134
Hunter, G. M., 200
Hurley, R. W., 116
Hurteau, W. W., Jr., 31

Iakovleva, I. I., 48
Il'in, E. A., 28
Il'in, Ye. A., 34
Ioseliani, K. K., 79,132,210,211
Isaac, W., 69
Isay, R. A., 298
Issekutz, B., Jr., 21
Izard, Carrol E., 221

Jackson, D. D., 230
Jackson, J. K., 62
Jackson, J. M., 118,119
Jackson, M. M., 27
Jacobs, S., 248
Jacobs, T. O., 167
Jacobson, G., 252
Janis, I. L., 174
Jarrett, F., 140
Jenney, L. L., 125
Jennings, C. L., 21,23
Jenson, R. L., 21
Jex, H. R., 23
Joern, A. T., 11,13
Johansen, R., 197,202
Johnson, C. C., 63,73
Johnson, P. C., 21
Johnson, R. L., 20,21,26,27
Johnson, W. H., 33
Johnston, R. S., 20,31
Jones, D. R., 49,50
Jones, E. E., 91,93
Jones, F. N., 69
Jones, G., 72
Jones, G. B., 223,226,254
Jones, S., 225
Jonsson, A., 73
Jordan, D. P., 20
Jordan, J. P., 20
Jourand, S., 91,92
Jovy, D., 26
Julian, J. W., 263

Kadyskin, A. V., 71
Kahn, G. H., 91
Kahn, R. L., 4,165,166,187,206,
    207,264,266,270,271,272,
    278,283
Kajiyama, S., 43
Kamp, D., 87
Kanas, N. A., 5,9,13,14,15,145,
    146,151,269,280,282,289

Kaplan, I. T., 128
Kaplan, R., 78
Karlin, R. A., 87,89
Karnes, E. W., 77
Kasian, I. I., 36
Katlin, E. S., 225
Katz, D. L., 4,165,166,187,206,
 207,264,266,270,271,272,278
Kearney, O. F., 129
Keith, J. F., Jr., 161,164
Kellogg, R. S., 42
Kelly, G. F., 107,116,137
Kelly, S., 74
Kelman, H. C., 91
Kendall, D. A., 69
Kennedy, R. S., 34,42,46
Kerckhoff, A. C., 232
Kerwin, J. P., 36
Kidera, G. J., 42
Kiesler, C. A., 92
Kiesler, S., 92
Kikoshima, A., 23
Kimmel, D. C., 150
Kimzey, S. L., 20
Kincaid, D. L., 207
Kinsey, J. L., 79
Kiritz, S., 87
Klapp, O. E., 232
Kleiber, D. A., 100
Klein, D. C., 236
Klein, K. E., 26,27,133,134
Klein T. J., 125
Klein, W. J., 11,12,13,237
Kleinhans, B., 168
Kleitman, N., 136,139
Kletti, R., 222,229,253
Kline, N. S., 90
Knapik, J. J., 134
Knapp, R. J., 223
Knauth, P., 133
Kobasa, S. C., 238
Kohl, R. L., 33
Kolosov, I. A., 36
Kopanev, V. I., 36

Korchin, S. J., 223
Kornilova, L. N., 35,41,48
Korol'kov, V. I., 34
Kosmalinskiy, F. P., 30,126
Kotovskaya, A. R., 34
Kottenhoff, H., 44
Kozerenko, O. P., 79,132,210,
 211
Krantz, D. S., 74,96
Krasnykh, I. G., 20
Kristen, C., 202
Krylov, Yu. V., 73
Kryter, K. D., 72
Kubis, J. F., 2,5,12,13,118,119,
 131,145,146,147,165,171,
 270,271,282,289
Kuklinski, P., 27
Kurash, S., 20
Kushner, E. N., 11
Kuznetsov, V. S., 73

Lackner, J. R., 50
Lamb, L. E., 20,21,26,107
Lamberth, J., 158
Landaw, S. A., 34
Landsman, M., 92
Lang, G., 88
Langdon, D. E., 139
Langdon, L. E., 73
Lange, R. D., 20
Langer, E. J., 78
Lankford, H. G., 129
Laporte, A., 132
Latané, B., 5,10,13,145,146,147,
 158,176
Lathrop, R. G., 123
Lauber, J. K., 230
Laufer, R. S., 83
Law, P., 12,13,153,211,235,278,
 282
Lawler, E. E., III, 176,279,280
Lawton, M., 69
Lazarus, A. A., 236,237

Lazarus, R. S., 237
Leach, C. S., 47,80
Leach, L. S., 33
Lebedev, V. I., 2,36,40,74,76,78, 79,90,94,147,154,160,164, 165,168,230,243,264,271, 282,289,293
Lederer, L. G., 42
Lee, G. F., 231
Lefcourt, H. M., 72
Lehr, D. J., 129
Lentz, J. M., 43,44,45
Leon, A. S., 81
Leon, H. A., 34
Leonard, J. I., 20
Leonov, A. A., 2,40,74,76,78,79, 90,147,154,160,164,165,168, 230,243,264,271,282,289,293
Leoppky, J. A., 26
Lessard, C. S., 22,23
Lester, D., 241
Lester, J. T., 225
Levanthal, G. S., 275
Levanthal, H., 224,225,254
Levin, H. A., 84
Levine, R. B., 114
Levy, A. S., 89
Levy, E. E., 226,239
Levy, E. Z., 13
Levy, R. A., 49,50,51
Lewin, K., 168
Lewinsohn, P. M., 236
Lewis, H. B., 46
Lewis, O. F., 21,22,23
Lidvall, H. F., 42
Likert, R., 208
Lindahl, L. E. H., 44
Lindemann, E., 240,247,248,249
Lindsley, D. B., 225,254
Lindvall, T., 69
Link, M. M., 115
Lippitt, R., 168
Llano, G. A., 231
Lockhart, J. M., 66

Loiko, M. A., 226
Loo, C., 87
LoScinto, L. A., 77
Lott, A., 169,173
Lott, B., 169,173
Loudis, L. A., 77
Lovelace, W. R., II, 29,107
Lubin, A., 129
Luft, U. C., 26,29,107
Lugg, D. J., 13,153
Lyman, J., 47
Lynch, K., 62
Lynch, T. N., 21

McBride, G. H., 118,119
McCally, M., 95
McCarthy, J. J., 264
McCauley, M., 66
McClintock, C. G., 155
McCormack, P. D., 128
McDonnell Douglas Aeronautics Company, 64
McGaffey, C. N., 145
McGrath, J. E., 5,10,13,145,146, 147,158,176
McGregor, P., 123
McGuire, F., 211
Machell, R. M., 116
Machover, K., 46
Mack, P. B., 20,21
McKenzie, R. E., 130
Mackie, R. R., 66
McKinney, W. T., 236
Mackworth, N. H., 110,128,129
McLaughlin, E. J., 12,118,119, 131,146
McLean, M. V., 76
McMillan, D. E., 20
McNaughton, G. B., 19
McNee, R. C., 239
Maddi, S. R., 238
Mahan, J. L., 14,147
Mahar, L., 167

405

Maki, J. E., 155
Maly, J., 140
Mandel, D. R., 89
Mandler, G., 45
Mandler, J. M., 45
Mandour, J. A., 116
Mann, L., 174,177
Mann, R. D., 165,166
Manno, B. R., 47
Manno, J. E., 47
Margolis, G. F., 31
Margulis, S., 83
Maris, R. W., 253
Marshall, E., 240
Marshall, N., 85
Mason, R. M., 63
Mateev, A. D., 48
Mathews, K. E., 71
Mathewson, G. C., 169
Matsnev, E. I., 48
Meddis, R., 44
Medina, M. A., 31
Mehrabian, A., 192
Meier, U., 26
Meissner, P., 46
Messick, J. M., 252
Metlay, W., 128
Michel, E. L., 122
Miles, S., 244
Miller, D. H., 250
Miller, D. R., 10
Miller, E. F., II, 35,36,41,42,46, 47
Miller, J. G., 4,196,273
Miller, L. M., 279,280
Miller, N., 226,230
Miller, P. B., 21,26
Miller, R. B., 109
Miller, R. H., 197
Miller, W. R., 236
Mills, J., 169
Mills, J. N., 132,134
Mintz, A., 229
Mire, J. L. C., 107

Misumi, J., 165
Mitchell, J. L., 222
Money, K. E., 42,50
Monk, T. H., 133
Monroe, J. T., 9,13,130
Montgomery, L., 28
Moore, H. G., 117
Moos, R. H., 87
Morey, E. R., 20
Morgan, B. B., Jr., 114,129,133, 134,136,139
Morgan, T. E., 130
Morgan, W. P., 101
Morley, I. E., 201,202
Morley, W. E., 252
Morris, C. G., 175,176,184
Morris, R. L., 132
Morse, M. L., 116
Morway, D. A., 123
Moss, M. S., 249
Moss, S. Z., 249
Mostert, N., 9,152,243,246,250
Motley, E. P., 244
Moylan, J. A., 140
Mueller, M., 21
Mullin, C. J., 9,11,12,13,65,130
Munter, P. K., 298,299
Murakawa, N., 158
Murphy, M. R., 125,126,175
Murray, R. H., 95
Myasnikov, V. I., 30,79,132,189, 192,210,211
Myers, T. U., 146,147,159
Myhre, L. G., 26

Nagatsuka, Y., 130
Nandy, K., 22
Nardini, J. E., 16
NASA, 27
NASA, Office of Manned Spaceflight, 70

Natani, K., 11,12,13,16,136,146, 147,153,170,171,245,271, 275,276,278,280,282,285
National Academy of Sciences, 71
Nedzel, A. J., 31
Nelson, P. D., 12,13,14,16,130, 147,154
Neuringer, C., 251
Newman, P. P., 130
Newsom, B. D., 34
New Yorker Magazine, 62
Nichols, G., 239
Nicogossian, A. E., 19,27,31,52, 116,122
Nixon, C. W., 73
Nogeire, D., 123
Nordlie, P. G., 197
Novikov, V. E., 28
Noyes, R., 222,229,253
Nyberg, J. W., 34

Oberg, J. E., 64,67,78,81,137, 148,152,243,257
O'Brien, J. F., 71,72
Ochsman, R. B., 197
O'Donnel, R. D., 42
O'Hanlon, J. F., 66
Olasky, C. C., 107
Older, H. J., 125
Oldham, G. R., 176,283
O'Leary, B., 228
O'Lone, R. G., 51
Olree, H. D., 80
Olsen, L., 27
Olson, R. M., 140
O'Neal, E. C., 71
Ordiway, V., 235
Orne, M. T., 242
Osborne, R. T., 20,161,164
Osborne, W. Z., 53
Osgood, C. E., 297

Ostfeld, A., 248
Ostfeld, A. M., 88

Page, R. N., 70
Palamarek, D. L., 95
Pallak, M., 92
Palmai, G., 13,15
Panchenkova, E. F., 189,192
Panko, R. R., 197,202
Papov, N. I., 36
Parin, V. V., 30,126
Parker, D. M., 45
Parker, J. F., Jr., 19,31,52,75, 110,113,116
Parkes, C. M., 248
Parrish, R. N., 197
Parsi, R., 140
Patterson, C. H., 241
Patterson, P. H., 227
Patton, J. F., 134
Paykel, E. S., 237
Pearlman, C. A., 298
Pearson, D. W., 221
Pearson, R. G., 71,72
Pearsons, K. S., 72
Pepler, R. D., 66
Peraky, H., 223
Perea, A., 140
Perlow, M., 123
Perrow, C., 259,260,261,266, 268,272
Perry, C. J. C., 153,178,239,266, 268
Perry, S., 235
Peters, R. J., 23
Petrov, Y. A., 68
Phillips, S., 59,72
Pierce, B. F., 61
Pierce, C. M., 136
Pinner, B., 89
Pinsky, L. S., 53
Pishchik, V., 52

407

Pond, J. I., 65,163-164
Pope, F. E., 12,13,211,240
Popov, V. A., 52
Popp, R., 28
Portugalov, V. V., 34
Poulton, E. C., 66
Poza, F., 72
Prasad, A., 158
Prasad, M., 158
Preber, L., 42
Price, J., 89
Prim, J. W., 116
Proshansky, H. M., 83
Proulx, P., 191
Pushkar, D., 130

Quarantelli, E. L., 231
Quilter, R. W., 128

Rabbie, J. M., 226,230
Radloff, R., 5,7,10,13,64,145,
 146,147,150,158,165,170,
 176,211,222,225,281,301
Rahe, R. H., 227,237
Rall, M., 89
Rambaut, P. C., 80
Ramirez, M., 158
Ramsey, R. W., 247,249
Randall, L. S., 200
Rasmussen, E., 49,50
Rasmussen, J. E., 16,65,67
Raven, B. H., 262,300
Rawls, D. J., 152
Rawls, J. R., 145,152
Rayman, R. B., 19
Raymon, R. S., 235
Razumeev, A. N., 30
Reason, J. T., 35,38,40,42,43,
 44,48,50
Redetzki, H. M., 47
Redgrove, J. A., 30

Rehme, H., 133
Reilly, R., 71
Reilly, R. E., 110
Reinberg, A., 133
Reitman, E. E., 12,13
Remek, V., 191
Reschke, M. F., 33,41
Ricciuti, E. A., 36
Richman, J., 253
Rickarby, G. A., 247
Rimpler, A., 26
Rippere, V., 243
Rivolier, J., 95
Robbins, E., 251
Robbins, K., 285
Robinson, S. M., 31
Rodahl, K., 21
Rodgers, W., 60
Rodin, J., 78
Rogers, E. M., 192,205,207,208
Rogers, J. G., 14,65,67,90,246
Rogers, T. A., 12,13,211,240
Rohrer, J. H., 13,14,79,130,174,
 237,282
Rose, C. L., 81
Rosenbaum, M. A., 253
Rosenblatt, L. S., 47
Rosengren, K. E., 229
Rosenkrantz, P. S., 156
Rositano, S. A., 27
Ross, S., 128
Roth, E. M., 66
Rothstein, L. 23
Rowell, L. B., 21
Rowes, B., 29,30
Rubin, R. T., 227
Ruff, G. E., 13,112,226,239
Ruffell Smith, H. P., 230
Rule, B. G., 95
Rummel, J. A., 25,120,122
Rumyantseva, M. P., 30
Runge, T. E., 145,155,156,157,
 270,271,281,301
Rushing, W. A., 253

Rusnak, R., 118,119
Rutenfranz, J., 133
Ryan, M. G., 202
Ryback, R. S., 21,22,23

Salamy, J. G., 137
Sampson, J. B., 134
Samuel, J. A., 128
Sanders, M., 69
Sandler, H., 20,22,27,28,29,30, 133,213
Sansom, W., 130
Sarnoff, I., 221,230
Sassin, J., 123
Sawin, C. F., 27,122
Saxon, S. C., 118,119
Scales, J. T., 140
Scarr, S., 85
Schachter, S., 158,173,225,230, 237
Schalling, D., 256
Schatte, C. L., 20
Schludermann, E., 130
Schopler, J., 88,89
Schork, M., 88
Schotte, J., 26
Schreuder, O. B., 126
Schroder, H. M., 191
Schull, W., 88
Schultze, D. P., 65,163-164
Schwartz, G. E., 242,256
Schwartz, S., 231
Schweickart, R., 65
Schwichtenberg, A. H., 29,107
Scott, T. H., 129
Secrest, R. R., 29,107
Seeman, J. S., 62,76
Seibold, D. R., 92
Seligman, M. E. P., 236
Sells, S. B., 2,4,5,145,146,170
Selye, H., 96,237,238
Sengel, R., 147,171,245,276,285

Serova, L. V., 34
Serxner, J. L., 9,13,235,242
Shackel, B., 107
Shafer, W. A., 34
Shannon, C. E., 195
Shapland, D. J., 239
Sharpe, M. R., 242
Shashkov, V. S., 30
Shaw, M. E., 173,208
Sheard, J. H., 65,163-164
Shearer, J. W., 110
Shears, L. M., 154
Sheer, D., 11,12,13
Shepard, H. A., 177
Sherif, C. W., 139,170,295
Sherif, M., 139,170,295
Sherrod, D. R., 88
Shestkov, B. P., 30
Shipka, W., 26
Shirakashi, S., 165
Shorr, E., 20
Short, J., 197,198,201,202
Shul'Zhenko, Ye. B., 20
Shumate, W. H., 137
Shurley, J. T., 9,11,12,13,136, 147,170,171,245,248,271, 275,276,278,280,282,285
Shyken, N. P., 116
Sieber, J. E., 70,195,210,271, 274,289
Simmons, J. B., II, 20
Simons, J. C., 110,121,122,123
Sinaiko, H. W., 202
Singer, J. E., 72,88,96
Singer, R. V., 76
Singer, S. F., 34
Sistrunk, F., 197
Smelser, N. J., 232
Smith, C. W., 80
Smith, D., 202
Smith, M. B., 151
Smith, P. K., 42
Smith, S., 8,9,11,15,61,67,70, 95,160,161,236

Smith, W. M., 13,229,253,271, 272
Solodovnik, F. A., 36
Solomon, R. L., 226
Sours, J. A., 244
Space Sciences Board, 2,4
Spangler, K., 197
Sparvieri, F., 31
Spence, J. T., 156
Spielberger, C. D., 220,221
Stager, P., 191
Stapantsev, V., 80
Stapley, B., 199
Stapp, J. P., 34
Steadman, J., 132
Stegemann, J., 26
Stein, D. D., 151
Steiner, I. D., 168
Steinhoff, W. D., 133
Steinmann, D. O., 291
Stephenson, G. M., 201,202
Stern, M., 251
Stevens, P. M., 20,21
Stewart, G. T., 223
Stewart, J. D., 44
Stewart, T. R., 291
Sticha, P. J., 200
Stockton, W. S., 8,145
Stokols, D., 74,87,88,89,96,104
Stolze, H., 133
Stone, H. L., 20
Storm, W. F., 23
Strange, R. E., 11,12,13,237
Straumfjord, A., 92
Streimer, I., 123
Stretton, M., 139
Streufert, S., 191
Strickland, D. A., 90
Strope, W. E., 65,163-164
Strughold, H., 134,137
Stupakov, G. P., 34
Sturesson, D., 229
Suedfeld, P., 70
Sullins, W. R., 65

Sundstrom, E., 84,87,89
Surgeon General's Scientific Advisory Committee on Television and Social Behavior, 77
Surwillo, W. W., 128
Suvorov, G. A., 71
Suzuki, Y., 130
Swenson, L. S., 116
Sykes, H. A., 20
Syrykh, G. D., 41

Talmon, Y., 149
Tanner, T. A., 70,195,210,271, 274,289
Tarasov, I. K., 35,41,48
Taylor, D. A., 10,11,147,159, 168,178,193
Taylor, J. H., 52
Teichner, W. H., 66,71
Tennent, C., 238,255
Teoh, K., 20
Thackray, R. I., 221
Thaler, E. H., 13
Thomas, J. K., 77
Thomas, K., 293,294,295,296, 297,302
Thompson, L. F., 53
Thompson, T. L., 92
Thomson, R. M., 244
Thorngate, W. B., 155
Thornton, C. L., 40,45
Thornton, W. E., 25,121
Thurmond, J. B., 109
Tickner, A. H., 223,226
Tiebes, U., 26
Tigranyan, R. A., 34
Tobias, C. A., 19
Tolchin, S., 211
Toscano, B. W., 44,49,50
Toufexis, A., 20,25,36
Trego, R. E., 145

Trembly, G., 225
Triebwasser, J. H., 27
Trimble, R. W., 21,23
Truchaud, M., 31
Tuckman, B., 177
Turner, D. P. W., 123
Turner, R. T., 34
Tyler, D. B., 42

Ulvedal, F., 130
Ushakov, A. S., 30
Uskov, F. N., 79,132,189,192, 210,211
Uviller, E. T., 45

Vachon, M. L. S., 247
Vaillant, G. E., 238,240,255
Vallee, J., 197,202
Vallucki-Morf, M., 21
Van Bockel, J. J., 107
Van Huss, W. D., 65,80
Vasil'yev, P. V., 36
Veitch, R., 66
Vekovius, W. A., 47
Veltfore, H. R., 231
Venters, M. D., 26
Vernikos-Danellis, J., 22,47,80, 132,213
Vieux, N., 133
Vinogradov, V. M., 30
Vinsel, A., 92
Voas, R. B., 107
Vogel, S. R., 156
Voget, J. A., 134
Vogt, F. B., 21
Vogt, L., 26
Volkmer, K., 123
Volynkin, U. M., 36
Von Gierke, H. E., 73
von Restorff, W., 33
Vroom, V. H., 168

Wagner, C. M., 65,67
Waligora, J. M., 19
Wallis, D., 128
Walsh, B., 191
Walster, E., 149,275
Walster, G. 275
Walton, M., 88
Wamsley, J. R., 62
Warfield, J. T., 116
Watson, J. A., 161,164
Watters, H., 132
Weaver, W., 195
Webb, W. B., 123,133,139
Weber, M., 261
Weeks, G. D., 197
Wegmann, H. M., 26,27,133,134
Weick, K. E., 260,279,290
Weigel, R. G., 92
Weinstein, N. D., 72
Weitzman, E. D., 123
Welch, B. E., 20,130
Welford, A. T., 125,126,127
Wendt, G. R., 36
Wenger, D. E., 231
Wenger, M. A., 45
Westin, A., 85
Weston, J. R., 202
Westrum, R., 197
Wexley, K. N., 265,274
Weybrew, B. B., 12,257
Whedon, G. D., 20
Wheeler, L., 11
Wheeless, L. R., 92
Whitaker, L. A., 140
White, B., 170,295
White, P. D., 34
White, R. K., 168
White, W. J., 34,52
Whittenburg, J. A., 128
Whyte, W. F., 262,268,269
Wicker, A. W., 171,172
Wilding, J. M., 44
Wilford, J. N., 8,145
Wilgosh, L., 130

Wilhelm, J., 70,145,155,156, 157,195,210,270,271,274, 281,289,301
Wilkinson, R. T., 128,138,139
Williams, E., 197,198,201
Williams, H. L., 129
Williams, W. K., 107
Wilson, C. L., 107
Wilson, D. W., 71
Wilson, M., 84,93
Wilsoncroft, W. E., 45
Winget, C. M., 22,24,47,80,133, 213
Winocur, G., 130
Winter, D. L., 29,30
Winter, G. D., 140
Witkin, H. A., 46
Wolfe, M., 83,92
Wolfe, T., 8,131,152,156,275
Wolthuis, R. A., 27
Wood, C. D., 46,47
Wood, D. T., 80
Wood, M., 47
Woodhead, M. M., 71
Woodling, C. H., 107
Woodson, W. E., 107
Wopner, S., 46

Worthy, M., 91
Woskow, M. H., 69
Wright, J. E., 134
Wrightsman, L. S., Jr., 230
Wurster, W. H., 33

Yakovleva, I. Ya., 35,41
Yanowitch, R. E., 175
Yegorov, A. D., 25
Yegorov, B. B., 30
Yeremin, A. V., 108
Yetton, P., 168
Yingling, R. W., 74
Yoesting, D. R., 77
Youngblood, D., 223
Yuganov, E. M., 36
Yuganov, Ye. M., 73
Yuki, G. A., 265,274

Ziller, R. C., 178
Zimbardo, P. G., 221,226,230
Zimberg, S., 251
Zubek, J. P., 9,129,130

SUBJECT INDEX

Acceleration — see "Centrifugation"
Accidents, 175
Acting out, 92
Acuity — see "Visual changes"
Affiliation, need for, 158-159
Aftereffects of stress, 96
Age range, space travelers, 150-151
Alcohol, 9,126,245-247,248
Ambiversion, 157,317
Ames Research Center, 21,22, 24,30,44,273
Androgeny, 156-157,317
Annoying characteristics, 153
Anxiety, 12-14,23,80,220-221, 232,233,235
Artificial gravity, 33-35
Athletes — see "Exercise"
Atmospheric pressure, 73,189
Attitudes, 148-149,151-152, 157-158
Audio systems — see "Mediated communication"
Authority/Control, 210
Autonomic dominance, 45-46
Autonomic response control — see "Biofeedback"
Autonomy, 83
Awareness training, 241

Bad news, 212
Barriers to communication, 206-207

Bathing — see "Hygiene"
Bedrest, 21-25,213-214,329
Bedrest, individual differences, 24
Bereavement — see "Grief"
Biofeedback, 49-51,314
Biological rhythms — see "Circadian rhythm" and "Desynchronosis"
Blaming, 117,231-232,316
Boredom, 323
  (See also "Monotony")
Boundary role, 208,288-289, 294
Breakoff, 243-245,330
Bridging — see "Boundary role"
Broadband communication, 198,202,205

Canal-Otolith Conflict — see "Sensory Conflict Theory"
CAPCOM, 210
Centralization of authority, 265-266
Centrifugation, 26,27,29,37, 49-50, 51,115
  Effects of, 33-35
Channel hypothesis, 198,199
Chronobiology — see "Biological rhythms"
Chronopharmacology — see "Drugs, time of administration"

Circadian rhythm, 32,135–136, 139
  (See also "Desynchronosis")
Cliques, 149,152,164,208,275
Clothing, 86,92,193
Cocooning, 13
Cognitive conflict, 291–292
Cohesiveness, 169–173
Compatibility, 146–162
Compatibility, size effects, 160–161
Competence, 154
  Social, 154,164–165
  Task, 154,164–165
Competition, 78,81,155–156
Compliance — see "Conformity"
Computer games, 78
Computers, 194–195
Conflict, 188,209–210
  (See also "Hostility")
Conflict management, 293–297
  Process, 296–297
  Structural, 293–296
Conformity, 173–175
Contingency theory, 166–167
Convents and seminaries, 330
Cooperation, 155–156
Crew rotation, 132
Crisis intervention, 252–253
Crowding, 87–89
Crowding, gender differences, 89
Cuban boots, 49

Effects on cardiovascular system, 20,21
Effects on musculoskeletal system, 20,21,25–26
Gender effects, 28–30
Decor, 67–68,92
Defense Advanced Research Projects Agency, 200
Density — see "Crowding"
Depersonalization, 222
Depression, 13,23,79,137, 235–237,247,249,251,253, 293,298
Desynchronosis, 114,118, 132–136,308,322
Diet — see "Food"
Digestive problems, 12
Direct communication, 188–194
  (See also "Face-to-face communication")
Direction of gaze, 200
Discrete task assessment, 110–113,120
Distance, interpersonal, 193
Diving bells, 329
Drugs, 30–33,46–47,126, 245–247, 248,251,286
  In weightlessness, 31–32
  Time of administration, 32
Duration, research issues, 322–323
Dyads, 145,159,161

Danger control, 224–225
Decentralization of authority, 265–266
Deconditioning in weightlessness, 35
  Age effects, 27–28
  Countermeasures, 30–35

Electronic communication — see "Mediated communication"
Emergencies, 124,175
Emergencies in space, 218–220
Emotional stability, 153–154
Ergonomics, 108
  (See also "Work capacity")
Ethnicity, 151–152

European Space Agency, 152,330
Excitement/novelty, 117,120,131
Exercise, 25-27,80-82,242
Exercise, motivation to, 31-32
Exit event, 237
Extravehicular activity, 119
Extroversion-introversion, 44,46, 85-86,157,165
Eysenck Personality Inventory, 44

Face-to-face communication, 188-192,197-198, 199-202(passim),212
Facial expression, 193-194
Factions — see "Cliques"
Families, 298-299
Fatigue, 112,125-131,132,134
  (See also "Desynchronosis," "Monotony," "Sleep")
Fear, 72
  Birth-order effects, 225-226
  Control, 224-225
  Gender effects, 225
  Inhibition, 224
  Research results, 221-224
  Time effects, 223,224,229
Field dependence/independence, 46
Followership, 326
  (See also "Leadership")
Food, 8-9,12,30,63-65
Fright, 221

Gatekeeping, 197,206,288,302, 310
Gender effects, 147-149, 156-157
Goals, 169-171

Grief, 247-252
Group performance, 175-177
  Long-term, 177-180
  Short-term, 177-180
Groups as groups, 327
"Groupthink", 174,175
Guilford-Zimmerman Temperament Survey, 44
Guilt, 298

Head restraint, 48-49
Head schedule movements, 47-48
Headaches, 12,235
Heterogeneity, research issues, 321-322
Heyerdahl, Thor, 165
Homeostasis, 84,160,161,230, 316
Homicide/suicide, 250-252
Horizontal communication, 207
Hostility, 13-14,23,156,158
  (See also "Conflict")
Hostility to outsiders, 14,326
Human performance, 109-121
  Abilities, 109-110
  Assessment, 110-121
Human relations — see "Training"
Humidity, 66-67
Hygiene, 65-66,81
Hypnosis, 242
Hypokinesis, 20,23,29
  (See also "Bedrest")
Hysterical contagion, 232-233

Idiosyncrasies, 93
Illusions — see "Vestibular changes"
Incentives, 264,279
  (See also "Rewards" and "Sanctions")

415

Independence — see "Conformity"
In-flight performance assessment, 118–120
Information, 195–196
Insomnia — see "Sleep, disturbances"
Intellectual functioning, 11–12
Intergroup relationships, 287–299
International crews, 152, 190–192,194
Interpersonal friction, 10,13 (See also "Conflict")
Introversion — see "Extroversion-introversion"
Isolation and confinement, 5–15
   Application to space, 5–11
   Research results, 11–15
   Temporal effects, 14–15

Johnson Space Center, 116,273

Kinesic cues — see "Non-verbal communication"

Laboratory in space, 141
Laws — see "Rules and laws"
Leadership, 162–169
   Autocratic-democratic, 167–169
   Levels of, 266–268
   Personality characteristics, 165–167
   Socioemotional, 164–165, 166–167
   Task, 164–165,166–167
Leisure, 75–82
Liaison, 208

Light Flashes — see "Visual changes"
Lighting, 67,68–69,87
Linking pin, 208
Living space, 60–63
Living space, personal, 86
Locus of control, 88–89
"Long Eye", 174,237

Manning Theory, 171–172
Meals — see "Food"
Mediated communication, 194–205, 328
   Audio systems, 200–201
   Computer/telegraphic systems, 201–202,205
   Versus face-to-face, 197–198
   Video systems, 199–200,201, 202,210
Mediated counseling, 241
Meditation, 242
Metabolic costs, 122
Military bases, 329
Mixed crews — see "Gender effects"
Monotony, 112,128,131–132, 134
Morale, 12,14–15,79,130,147, 164,173,279,281,282,299
Motion sickness — see "Space sickness"
Motivation — see "Morale"
Mourning — see "Grief"
Multiple stressors, 95–96
Multiple task batteries, 114–115
Musgrave, Story, 62,135
Music, 74,246

Narrowband communication, 199,202,205

Networks, 205-212
    Centralized, 208-209
    Decentralized, 208-209
    Formal, 206-207
    Informal, 207-208
Newcomers, 178-179
Nitrogen narcosis, 244-245
Noise, 70-75,113,189
    Effect on performance, 71
    Levels, 71-73
Non-verbal communication, 192-194
Norms, social, 277-278,288, 294
Norms, social, reciprocity of, 297

Odor, 69-70
Off-board communication, family and friends, 211-212
Off-board communication, ground control, 209-211
Oil rigs, tankers, 329
On-board training, 131,179, 264,328
Opinion leader, 208
Opponents' processes, 226-227
Ostracism, 173,237,278,318
Othostatic tolerance — see "Exercise"
Out-of-body experiences, 222
Overloading — see "Fatigue" and "Work roles"
Overmanning — see "Manning theory"

Panic, 228-229
Parachuting, 223
Paralinguistics — see "Non-verbal communication"

Performance adaptation, 118-120,121
Performance assessment — see "In-flight performance assessment"
Person-environment fit, 59-60, 103-104
Polaris, 257
Possessions, 86
Power — see "Social power"
Prejudice — see "Attitudes"
Primary/secondary tasks, 113
Prisons, 330
Privacy, 82-94,209-210
Privacy, regulation, 86-87
Professionalism, 259-260
Proxemics — see "Non-verbal communication"
Psychiatric disturbances, 13 (See also "Anxiety," "Depression," "Psychological episodes")
Psychological episodes, 234-243,244
Punishment — see "Sanctions"

Readaptation/reassimilation, 138,298-299
Recreation, 76-79
Recreation, passive, 76-78
Redundancy, 190
Rejection — see "Ostracism"
Rewards, 280-284,294,295
    Extrinsic, 280-282
    Intrinsic, 280,282-284
Rules and laws, 285-286,295
"Ruston", 190

Sanctions, 284-286
Scapegoating — see "Blaming"

Scheduling, 118
Sealab, 222
Selecting in, 153,327
Selecting out, 284
Selection, 146-147,161-162, 238-240
Self-designing systems, 260-261
Self-disclosure, 91-93
Sensory conflict theory, 37-41
Sensory deprivation, 70,129
   (See also "Uniform Environments")
Simulation, 20-25,25-30(passim),49-51(passim),329
Simulation assessment, 115-118
Size, research issues, 320-321
Sleep, 25,107,114,118,123-125, 132, 136-140,247,308
   Disturbances, 12
   Sudden awakening, 139-140, 308
Social comparison, 188
Social organization, 261-278
   Bureaucratic/hierarchical model, 262,266-268
   Community democracy model, 268-269
   Labor relations model, 268, 269
Social power, 262-269
   Bases of, 262-265
   Coercive, 264
   Expert, 264-265
   Legitimate, 262-264,265
   Referrent, 265
   Reward, 264,265
   Structure, 262-269
Social presence hypothesis, 198,199
Socioemotional function, 188,209
Socioemotional requirements, 272

Sound, 74-75
   (See also "Noise")
Space adaptation syndrome — see "Space sickness"
Space environment, 7-15
Space Shuttle, 330
Space sickness, 36-41, 41-51(passim),120,307, 314
   Countermeasures, 46-51
   Personality correlates, 44-46
Space Station, 330
Stress, performance under, 124-125
Strong-Campbell Interest Inventory, 283,287
Suicide — see "Homicide/ suicide"
Synthetic work methodologies — see "Multiple task batteries"
System acceptance, 203-204
Systems approach, 312,313
Systems approach, Open Systems Theory, 4

Task function, 188,209
Task requirements, 272
Task/socioemotional hypothesis, 198,199
Tektite, 173
Temperature, 66-67
Tereshekova, Valentina, 148
Territoriality, 89-90
Three-dimensional space, 62
Time effects — see "Duration, research issues"
Training, 171
   For danger, 226-228
   Interpersonal relations, 210
   On-board — see "On-board training"
   Survival, 227-228

Transderm-V system, 32–33
Transmeridian flights, 133–135
Triads, 145,161

Underloading — see "Fatigue" and "Uniform environments"
Undermanning — see "Manning theory"
Uniform environments, 129–131
"Up" and "down" — see "Vertical, true"

Validity, 120
Verbal communication, 188–192
Vertical communication, 206–207
Vertical, true, 38–40
Vertigo — see "Vestibular changes, vertigo"
Vestibular changes, 35–51
    Age effects, 42–43
    Gender effects, 43–44
    Resistance to, 41–46
    Spatial disorientation, 35–36
    Vertigo, 35–36,41
      (See also "Space sickness")
Vibration, 53
Visual changes, 51–53
Visual-vestibular conflict theory —see "Sensory conflict theory"

Vocational interests, 283

Waste management — see "Hygiene"
Water immersion, 26–27,40, 116,122–123
Weightlessness, 19–56(passim), 145,192–193,279
    (See also "In-flight performance assessment" and "Work capacity")
Withdrawal, 9–10,13–14,15,76, 84,92,137,307
Women in space — see "Gender effects"
Work capacity, 121–123
Work categories in space, 108–109
Work emphasis, 75–76
Work modules, 284
Work roles, 269–277
    Conflicts, 274–276
    Emergent, 271–273
    Overload, 260–261,273–274, 292
    Prescribed, 271–273
    Rotation, 276
Work schedules, 123–125,127
Work/Efficiency ratio, 119–120
Workload, 117,125–132
Work-rest cycles, 118
Written communication — see "Mediated communication"

Zeitgebers, 135,314
Zero-g — see "Weightlessness"